Bioinvasions and Globalization

Ecology, Economics, Management, and Policy

Edited by

Charles Perrings, Harold Mooney, and Mark Williamson

OXFORD
UNIVERSITY PRESS

OXFORD
UNIVERSITY PRESS

Great Clarendon Street, Oxford OX2 6DP

Oxford University Press is a department of the University of Oxford.
It furthers the University's objective of excellence in research, scholarship,
and education by publishing worldwide in

Oxford New York

Auckland Cape Town Dar es Salaam Hong Kong Karachi
Kuala Lumpur Madrid Melbourne Mexico City Nairobi
New Delhi Shanghai Taipei Toronto

With offices in

Argentina Austria Brazil Chile Czech Republic France Greece
Guatemala Hungary Italy Japan Poland Portugal Singapore
South Korea Switzerland Thailand Turkey Ukraine Vietnam

Oxford is a registered trademark of Oxford University Press
in the UK and in certain other countries

Published in the United States
by Oxford University Press Inc., New York

© Oxford University Press 2010

The moral rights of the authors have been asserted
Database right Oxford University Press (maker)

First published 2010

All rights reserved. No part of this publication may be reproduced,
stored in a retrieval system, or transmitted, in any form or by any means,
without the prior permission in writing of Oxford University Press,
or as expressly permitted by law, or under terms agreed with the appropriate
reprographics rights organization. Enquiries concerning reproduction
outside the scope of the above should be sent to the Rights Department,
Oxford University Press, at the address above

You must not circulate this book in any other binding or cover
and you must impose the same condition on any acquirer

British Library Cataloguing in Publication Data
Data available

Library of Congress Cataloging in Publication Data
Data available

Typeset by SPI Publisher Services, Pondicherry, India
Printed in Great Britain
on acid-free paper by
CPI Antony Rowe, Chippenham, Wiltshire

ISBN 978–0–19–956015–8 (Hbk.)
       978–0–19–956016–5 (Pbk.)

1 3 5 7 9 10 8 6 4 2

Preface

This volume is motivated by a concern expressed at the 8th Conference of the Parties to the Convention on Biological Diversity (CBD) over "gaps and inconsistencies" in the international regulatory framework (Decision VIII/27 of COP 8: Alien species that threaten ecosystems, habitats or species (Article 8 (h)). At least part of the problem identified by the CBD lies in the fact that there are similar "gaps and inconsistencies" in our knowledge of biological invasions. As with many anthropogenic environmental problems, the social and natural science of biological invasions are still poorly articulated. There is, for example, almost no research into interactive effects between the main social and environmental drivers of change in the global distribution of species. This volume explores the current state-of-the-art in the social and ecological science of invasive species, and draws out the implications for the national and international regulation of the problem. It is organized into three parts.

Part 1 identifies the major drivers behind the problem: globalization (the ever closer integration of the world system), climate change, and land use change—and explores the extent to which an understanding of these drivers can help us predict the introduction, establishment, and spread of species that have the potential to impose significant harm. While the individual chapters focus on distinct drivers—Thomas and Ohlemuller and New and McSweeny on climate, Perrings et al. on trade, and Williamson and Pyšek et al. on landscape and land use change—together they establish two of the key messages of the book. One is that invasive species risks are endogenous: they are the product of human decisions that determine the direction and volume of trade, the level of sanitary and phytosanitary effort, and the way in which landscapes are structured. This makes it important for policy and management to address the factors that determine those decisions. The other is that the risks associated with different species are interdependent. Many human and animal pathogens, for example, not only share the same origins but also the same pathways and vectors. Equally, vulnerability to one invasive species may be determined by the success of others. This matters precisely because it is ignored by the very fragmented regulatory and institutional environment within which the invasive species problem is addressed. Indeed, this is where many of the gaps and inconsistencies identified by the CBD come from. Aside from these two messages, Part 1 also underlines the limitations on our capacity to predict biological invasions from the data on drivers, and hence the importance of learning—a theme picked up in subsequent chapters.

Part 2 comprises a set of five chapters that explore the economics of the problem, focusing on the main elements of the decision problem facing environmental authorities at the national level. A dominant theme of these chapters is the treatment of uncertainty. Given the fundamental uncertainty associated with novelty in trade goods, trade routes, landscape change, and so on, several chapters ask what the right way to deal with the problem may be. Simpson argues forcefully that reversing global integration is not desirable even if it were feasible, and that a better option is to control the high-risk species introduced alongside trade goods. This raises a series of questions about the optimal control of such species. The question of how to identify problematic species at the port of entry is explored by Springborn et al. The question of whether to control an invasive species at the port of entry or beyond, and what resources to commit to control, is explored by Polasky. The question of whether control is better directed to the structure of the landscape is explored by Touza *et al.*, and the question whether to target the current or steady-state abundance of invasive species is addressed by Finnoff *et al.*

One key message coming out of Part 2 is that the rate of growth and spread of an invasive species are critical parameters in the development of efficient (or cost effective) control strategies. Finnoff *et al.*, for example, show that control at the steady-state abundance of the invasive species is a "good enough" strategy for fast growing species, whereas for slow-spreading species it is necessary to understand the transition dynamics. At the same time, as Perrings *et al.* show in Part 1, inspection and interception effort will be decreasing along an optimal trajectory wherever species are "slow-growing" relative to the economy. A second key message is that while prevention is almost always better than cure, there exist a range of possible strategies that, at least in Part 2, extends from inspection and interception all the way to control of the spatial structure of an invaded landscape. Just what strategy dominates depends on the costs and benefits of different forms of intervention.

Part 3 comprises a set of papers on different aspects of the management and policy problems. All are informed by just this point: that strategy choice depends on the costs and benefits of different forms of intervention. Indeed, as Pejchar-Goldstein and Mooney point out, identifying the costs and benefits of different options is an essential first step. As in Part 2, the range of options considered varies widely. Keller and Lodge, Uma Shaanker *et al.*, and van Wilgen and Richardson all consider the options open to the nation state. Keller and Lodge focus on species introductions and take the cost-effectiveness of preventive measures as their starting point. Uma Shaanker *et al.* focus on established invasive species, and assume that the costs of eradication are prohibitive. Van Wilgen and Richardson are in between. They pose the following question: why should control measures be delayed until the costs become prohibitive? They identify the problem as lying in the perceived costs and benefits of control options, but connect the decisions made at each stage with the quality of information available to the decision-maker. At the point where rapid intervention and eradication might be efficient, decision-makers frequently do not have information on either the ultimate damage or control costs of invasive species and hence choose to do nothing. The authors observe that once a species has established and spread, control will not be selected unless the damage costs are understood to be extremely high—as with the problem being tackled by the Working for Water project in South Africa.

In the final chapter in Part 3 we return to the original motivation for this volume, the gaps and inconsistencies in the international regulatory regime identified by the CBD. Most chapters focus on national defensive measures. This is a default strategy. It is what is allowed under Article XX of the General Agreement on Tariffs and Trade (GATT) and the Sanitary and Phytosanitary (SPS) Agreement (1994). Both agreements authorize unilateral actions to address national risks associated with imports from individual countries, but ignore both the fact that many countries lack the information or capacity to take advantage of that authority, and also the effect of those actions on third countries. We contrast this position with the International Health Regulations (2005) (IHR) which directly focus on global public health risks. The IHR not only authorize collective action to mitigate such risks at the global level, they also provide for resource flows to low income countries to help build capacity to implement the regulations. The fact that the main authorizing agreement for action on trade-related invasive species risks ignores the global public good is the most fundamental gap in the international regulatory regime. Indeed, two key messages from Part 3 are that the SPS Agreement be brought into conformity with the IHR, and that a body be established to undertake the monitoring and risk assessment needed to enable all countries to take advantage of the provisions of the SPS Agreement.

Despite the growing focus on climate change, it is hard to believe that there is any other environmental problem of greater importance than the one addressed in this volume. It may not be as visible as many other problems, in part because expenditures on invasive species and their wider impacts are spread across a large number of sectors—health, agriculture, fisheries, forestry, and conservation to name but a few. But it is likely that the total impact of invasive pests and pathogens on human well-being is greater than the impact of almost any other environmental issue. There is certainly no greater global environmental threat than that posed by emerging zoonotic diseases. Nor is there another environmental issue where persistent failure to address the global dimensions of the problem carries a higher cost. This should be sufficient to trigger both the international coordination needed to resolve the problem, and the science needed to inform that effort.

The preparation of this volume was supported by the NSF-funded research coordination network, Biodiversity and Ecosystem Services Training Network (BESTNet) and the ecoSERVICES Core project of DIVERSITAS. We gratefully acknowledge that support. We also acknowledge the contributing authors and external reviewers, the support of Ian Sherman and Helen Eaton of Oxford University Press, and research assistance provided by Gustavo Garduno and Qiong Wang.

March 14, 2009
Charles Perrings
Harold Mooney
Mark Williamson

Contents

Acronyms		xv
List of Editors and Contributors		xvii
1	**The Problem of Biological Invasions**	**1**
	Charles Perrings, Harold Mooney, and Mark Williamson	
	1.1 Introduction	1
	1.2 The ecological dimension	3
	1.3 The economic problem	7
	1.4 Policy and management options	9
	1.5 Recognizing the global dimension of the problem	12

PART I THE DRIVERS OF BIOLOGICAL INVASIONS

2	**Climate Change and Species' Distributions: An Alien Future?**	**19**
	Chris D. Thomas and Ralf Ohlemüller	
	2.1 Introduction	19
	2.2 Natives and aliens	19
	2.3 Future distributions will be very different	20
	2.4 What are "invasive species"?	21
	2.5 Climatic invasibility	22
	2.6 The invasion–climate change interaction	23
	2.7 Love thine enemies—but not all of them!	26
3	**Climate and Invasive Species: The Limits to Climate Information**	**30**
	Mark New and Carol McSweeney	
	3.1 Introduction	30
	3.2 Climate observations	31
	3.3 Climate models	33
	3.4 Emissions and other forcing of climate	35
	3.5 Approaches to the use of uncertain climate information	35
	3.5.1 Scenarios	36
	3.5.2 Probabilistic projections	36
	3.5.3 A middle ground?	36
	3.6 Discussion	37

4 Globalization and Invasive Alien Species: Trade, Pests, and Pathogens — 42
Charles Perrings, Eli Fenichel, and Ann Kinzig

4.1 Identifying the problem — 42
4.2 Species dispersal and inspection and interception effort — 44
4.3 Data and methods — 46
4.4 Empirical results — 49
4.5 Discussion — 52

5 Variation in the Rate and Pattern of Spread in Introduced Species and its Implications — 56
Mark Williamson

5.1 Introduction — 56
5.2 Variation in spread rates within species — 56
5.3 Variation in pattern and rate of spread between species — 59
5.4 Time to complete spreading — 61
5.5 Implications and conclusions — 63

6 Habitats and Land Use as Determinants of Plant Invasions in the Temperate Zone of Europe — 66
Petr Pyšek, Milan Chytrý, and Vojtěch Jarošík

6.1 Introduction — 66
6.2 Overview of studies on the level of invasion in habitats — 66
    6.2.1 Representation of alien species in habitats — 67
    6.2.2 Which habitats are most invaded? — 67
    6.2.3 Importance of scale — 68
    6.2.4 Native–alien relationship — 68
6.3 Theoretical background of community invasibility — 69
6.4 Separating the level of invasion from invasibility — 70
    6.4.1 Comparison of the levels of habitat invasions with their invasibility — 70
    6.4.2 Relative importance of factors determining the level of invasion — 74
    6.4.3 Habitat vs. propagule limitation and methodological pitfalls — 75
6.5 Habitat-based mapping of plant invasions in Europe and prediction of future trends: the next step? — 75
6.6 Conclusions — 76

PART II ECONOMICS

7 If Invasive Species are "Pollutants", Should Polluters Pay? — 83
R. David Simpson

7.1 Introduction — 83
7.2 Assumptions and literature — 85
7.3 Some revealing figures — 88
7.4 A simple model of invasive species with nonconvex damages — 90
7.5 Elaborations and extensions — 93
7.6 Implications for policy — 95
    7.6.1 When is aggressive policy appropriate? — 95
    7.6.2 Are there any easy diagnostics? — 95

	7.6.3 How can abuse be prevented?	96
7.7	Conclusion	97

8 A Model of Prevention, Detection, and Control for Invasive Species — 100
Stephen Polasky

8.1 Introduction	100
8.2 The model	102
8.3 Analysis	104
8.3.1 General solution	105
8.3.2 Numerical example	105
8.4 Discussion	107

9 Second Best Policies in Invasive Species Management: When are they "Good Enough"? — 110
David Finnoff, Alexei Potapov, and Mark A. Lewis

9.1 Introduction	110
9.2 Invasion dynamics	111
9.3 Decision models	112
9.3.1 Second best rule	113
9.3.2 First best rule	117
9.3.3 Comparison of the two rules	119
9.4 Discussion	124
9.5 Conclusions	124

10 Optimal Random Exploration for Trade-related Nonindigenous Species Risk — 127
Michael Springborn, Christopher Costello, and Peyton Ferrier

10.1 Introduction	127
10.2 A Bayesian learning model of trade-related NIS risk	129
10.3 The inspection allocation decision problem	130
10.4 Allocation method and results	132
10.4.1 The tractable inspection allocation decision problem	132
10.4.2 A fully dynamic model and non-constant exploration policy	135
10.5 Discussion	142

11 The Role of Space in Invasive Species Management — 145
Julia Touza, Martin Drechsler, Karin Johst, and Katharina Dehnen-Schmutz

11.1 Introduction	145
11.2 Spatial aspects of the economics of biological invasions	146
11.2.1 Spatially implicit models	146
11.2.2 Spatially differentiated models	148
11.2.3 Spatially explicit models	149
11.2.4 Spatially explicit and differentiated models	149
11.3 Modeling invasions in spatially heterogeneous systems	150
11.3.1 The management problem	150
11.3.2 The model	151
11.3.3 Analysis and results	152
11.4 Discussion and conclusions	154

PART III MANAGEMENT AND POLICY

12 The Impact of Invasive Alien Species on Ecosystem Services and Human Well-being — 161
Liba Pejchar and Harold Mooney

- 12.1 Introduction — 161
  - 12.1.1 Invasive alien species and global change — 161
  - 12.1.2 Quantifying impacts of invasive alien species — 161
  - 12.1.3 Defining ecosystem services — 162
- 12.2 Mechanisms — 162
- 12.3 Ecosystem services — 163
  - 12.3.1 Provisioning — 163
  - 12.3.2 Regulating Services — 166
  - 12.3.3 Cultural Services — 171
- 12.4 Case studies — 173
- 12.5 Research and policy recommendations — 173

13 Current and Future Consequences of Invasion by Alien Species: A Case Study from South Africa — 183
B.W. van Wilgen and D.M. Richardson

- 13.1 Introduction — 183
- 13.2 Invasive species in South Africa — 184
  - 13.2.1 Plants — 184
  - 13.2.2 Mammals — 184
  - 13.2.3 Birds — 186
  - 13.2.4 Reptiles and amphibians — 186
  - 13.2.5 Terrestrial invertebrates — 186
  - 13.2.6 Freshwater biota — 187
  - 13.2.7 Marine organisms — 187
  - 13.2.8 Pathways of invasion — 187
- 13.3 Current impacts of invasive species in South Africa — 188
  - 13.3.1 Impacts on water resources — 188
  - 13.3.2 Impacts on grazing resources — 189
  - 13.3.3 Impacts on biodiversity in terrestrial ecosystems — 189
  - 13.3.4 Impacts on fire regimes and erosion — 190
  - 13.3.5 Impacts on human health and safety — 190
  - 13.3.6 Impacts on freshwater aquatic ecosystems — 190
  - 13.3.7 Economic assessments of impact — 191
- 13.4 Future impacts — 193
  - 13.4.1 Estimates of magnitude of future impacts — 193
  - 13.4.2 Motivating for control on the basis of expected benefits — 193
- 13.5 Conclusions — 197

14 Invasive Plants in Tropical Human-Dominated Landscapes: Need for an Inclusive Management Strategy — 202
R. Uma Shaanker, Gladwin Joseph, N.A. Aravind, Ramesh Kannan, and K.N. Ganeshaiah

- 14.1 Introduction — 202

	14.2 Dynamics of invasive species and management scenarios	204
	14.2.1 How does management scale with the dynamics of invasive species?	204
	14.2.2 When control of invasive species fails: three examples	207
	14.3 The specific case of *Lantana* in India	208
	14.3.1 Invasion, spread, and ecosystem impacts	208
	14.3.2 When you cannot break it, at least bend it!	212
	14.4 The management of uncontrollable invasive species: from exclusion to inclusion	214

15 Prevention: Designing and Implementing National Policy and Management Programs to Reduce the Risks from Invasive Species   220
Reuben P. Keller and David M. Lodge

	15.1 Introduction	220
	15.1.1 Economic and ecological context for invasive species prevention	220
	15.2 Prevention	222
	15.2.1 Risk assessment for prevention of introductions via commerce in live organisms	223
	15.2.2 Risk assessment for prevention of transportation-related introductions	228
	15.3 Preventing invasive species: opportunities for improved risk assessments	229
	15.3.1 Quantitative vs. qualitative approaches	229
	15.3.2 What is an appropriate risk threshold?	230
	15.3.3 Pre-border vs. border prevention	230
	15.3.4 Resource availability	231
	15.3.5 Gaps and overlaps among government agencies	231
	15.3.6 Species spread across borders	232
	15.4 Conclusions	232

16 Globalization and Bioinvasions: The International Policy Problem   235
Charles Perrings, Stas Burgiel, Mark Lonsdale, Harold Mooney, and Mark Williamson

	16.1 Dimensions of the policy problem	235
	16.2 The institutional environment	237
	16.3 Provide the public good at the right geographical scale	239
	16.4 Precautionary action should be targeted, and support learning	241
	16.5 Many invasive species problems may be most effectively managed at the regional scale	243
	16.6 Pay special attention to the impacts of invasive species on the poor	245
	16.7 Conclusions	247

Appendix 1: Agreement on the Application of Sanitary and Phytosanitary Measures (1995), Articles 1–11   251

	Article 1	General provisions	251
	Article 2	Basic rights and obligations	251
	Article 3	Harmonization	251

Article 4	Equivalence	252
Article 5	Assessment of risk and determination of the appropriate level of sanitary or phytosanitary protection	252
Article 6	Adaptation to regional conditions, including pest- or disease-free areas and areas of-low pest or disease prevalence	253
Article 7	Transparency	254
Article 8	Control, inspection and approval procedures	254
Article 9	Technical assistance	254
Article 10	Special and differential treatment	254
Article 11	Consultations and dispute settlement	254

Appendix 2: International Health Regulations (2005) Articles 2, 5–13 **256**

Article 2	Purpose and scope	256
Article 5	Surveillance	256
Article 6	Notification	256
Article 7	Information-sharing during unexpected or unusual public health events	257
Article 8	Consultation	257
Article 9	Other reports	257
Article 10	Verification	257
Article 11	Provision of information by WHO	258
Article 12	Determination of a public health emergency of international concern	258
Article 13	Public health response	259

Index **261**

Acronyms

AFTA	ASEAN Free Trade Area
ALARM	Assessing large-scale risks for biodiversity with tested methods
APHIS	Animal and Plant Health Inspection Service
AQIM	US Agricultural Quarantine Inspection Monitoring
ASEAN	Association of South East Asian Nations
BSE	Bovine spongiform encephalopathy
CARICOM	Caribbean Community
CBD	Convention on Biological Diversity
CDC	Center for Disease Control
CITES	Convention on International Trade in Endangered Species
CMS	Convention on Migratory Species
COMESA	Common Market for Eastern and Southern Africa
ENRM	Environmental and natural resource management
EPA	United States Environmental Protection Agency
EU	European Union
FAO	Food and Agriculture Organization
FMD	Foot and mouth disease
GATT	General Agreement on Tariffs and Trade
GCM	General Circulation Model
GCVE	Generalized Cross Validation Error
HESCO	Himalayan Environment Studies and Conservation Organization
IAS	Invasive Alien Species
IATA	International Air Transport Association
ICAO	International Civil Aviation Organization
ICRAF	International Center for Research in Agroforestry
IDRC	International Development Research Centre
IHR	International Health Regulations
IIP	Inspection and interception programs
IMO	International Maritime Organization
IPCC	Intergovernmental Panel on Climate Change

IPPC	International Plant Protection Convention
IUCN	International Union for Conservation of Nature
MA	Millennium Ecosystem Assessment
MERCOSUR	Southern Common Market (Argentina, Brazil, Paraguay and Uruguay)
NAS	National Academy of Sciences
NAFTA	North American Free Trade Association
NEPAD	New Partnership for Africa's Development
NIS	Non-indigenous species
NISC	National Invasive Species Council
NRC	National Research Council
NSF	National Science Foundation
OIE	World Organisation for Animal Health
OTA	Office of Technology Assessment
PCSEs	Prais–Winsten regression with corrected standard errors
PII	Pacific Invasives Initiative
PILN	Pacific Invasives Learning Network
RCD	Rabbit Haemorrhagic Disease
RTA	Regional Trade Agreement
SADC	Southern African Development Community
SCOPE	Scientific Committee on Problems in the Environment
SITC	Standard International Trade Classification
SPC	Secretariat of the Pacific Community
SPREP	South Pacific Regional Environment Programme
SPS	Sanitary and Phytosanitary
TBT	Technical Barriers to Trade
TRMM	Tropical Rainfall Monitoring Mission
UN	United Nations
UNCLOS	UN Convention on the Law of the Sea
UNEP	United Nations Environment Program
USDA	United States Department of Agriculture
USDOC	United States Department of Commerce
WHO	World Health Organization
WRA	Weed Risk Assessment
WTO	World Trade Organization

List of Contributors

Aravind, N. A. Ashoka Trust for Research in Ecology and the Environment (ATREE), #659, 5th A Main, Hebbal, Bangalore 560024, India.

Burgiel, Stas. Global Invasive Species Team, The Nature Conservancy, Arlington, VA 22203-1606, USA.

Chytrý, Milan. Department of Botany and Zoology, Masaryk University, Kotlářská 2, CZ-611 37 Brno, Czech Republic.

Costello, Christopher. Bren School of Environmental Science and Management, University of California, Santa Barbara, CA 93106-513, USA.

Dehnen-Schmutz, Katharina. University of Warwick, Warwick HRI, Warwick, CV35 9EF, UK.

Drechsler, Martin. Department of Ecological Modelling, Helmholtz Centre of Environmental Research – UFZ, Permoserstr. 15, Leipzig 04318, Germany.

Fenichel, Eli. School of Life Sciences, Arizona State University, Tempe, AZ 85287, USA.

Ferrier, Peyton. Economic Research Service, USDA, 1800 M Street, Washington, DC 200365831, USA.

Finnoff, David. Dept. of Economics and Finance, University of Wyoming, 1000 E. University Avenue, Laramie, WY, 82071.

Ganeshaiah, K. N. Forestry and Environmental Sciences and School of Ecology and Conservation, University of Agricultural Sciences, GKVK Campus, Bangalore 560065, India.

Jarošík, Vojtěch. Department of Ecology, Faculty of Science, Charles University, Viničná 7, CZ-128 01 Praha 2, Czech Republic.

Johst, Karin. Department of Ecological Modelling, Helmholtz Centre of Environmental Research – UFZ, Permoserstr. 15, Leipzig 04318, Germany.

Joseph, Gladwin. Ashoka Trust for Research in Ecology and the Environment (ATREE), #659, 5th A Main, Hebbal, Bangalore 560024, India.

Kannan, Ramesh. Ashoka Trust for Research in Ecology and the Environment (ATREE), #659, 5th A Main, Hebbal, Bangalore 560024, India.

Keller, Reuben P. Department of Biological Sciences, University of Notre Dame, Notre Dame, IN 46556, USA.

Kinzig, Ann. School of Life Sciences, Arizona State University, Tempe, AZ 85287, USA.

Lewis, Mark A. Department of Mathematical and Statistical Sciences and Centre for Mathematical Biology, University of Alberta, Edmonton, AB, T6G 2G1 Canada.

Lodge, David M. Department of Biological Sciences, University of Notre Dame, Notre Dame, IN 46556, USA.

Lonsdale, Mark. CSIRO Entomology, Black Mountain Laboratories, GPO Box 1700, Canberra ACT 2601, Australia.

McSweeney, Carol. School of Geography and the Environment, University of Oxford, Oxford, OX1 3QY, UK

Mooney, Harold. Department of Biological Sciences, Stanford University, Stanford, CA 94305, USA.

New, Mark. Tyndall Centre for Climate Change Research and School of Geography and the Environment, University of Oxford, Oxford, OX1 3QY, UK

Ohlemüller, Ralf. Institute of Hazard and Risk Research (IHRR) and School of Biological and

Biomedical Sciences, Durham University, South Road, Durham DH1 3LE, UK.

Perrings, Charles. School of Life Sciences, Arizona State University, Tempe, AZ 85287, USA.

Pejchar, Liba. Department of Fish, Wildlife and Conservation Biology Campus Delivery 1474 Colorado State University Fort Collins, CO 80523, U.S.A.

Polasky, Stephen. Department of Applied Economics and Department of Ecology, Evolution and Behavior, University of Minnesota, St. Paul, MN 55108, USA.

Potapov, Alexei. Dept. of Mathematical and Statistical Sciences and Centre for Mathematical Biology, University of Alberta, Edmonton, AB, T6G 2G1 Canada.

Pyšek, Petr. Institute of Botany, Academy of Sciences of the Czech Republic, CZ-252 43 Průhonice, Czech Republic.

Richardson, David M. Centre for Invasion Biology, Department of Botany and Zoology, Stellenbosch University, Private Bag X1, Matieland, 7602, South Africa.

Shaanker, R. Uma. Department of Crop Physiology and School of Ecology and Conservation, University of Agricultural Sciences, GKVK Campus, Bangalore 560065, India.

Simpson, R. David. National Center for Environmental Economics, US Environmental Protection Agency, 1200 Pennsylvania Ave, N.W., Washington, DC 20460.

Springborn, Michael. Department of Environmental Science and Policy, University of California, 1 Shields Avenue, Davis, CA 95616, USA.

Thomas, Chris D. Department of Biology, University of York, York YO10 5DD, UK.

Touza, Julia. Department of Applied Economics, University of Vigo, Lagoas-Marcosende s/n, Vigo 36310, Spain.

van Wilgen, Brian W. CSIR Natural Resources and the Environment, PO Box 320, Stellenbosch, 7599, South Africa.

Williamson, Mark. Department of Biology, University of York, York YO10 5DD, UK.

CHAPTER 1

The Problem of Biological Invasions
Charles Perrings, Harold Mooney, and Mark Williamson

1.1 Introduction

Invasive species are defined by the Convention of Biological Diversity as "...those alien species which threaten ecosystems, habitats or species". From biblical times biological invasions have been an ever-present threat to human life and livelihood. In many cases they have been shattering in their impact. They matter. The bubonic plague that swept through Europe in the fourteenth century, the Black Death, caused human populations to decline for more than a hundred years, and outbreaks of plague continued to take a toll until the eighteenth century. For 300 years it was "normal" for outbreaks to kill up to half the population of infected cities. Outbreaks in Venice in 1575–77 and 1630–31, for example, killed more than a third of the inhabitants, as did outbreaks in Seville in 1647–1652, London 1664–65, Vienna in 1679 and Marseilles in 1720–21. The first global pandemic, the "Spanish flu" of 1918–20 killed up to 100 million people, more than twice the number killed in the whole of the 1914–1918 World War (McNeill 1977). More recently, HIV AIDS has killed around 22 million people worldwide since the 1980s.

Writing at the end of the eighteenth century, Malthus observed that a combination of war and disease was the dominant check to population growth in most parts of the world:

The frequency of wars, and the dreadful devastations of mankind occasioned by them, united with the plagues, famines and mortal epidemics of which there are records, must have caused such a consumption of the human species that the exertion of the utmost power of increase must, in many cases, have been insufficient to supply it... (Malthus 1970: 253)

His *Essay on the Principles of Human Population* was first about the role of the environment in limiting population growth. But it was also about the interactions between both drivers and checks on population. Others have since noted the close coincidence of these causes. War and disease naturally go together (Diamond 1997, 2005). War and conflict may be either a cause or an effect of human population movement, but it has always been a theater in which people have been brought into direct contact with one another, and so a theater for the transmission of human pathogens. The fact that war is also a time of stress has increased the interdependence.

Invasive species have had much more than a demographic impact, though. The Columbian exchange—the transmission of species across the Atlantic first through Spanish conquistadores and later through the slave trade in the sixteenth and seventeenth centuries—certainly had profound demographic effects. In the century after Cortez had initiated the Columbian exchange the population of central Mexico was reduced to little more than 5 per cent of its pre-Columbian levels due to the effects of a suite of European diseases, especially smallpox, influenza, measles and typhus. At the same time, however, the Columbian exchange completely disrupted pre-Columbian societies, and their capacity to resist the conquistadores. Many of the social and political systems of the New World were effectively destroyed as populations collapsed. So too were many of the ecological systems.

Human pathogens were not the only species involved in the exchange. The introduction of many European plants and animals, along with their pathogens, frequently transformed native American landscapes and displaced native American species, fundamentally changing both the character of the physical environment and the ways in which people interacted with it (McNeill 1977).

Invasive human pathogens are the species that have had the most profound effects on human well-being, but other species have not been far behind. Pathogens that affect the production of crops or livestock have been particularly costly. Although rinderpest (cattle plague) may now have been driven to extinction (Normile, 2008), it has historically been especially destructive. Three protracted epidemics in the eighteenth century, in 1709–20, 1742–60, and 1768–86, severely affected meat production in Europe—mortality approaching 100 per cent in immunologically naive populations. Nor have other regions been spared. A major epidemic in Africa in the 1890s is thought to have killed up to 90 per cent of all cattle as well as a number of other ruminants in Southern Africa and the Horn of Africa. The dollar value of the damage done by such outbreaks reflects prevailing conditions, including the incomes of the people most directly affected. Whereas a rinderpest epidemic that affected many countries in Africa in 1982–84 led to stock losses valued at only US$500 million, the 2001 outbreak of foot and mouth disease in the UK is estimated to have cost in the order of US$16 billion (both measured in 2004 dollars). But the impact on livelihoods reflects the extent to which people depend on livestock production. In many of the poorer regions of the world 70–90 per cent of the population still depend on agriculture for their livelihood.

Countering pests of one kind or another has historically been a major part of the cost of agriculture. In some cases there has been no effective response, and losses have—at least for a time—been total. In the late nineteenth century, for example, Phylloxera (*Daktulosphaira vitifoliae*), an insect similar to the aphid and of North American origin, destroyed most of the vineyards of Europe. While it is extremely difficult to estimate the cost of damage due to invasive agricultural pests and pathogens, it is clear that they remain substantial. Part of this is a direct effect. One estimate of the effect of invasive species on crop yields in poorer countries in the 1990s suggested that they may be around 50 per cent (Oerke *et al.* 1994). Invasive pests and pathogens that have had particularly severe effects on crop yields in the world's poorest region, Sub-Saharan Africa, include witchweed (*Striga hermonthica*), grey leaf spot (*Circosporda zeae-maydis*), the large grain borer (*Prostephanus truncatus*), cassava mealybug (*Phenacoccus manihoti*), and the cassava green mite (*Mononychellus tanajoa*) (Rangi 2004).

In fact, if we take the expenditure to counter pests in agriculture, forestry, and fisheries along with the output lost through invasive pests and pathogens in all sectors, and add to that the costs of both emergent and recurrent human diseases of international origin, the problem of invasive species makes all other environmental problem pale into insignificance. We collectively commit more resources to protecting ourselves against invasive pests and pathogens, and we lose more in productivity, health, and security from the same sources, than from any other environmental stress. To be sure, the various dimensions of the cost of invasive species are seldom put together—in large part because they are institutionally separate in modern governance structures. Agricultural pests and pathogens are addressed separately from human pathogens, or from plant or animal species that invade wider ecosystems. Yet the evidence is mounting that these things are not unconnected. SARS emerged as novel zoonotic disease of humans, for example, as a result of closer interactions between bats and livestock on Chinese farms (Dobson 2005).

Whilst military campaigns may not have the role in the transmission of human diseases that they once did, disease transmission still depends on the movement of people and goods. Trade, transport, and travel are the mechanisms that drive species introductions (Perrings *et al.* 2005). This book focuses on the global changes that lie behind a dramatic acceleration in the number of species introductions, the consequences of this trend for human livelihoods and well-being, and the options for dealing with it. The global changes we consider are both environmental (changes climate, land use, and land cover) and economic (changes in the volume, composition, and direction of trade). In recent years the growth of world trade relative to output indicates an increasingly integrated global economy. Indeed, the closer integration of the global economy is the defining characteristic of the process of globalization. It is also one of the main drivers of species dispersal. While climate change alters the natural range of species, the growth of trade affects the number and frequency of new introductions, and hence the likelihood that species will establish and spread (Enserink 1999). The link between trade and the introduction of new pest species is a long-standing problem. The development of the nineteenth century wool trade between the UK and Australia, for example, introduced a set of over 300 passenger species to the River Tweed in the UK (Salisbury, 1961)—the so-called "wool species"—three of which still survive and one of which remains a pest (M. Braithwaite *pers.com.*). What is new is the rate at which new introductions are occurring (McNeeley 2001).

This volume considers both the factors behind the growth of invasive species as a global problem, and the options for addressing that problem. These options are limited by the terms of the various international agreements governing trade, health, and biodiversity, and especially by the agreement that most directly deals with trade-related invasive species risks, the Sanitary and Phytosanitary Agreement. The default strategy in most cases has two parts: border protection and the control of, or adaptation to, introduced species that have escaped detection at the border. For that reason, most invasive species policy involves unilateral national defensive action as opposed to coordinated international action, and that is reflected in the chapters of this book. We conclude that this is not enough. The solution to a problem that is driven by globalization ultimately lies in global coordination and cooperation in the management of both pathways and sanitary and phytosanitary risks at all scales. It is still important to improve national capacity to defend against the threats posed by introduced species, but we need to do more.

1.2 The ecological dimension

The definition of invasive species embedded in the Convention on Biological Diversity emphasizes the harm that they do. This is partly because the signatories to the convention see no benefit in countering invasive species that have no discernible impact on either ecological functioning or human well-being. But it also serves as a reminder that inspection, interception, quarantine, eradication, and control all have an opportunity cost that needs to be balanced against the gains of those actions—the damage they avoid. That is, the problem of invasive species is as much an economic problem as an ecological problem.

Taking the two separately for the moment, since Elton (1958) launched the ecological literature on invasive species, the field has tended to focus on two things: the factors involved in the risk that introduced species will establish and spread, and the ecological consequences if they do. Risk factors in the spread of introduced species include the characteristics of both the introduced species themselves and of the source and sink systems: where they have come from and where they are going to. The first includes the set of traits—plasticity, generalism, and so on—which make species more or less invasive. The second includes the characteristics of the ecosystems into which species are being introduced—the effects of fragmentation, disturbance, biodiversity loss, bioclimatic distance from the source country, as well as the existence of predators or competitors of the introduced species—which make those systems more or less invasible.

Of these, it is the characteristics of ecosystems and not traits that are most subject to change, and that have been most affected by global economic and environmental trends. The biodiversity in host ecosystems, the degree to which those systems have been fragmented or transformed, and the range of anthropogenic stresses on them are all changing rapidly. The Millennium Ecosystem Assessment (MA) concluded that:

"...human actions are fundamentally, and to a significant extent irreversibly, changing the diversity of life on earth and most of these changes lead to a loss of biodiversity. Changes in important components of biological diversity were more rapid in the past 50 years than at any time in human history." (MA 2005)

This is partly due to the accelerating conversion of ecosystems for the production of foods, fuels, and fibers, but also to the more generalized consequences of anthropogenic stress. The MA noted, for example, that more land was converted to cropland in the 30 years after 1950 than in the 150 years between 1700 and 1850; that less than 20 per cent of many biomes remained as viable habitat for species; that the loss of habitat had led to increases in species extinction rates by as much as 1000 times over background rates; and that the capacity of ecosystems to buffer shocks and stresses had been systematically reduced through loss of wetlands, forests, and mangroves. The net effect has been a systematic reduction in the resilience of ecosystems, particularly with respect to the shocks posed by invasive species.

Moreover, while the effects of anthropogenic stress have been especially acute in terrestrial systems, marine systems have also been severely impacted. There is strong evidence that the resilience of coral reefs (Hughes *et al.* 2003; Hughes *et al.* 2005) and kelp forests (Steneck *et al.* 2004) has been affected in ways very similar to many terrestrial systems, and for similar reasons. One common factor across both marine and terrestrial systems is, for example, the impact of changes in species diversity on the level of functional redundancy (Diaz *et al.* 2003). The most frequently cited cause of marine stress in pelagic and epipelagic systems is overfishing (Pauly *et al.* 2002; Myers and Worm 2003; Hughes *et al.* 2005), with bycatch (Lewison *et al.* 2004), loss of habitat (Pandolfi 2003), and climate change being contributory factors (Hughes *et al.* 2003). The MA (2005) noted that the number of invasive marine species has been increasing exponentially, and the spread of pathogens has been identified as a source of stress in marine and terrestrial systems alike (Harvell *et al.* 2004).

The dimensions of this problem most closely investigated in this volume are the impacts on the invasibility of ecological economic systems of (a) climate change (Thomas and Ohlemüller, Chapter 2 this volume) and (b) the relationship between ecosystem type and anthropogenic disturbance (Pyšek *et al.*, Chapter 6 this volume). Climate change is important for a number of reasons. First, it alters the "natural" distribution of species, their predators, and their pathogens. Second, it alters the ways in which humans interact with and transform their physical environments. Third, it alters the range of conditions over which ecosystems function—since changes in mean precipitation and temperature are expected to be accompanied by changes in both the second moment (the variance) and the tails of their distribution. A common projection in areas that are expected to get warmer and drier, for example, is that both precipitation and temperature are also expected to become more variable, and that the incidence of extreme events is expected to increase (IPCC 2007). Using one set of climate projections from the Hadley Centre, for example, Thomas and Ohlemüller argue that there is likely to be a significant increase in the area that is vulnerable to invasive species. This is because of changes in the distribution of bioclimatically similar but geographically distant areas—the distance between areas that are bioclimatically similar being a proxy for the likelihood that a species moving between them will encounter the same set of predators, competitors, or pathogens.

The relationship between ecosystem type and anthropogenic disturbance is important because of the evidence that the invasibility of ecosystems responds to both things. Pyšek *et al.* observe that some ecosystems are intrinsically more vulnerable to invasions than others, but the vulnerability of all ecosystems tends to increase with anthropogenic fragmentation, disturbance, biodiversity change, and the like. It follows that the same activities undertaken in different ecosystems might be expected to have very different consequences for the invasibility of those systems. It is important to understand how management actions affect invasibility, and how this varies between ecosystem types. Habitat loss through land use change affects the invasibility of the impacted ecosystems (Polasky *et al.* 2004). But less intrusive habitat changes can have similar effects. Fragmentation through road construction, the elimination of buffer zones, or an increasing pollution burden can also affect the vulnerability of ecosystems, although this effect will vary between ecosystem types.

Intervention in ecosystems also has a variable effect on the kind of species to which the system becomes vulnerable. Interventions that create habitat suited to particular species increase the likelihood that those species will establish and spread if introduced. Symmetrically, interventions that create habitat unsuited to particular species decrease the likelihood that those species will establish and spread if introduced. Indeed, this is the basis not just for conservation, but also for all agriculture, forestry, aquaculture, and for *in situ* sanitary and phytosanitary measures. One reason for the increasing incidence of malaria, for example, is the abandonment in many countries of public health strategies aimed at reducing the area of suitable habitat for the vector *Anopheles* spp. Interventions designed to select for one species also make the system more vulnerable to other species—competitors, predators, pathogens, symbionts, or commensals—associated with the targeted species. So, for example, the production of particular crops makes agricultural systems more vulnerable to weeds (competitors) and to crop pests and diseases (predators).

Another dimension of the same problem is that anthropogenic changes to ecosystems affect not just their vulnerability to the establishment and naturalization of introduced species, but also the rate at which naturalized species spread. In Chapter 5 of this volume, Williamson draws attention to the wide variation in the rate of spread of different species, and while this partly reflects differences in the traits of the species themselves, it also reflects differences in the structure of the landscapes. There is no systematic relation involved, though. The existence of roads creates corridors for species able to exploit the conditions along roadsides, and so encourages their spread (Christen and Matlack 2009). But roads may also be barriers to the spread of other species (see Niemelä *et al.* 2007 for these effects in carabid beetles). More generally, the structure of human dominated landscapes favors the spread of some species whilst discouraging the spread of others. Williamson cites the example of the cane toad (*Chaunus marinus*) in Australia, which spread at approximately 10 kilometers per year in the sugar cane fields of Queensland, but much more rapidly in the less structured, more "natural" landscapes of the Northern Territories (Phillips *et al.* 2006).

A second focus of the ecological literature on invasive species has been the ecological consequences of their introduction. The damage due to invasive species is largely a function of their impact on the systems they invade. This involves both direct and indirect impacts. The damage due to pathogens, for example, involves both the harm suffered by infected species and the indirect effects of changes in their abundance or behavior. In many cases these direct impacts are substantial. Introduced pathogens, predators, or competitors have all been implicated in the loss of native species over a wide range of ecosystems (Williamson 1996). Indeed, invasive species are commonly cited as the second most important cause of biodiversity loss after habitat conversion (Glowka *et al.* 1994). A number of species of frogs, toads, newts, salamanders, and caecilians, for example, either have already been driven to extinction or are at risk of extinction due to the effects of disease (Daszak and Cunningham 1999). Indeed, a parasitic fungus, amphibian chytrid (*Batrachochytrium dendrobatidis*), is one of the most severe current threats to the survival of amphibians in many systems. In Australia, for example, the disease has been implicated in four frog extinctions, and has put many more at risk. It also involves both actual and potential impacts.

Even if disease has much weaker impacts on the abundance of infected species, however, it may still have substantial indirect impacts. Invasive species are important for the effect they have on the functioning of ecosystems, frequently via their capacity to displace existing species through competition or predation in particular environmental conditions. A reduction in the abundance of members of a particular functional group, for example, has implications for the sensitivity to environmental conditions of the supply of the ecosystem services associated with that function. More particularly, it affects the capacity of the system to absorb anthropogenic and environmental stresses and shocks without losing resilience (Kinzig *et al.* 2006; Scheffer *et al.* 2000; Walker *et al.* 2004; Walker *et al.* 2006). Maintenance of functional diversity, in particular, supports the provision of ecosystem services over a range of environmental conditions (e.g. Loreau *et al.* 2003; Naeem and Wright 2003; Reich *et al.* 2004; Hooper *et al.* 2005). While the introduction of invasive pests may not immediately impact system performance, it can make it less able to cope with future variation in environmental conditions.

In the language of the Millennium Ecosystem Assessment, this implies a reduction in the regulating services of ecosystems: that is, in the security with which they are able to supply the things that people care about—the provisioning and cultural services. Of course, invasive species are not the only source of loss of resilience, and loss of resilience is not the only impact of invasive species. But it may well be the most important ecological effect they have. Positive feedbacks operating through interactions between species can magnify the effect of introductions over time, increasing the probability that a system in one state can transition to another. Amongst human pathogens, one of the best known examples of this is the interaction between HIV AIDS and tuberculosis. The resurgence of tuberculosis is closely associated with the current HIV AIDS epidemic. Amongst other species, examples include interactions between invasive *Pinus*, *Hakea*, and *Acacia* species and species of the fynbos biome in South Africa. The water demands of the invasive species alters the water available both to other species in the system and to human users, making both the ecological and social systems more vulnerable to fluctuations in precipitation (Turpie and Heydenryc 2000; Le Maitre *et al*. 2002; van Wilgen and Richardson, Chapter 13 this volume).Other examples are the effects of buffelgrass invasions in the US and Mexico on the frequency and severity of fires (Arriaga *et al*. 2004).

The mechanism behind a reduction in system resilience in most cases is a reduction in the diversity of functional groups. The homogenization or simplification of affected systems is a frequently cited driver of reductions in system resilience. This, in turn, is attributed both to land use change and to the effects of invasive species (MA 2005). The variability of external environments, and the fact that the cyclical nature of succession allows local transients to persist, means that some species may be "redundant" at any one moment in time. But the "redundancy" of one species at a moment in time may simply be the expression of the complementarities between species through time. A function best performed by one species under one set of environmental conditions may be best performed by another species under a different set of environmental conditions. This is the insurance hypothesis. While experiments showing the functional significance of biodiversity have so far involved only small scale experiments on grasslands and streams, Loreau *et al*. (2003) conjecture that the insurance value of biodiversity may only be evident at much greater temporal and spatial scales.

The argument is that species or phenotypes may be complementary over space as well as time. They postulate a spatially heterogeneous environment comprising a meta-community comprising a set of communities linked by the dispersal of species between them. Environmental conditions are assumed to vary asynchronously across these sub-communities. Within each community, variation in species characteristics allows selection of the set of species best adapted to prevailing conditions. This in some sense optimizes the functioning of local ecosystems. Loreau *et al*. argue that the performance of each local community may be covariant with the performance of other local communities in the meta-community, and that the mechanism is the dispersal of species between communities.

Dispersion at either very low or very high rates leads to competitive exclusion, and hence a reduction in the diversity of species in each community. If species vary in their adaptation to particular environmental conditions and there is no dispersion (each community is a closed with respect to all other communities), then the best-adapted species in each community will competitively exclude all other species. The community will then become vulnerable to changes in environmental conditions. At the other extreme, if the rate of dispersion is very high, then the meta-community will begin to behave as if it were a single community. This has the same effect. The species that is best adapted to average conditions across the meta-community will out-compete all other species. That is, competitive exclusion will cause the loss of biodiversity and increasing sensitivity to variation in environmental conditions. The implication of this work is that productivity in individual communities and the stability of productivity over time are maximized at intermediate rates of dispersion.

1.3 The economic problem

The ecological and economic dimensions of the problem of invasive species are connected at different levels. The ecological changes that lead ecosystems to be more vulnerable to the impacts of invasions (the fragmentation and disturbance of habitats, loss of biodiversity, and increasing pollution burdens) are a direct consequence of economic behavior. The ecological mechanism connecting invasive species, functional diversity, and ecosystem resilience (the rate at which species are dispersed) is highly correlated with the growth of trade, transport, and travel. The main consequence of a loss of resilience, a reduction in the capacity of ecosystems to maintain functionality and the production of ecosystem services over a range of environmental conditions, has direct implications for the value of both output and the underlying ecological assets—the natural capital—of the system (Kareiva *et al.*, 2009). At every level, the ecological impacts of the economic activities involved are incidental to and are ignored by the actors concerned. They are said to be externalities of the market transactions involved, meaning that they are not taken into account by the people engaged in those transactions. The economic problem of biological invasions is first to understand the nature of invasive species externalities and why they occur, second to evaluate the consequences they have for human well-being, and third to develop policies and instruments for their internalization.

There exist several surveys of the economics of biological invasions. Lovell and Stone (2005) have reviewed the literature on the economics of aquatic invasive species, while Evans (2003), Eisworth and Johnson (2004) have considered the literature on terrestrial systems—the latter in the context of a paper developing a general model for the management of invasive species. Stutzman *et al.* (2004) offer an annotated bibliography of economics of invasive plants. This volume does not add to these, but it does address the particular challenges that invasive species externalities pose for decision-makers at different levels—international, national, and local.

The economic forces that drive the problem of biological invasions include trends in land use that affect the vulnerability or susceptibility of ecosystems, trends in trade, transport, and travel that affect the likelihood of species introductions, and trends in technology that affect species' impacts. While globalization has implications for all three, it is especially important in the second, influencing both the species involved in exchanges and the likelihood of their becoming invasive.

The evidence for the closer integration of the world economy is striking. The best measure of this is the ratio of world trade to global GDP, where trade is typically measured by either import or exports. Between 1970 and 2004 this ratio has increased from around 13 to 29 per cent. While this has had many beneficial consequences for human well-being, it is the primary cause of the increased rate of species dispersal. The development of new trade routes has led to the introduction of new species either deliberately or accidentally, while the growth in the volume of trade along those routes has increased the frequency with which introductions are repeated (Cassey *et al.*, 2004; Semmens *et al.*, 2004). As economies have become more open (with respect to imports and exports) they have also become more likely to experience the introduction of potentially invasive species (Dalmazzone, 2000; Vilà and Pujadas, 2001). As trade between bioclimatically similar regions has increased, the likelihood that introduced species will establish and spread has also gone up (Levine and d'Antonio, 2003). A study of invasive species risks in San Francisco Bay, for example, (Costello *et al.*, 2007) found that the risks associated with imports to the area differ depending on the source of the exports and the volume of imports, and that the cumulative number of introductions from a particular source is a concave function of the volume of imports. Both the source and volume of imports matter, as does the mode of transport. Introduction of most of the non-indigenous species discovered in San Francisco Bay—as with the poster-child of invasive aquatic species, the Zebra Mussel (*Dreissena polymorpha*) and the Asian Clam (*Corbicula fluminea*)—depended on ballast water exchange in ships.

There are by now a number of case studies of the relationship between particular invasive species and trade, many focusing on emergent or re-emergent pathogens. A study by Tatem *et al.* (2006) considered the role of air and marine transport in the spread of two malaria vectors: *Aedes albopictus* and *Anopheles gambiae*. They found that the growth of the range of *Aedes albopictus* away from South and South East Asia was well predicted by combining climate and trade data. In areas where the range of *Aedes albopictus* had expanded, export volumes from the source countries were more than twice as high as in climatically similar areas where it had not invaded. *Anopheles gambiae*, by contrast, has not spread from Africa—but exports from that region remain among the lowest in the world. A second example involves an analysis of the spread of the avian influenza virus HD5N1. Kilpatrick *et al.* (2006) found that while the spread of the H5N1 virus in Europe was dominated by movement of wild birds, its spread in Asia and Africa involved the poultry trade in addition to wild bird movements, and transmission to the Americas was due wholly to the poultry trade (Hubalek 2003; Daszak *et al.* 2000).

An important trend in world trade that affects the risks of invasive species is the development of regional trade agreements (RTAs): bilateral or multilateral trade agreements to reduce barriers to the movement of commodities and people within the area covered by the agreement. Such agreements include, for example, the European Union and the European Free Trade Association (EFTA), the North American Free Trade Agreement (NAFTA), the Southern Common Market (MERCOSUR), the Free Trade Area of ASEAN, the Association of Southeast Asian Nations (AFTA), and the Common Market of Eastern and Southern Africa (COMESA). By the end of 2008 more than 420 RTAs had been notified to the World Trade Organization (WTO), most of which are free trade and "partial scope" agreements. While many RTAs incorporate explicit environmental agreements, one consequence of most has been to reduce the effectiveness of import protection measures applying to trade between member states. Since most involve neighboring states, they facilitate the dispersal of species between those states (e.g. Perrault *et al.*, 2003).

There is, however, still a very high degree of uncertainty about trade-related species dispersal. At a general level we can say much about the role of the widening and deepening of international trade for species introductions. Moreover, for certain activities, commodity groups, and trade routes, we have enough data to predict the implications of change for biological invasions. Targeted inspections of regular shipments of many agricultural and horticultural commodities along particular trade routes have yielded reliable information about the likelihood of discovering particular species (e.g. Areal *et al.* 2008). In many other cases, however, there is as much uncertainty about the probability attaching to the introduction, establishment, and spread of a particular species as there is about the ecological and economic consequences if they do get in. Although it may be known that changes in the composition and volume of trade and land use will both affect the likelihood of species introductions, establishment, and spread, there are insufficient data to predict how—whether the risks are increased or decreased and by how much. This is particularly true of the development of new trade routes, the importation of new commodities, or the introduction of new shipping methods.

Species introductions are an externality of trade whose cost depends heavily on the way that ecosystems are exploited, both because that influences the vulnerability of those systems to invasion, and because it determines the value at risk from invasions. The introduction of any given species has very different consequences in landscapes exploited for distinct purposes. A plant pathogen specific to a particular cultivated crop, for example, may have no implications outside of agricultural areas, but may be devastating within those areas. Indeed, agro-ecosystems are typically the most vulnerable to invasive species, though also the most likely to be protected through controls. The first estimate of the costs of invasive species by the Office of Technology Assessment of the US Congress was concluded in 1993 (OTA, 1993). Since that time Pimentel and colleagues (Pimentel *et al.*, 2000, 2001, 2005) have updated the OTA estimates and extended them beyond the US. For the agricultural sector, for example, they conclude that invasive species cause damage costs equal to around 50 per cent of agricultural GDP in the USA and Australia, 30 per cent in the UK, but between 80 and 110 per cent of agricultural GDP in South Africa, India, and Brazil. While their estimates are only approximate, they indicate both the wide range in damage costs and their sensitivity to the importance of agriculture to employment and output (Perrings 2007). Pejchar-Goldstein and Mooney in Chapter 12 this volume review the very wide range of studies of the damage cost of invasive species that have since been undertaken.

To date, however, there are almost no studies of the implicit cost of habitat fragmentation, disturbance, or other landscape changes that affect the ease with which introduced species can establish and spread. Such changes imply a loss in the resilience of the system to invasive species shocks, the value of which is the value of the change in probability that the system will transition to a different state. A reasonable proxy for this is the value at risk of an invasion multiplied by the change in probability of an invasion, but we do not yet have estimates of this. Indeed, most work on the value of invasive species continues to focus on direct damage or control costs: on the cost of adaptation *ex post* as opposed to mitigation *ex ante*. The only exceptions to this are studies of the cost of inspection and interception at the border—the focus of chapters by Polasky (Chapter 8) and by Springborn, Costello and Ferrier (Chapter 10) in this volume. In many cases, the most cost-effective mitigation of invasive species risks is likely to be the protection of the characteristics that make ecosystems resilient to invasive species shocks, but in the current state of the art we are not yet able to show this.

The economic problem, where there is uncertainty about the likelihood of both introductions and the potential consequences of establishment and spread, resolves into an evaluation of the relative net benefits of mitigation versus adaptation strategies: of preventive action versus control after the fact. What makes the problem particularly difficult is precisely the uncertainty attaching to several aspects of the invasion process. The historic likelihood that any given introduced species will establish, spread, and inflict appreciable harm on the host system is low—the important point in Williamson's (1996) "tens rule". However, the few species that do turn out to be damaging can be very harmful indeed—as was the case with the plague in Europe, smallpox, measles, and typhus in the Americas, or the Spanish Flu worldwide. This has led to inconsistencies in the way that such problems have been addressed in the real world. It is well known that standard expected utility models of decision-making under uncertainty do not adequately capture people's responses to low-probability, high-cost events—they do not capture the way that people weight events in the tails of the distribution (Starmer 2000). But even though people have been shown to place a much higher weight on "dreadful" outcomes than is warranted by the objective probability of their occurrence, there is evidence that environmental regulatory authorities systematically underestimate the importance of new scientific information on emerging environmental threats. For example, a European review of regulatory responses to evidence of novel environmental threats, Harremoës *et al.* (2001), identifies a number of cases where early research results failed to stimulate any regulatory response at all—or even, in the case of antibiotic resistance in humans, the research required to corroborate those results. They also identified striking differences between the regulatory response to scientific results in different countries. An example of this is bovine spongiform encepalopathy (BSE). The disease fell within the remit of UK veterinary officials who regarded the "risk" of transmission to humans as "acceptably slight", and hence neglected that outcome. In the US, by contrast, the possibility of transmission of scrapie and Creutzfeldt-Jacob Disease (CJD) had led to a ban on infected animals being admitted to the human food chain as soon as it was recognized in the 1970s (Harremoës *et al.*, 2001). By one view, this is not unexpected. In Chapter 7 in this volume, for example, Simpson observes that where the expected costs of invasive species are not convex, policies can range between extreme (corner) solutions: either to prohibit trade entirely or to allow unrestricted free trade. Moreover, he argues that major changes in policy may turn on small changes in underlying parameters, and little difference to overall welfare. There is a need for more systematic evaluation of both the risks, and the most cost-effective way of containing risks.

1.4 Policy and management options

The policy and management implications of the analysis offered in this volume are reported in the final chapter (Chapter 16). To get a sense of the range of actions in play, the volume includes both theoretical analyses of the optimal controls for species spreading at different rates and case studies of adaptive responses to invasive species in different environments. Three related issues turn out to be important in the theory of invasive species control. The first is the relative costs and benefits of alternative strategies and, in particular, the relative costs and benefits of mitigation versus adaptation strategies. The second is the degree of uncertainty involved, and the third is the rapidity and spatial extent of the potential spread of the invasive species.

In the absence of reliable estimates of the net benefits of investment in the defensive capabilities of ecosystems themselves, the relative costs and benefits of alternative strategies generally involve a comparison of net benefits of inspection and interception, or detection and eradication, versus the control of established invaders. Simpson draws attention to the difference between taking specific measures to combat specific threats, versus taking general measures to combat general threats, and concludes that it will seldom be optimal to pursue the latter. While the choice between adaptation and mitigation is, in part, a distinction between specific and general threats, it is more about the *ex ante* management of the risk of invasions versus the *ex post* management of the consequences of realized invasions. Given the evidence for the cost advantage of prevention over later control in the case of most harmful species (Leung *et al.* 2002), many scientists working on the problem are pre-disposed towards preventive action (Keller and Lodge, Chapter 15 this volume). However, the optimal choice between mitigation and adaptation does depend on the costs and benefits of each strategy.

Polasky, in Chapter 8, addresses the choice between three strategies: inspection (to prevent introduction), detection (to identify and eradicate species that have got past the border but have not yet spread), and control (management of species that have established and spread). The first two correspond to the "prevention" and "early detection and rapid response" options of the US National Invasive Species Council, and both involve the mitigation of risk. He finds that inspection and detection are generally substitutes—that reducing the cost of one of these two strategies will increase the optimal effort devoted to it and reduce the optimal effort devoted to the other. Detection and eradication are complements, but both are substitutes for the control of established populations. As he puts it:

"...when control is impossible at large population levels, making invasions that reach a certain size irreversible, then the value of both inspection and detection increase. There is greater value in stopping the invasion from reaching the critical population level, either by preventing introduction or by early detection and control."

As with many other environmental issues, the timing of the costs or benefits (avoided damage) of alternative strategies matters. Many invasive plants, for example, may be present in an ecosystem for more than a hundred years before they become a problem (Williamson, 1996; Chapter 5). Discounting the costs of such invasions at any positive rate ensures that costly interception or eradication programs will be rejected, even though the current value of that cost may be low relative to the damage done.

The second important issue is the nature and degree of uncertainty associated with invasive species. Aside from the implications it has for the estimation of the net benefits attaching to any strategy, the nature of uncertainty influences the optimal inspection strategy. The lack of historical precedents for any emergent threat, including the threat posed by invasive species, means that there do not exist sufficient observations to calculate the probabilities attaching to the establishment, spread, and impact of new introductions. While it may be possible to go some way towards improving our understanding of the risks of emergent pests and pathogens using models, this ultimately depends on a Bayesian learning process. Applying this insight to the analysis of inspection and interception regimes, Springborn, Costello and Ferrier in Chapter 10 show that the optimal inspection strategy depends on the level of uncertainty attaching to commodity groups and trade routes. Where the risks associated with particular commodity groups or trade routes are known, then a targeted inspection strategy makes sense. But where the risks are not known, they show that it is optimal to adopt a random audit approach.

The important point made by both Springborn *et al.* and Polasky is that the primary value of inspection and interception regimes is in fact the information those regimes provide about the evolution of risk. And the greater the uncertainty associated with novel commodities or trade routes, the greater is this value. Since the uncertainty associated with species introductions is increasing in both the number of new trade routes and commodity groups traded, the importance of activities which provide an opportunity for Bayesian learning is also increasing. We return to this point in Chapter 16 of this volume. Here, we note that in an evolutionary system, where the capacity to predict the consequences of novel events is limited by the lack of historical precedents, building that capacity through both monitoring and experimentation is an essential part of the policy tool kit (Perrings 2007).

The third important issue relates to the speed with which invasive species are able to spread. This turns out to be important both for the choice of policy and for the way in which the problem is analyzed. Williamson, in Chapter 5, draws attention both to the fact that there is extreme variability in the rate at which different species spread, and that "managers attention is inevitably on those that spread fast". The point was made earlier that discounting the costs of the invasions of very slow-spreading species at any positive rate ensures that costly interception or eradication programs will be rejected, even though the current value of that cost may be low relative to the damage done. Given this, it is not surprising that there is a pronounced bias in favor of mitigation strategies for species with the potential both to spread rapidly and to impose significant damage (primarily pathogens), and adaptation strategies for species whose likely rate of spread is much slower whether or not they have the potential to impose significant damage (primarily plants).

Touza et al. in Chapter 11 show how landscape heterogeneity, spatial population dynamics, and cost effective control strategies are linked in the case of species that themselves spread through the landscape. In the case of pathogens, however, the agent of spread is generally a host species. The rapid spread of human infectious diseases, for example, is due to air travel. Rapidly spreading species affect the cost of biological invasions in two separate ways. First, the more widespread is a species, the greater the physical area and so the greater the population it is likely to affect. Second, the more rapid the rate of spread of invasive species, the earlier its impacts will be realized. Both things increase the present (discounted) value of the damage avoided by inspection and interception, or detection and eradication. An additional insight from Finnoff et al. in Chapter 9 is that the more rapid is the rate of spread of invasive species, the more effective is a rule-of-thumb policy that bases control effort on the costs associated with the final potential spread of those species. For more slowly growing species, by contrast, the only effective strategies are those that track the expansion of the species.

The case studies included in the volume, by Uma Shaanker et al. (Chapter 14) and van Wilgen and Richardson (Chapter 13) address problems of the second kind. They deal with species that are in many instances long-established and that continue to spread, that are damaging but not in the acute way that many pathogens are damaging. They also deal with the options for combining control of invasive species with their exploitation. Both involve examples of what is a very long list of established invasive species that are impossible or very difficult to eradicate. Other well-known examples of such species include fire ants in the USA (Buhs 2005) and rabbits in Australia (Johnson 2006). But even though there is considerable pessimism over the prospects for effectively dealing with such species on large land masses, there are a growing number of examples of management strategies that have either eliminated invaders or maintained them at tolerably low levels on islands. The elimination of rats (*Rattus spp.*, mostly but not only *R.norvegicus*) from islands is one such. Developing techniques first on very small islands, it is now possible to tackle successfully islands over 100 km^2 and larger mammals such as coypu, muskrat, goats, and pigs have been eliminated from much larger islands such as Britain or Isabela in the Galapagos (Simberloff 2009). There have also been some successes amongst marine organisms. *Caulerpa taxifolia* is a tropical marine alga. A strain tolerant of lower temperatures was developed inadvertently in one aquarium and released to the Mediterranean from another. It now covers about 10,000 ha of sea bottom in various parts of that sea with no effective control program. It was later introduced, again accidentally, to the Pacific off the coast of California, but was detected early enough that it could be successfully eradicated (Simberloff 2009).

1.5 Recognizing the global dimension of the problem

While there is much that can and should be done defensively by individual countries, the dispersal of species through international trade, transport, and travel is a global problem that needs to be addressed at a global level. Biological invasions are an externality of international trade, and the solution to the problem requires policies to address that fact directly. These imply international cooperation, which means collaborative action both in terms of the multilateral agreements governing trade (the General Agreement on Tariffs and Trade) and the effects of trade (the Sanitary and Phytosanitary Agreement, and the International Plant Protection Convention), and of the intergovernmental organizations established to address different dimensions of the invasive species problem: the World Health Organization (WHO), the World Animal Health Organization (OIE), the Food and Agriculture Organization, the United Nations Environment Program, and so on. More importantly, it requires coordination between these bodies.

Some of the issues involved are discussed in Chapter 16 of this volume. Here we draw attention to three dimensions of biological invasions as a global problem that pose particular challenges to science and policy alike. Thomas and Ohlemüller, in Chapter 2, point out that climate change is likely to alter the vulnerability of ecosystems to invasion in fundamental ways. Yet, as New and McSweeney (Chapter 3) indicate, our capacity to predict the likelihood of invasions from the climate data is quite limited. This accords with other observations on the predictability of invasions. Williamson (1999), for example, argued that because there are no general laws governing invasions it is difficult both to model the process (Kareiva *et al.* 1996) and to predict the population dynamics of a particular species in a particular habitats (Lawton 1999). He identified only two reasonable predictors of the invasiveness of particular species: (a) a previous history of invasions by the same species, and (b) propagule pressure. A later assessment of the predictive capacity of models of the invasions process by the NAS Committee on the Scientific Basis for Predicting the Invasive Potential of Nonindigenous Plants and Plant Pests in the United States suggested that there may be other "biological leads" that have the capacity to improve predictability of invasiveness (NAS 2002).

What is equally important, though, is that our capacity to predict the likelihood of invasions from other data is rather better. The probability that species from particular ecosystems will be introduced to bioclimatically similar but distant ecosystems depends on global trade patterns and volumes, both of which are predictable over a reasonable period (Williamson's propagule pressure). The same is true of the land use changes that determine both the invasibility of managed ecosystems and the value at risk. Of course there is uncertainty about the dynamics of many components of the global system. Many ecosystems are only partially observable and controllable, implying that the most that may be achieved is their stabilization—the regulation of stresses on the uncontrolled part of the system so as to maintain its stability (or at least resilience). But we do have sufficient information to identify the consequences of closer global integration for a number of ecosystem types, and to use this information to inform policy at many spatial scales. However, this does depend on the capacity to integrate modeling efforts. If propagule pressure, host system vulnerability, and value at risk all depend on economic factors, then models of invasion that ignore these factors are almost certainly inadequate to the task.

Second, fundamental differences in the pattern and rate of spread of invasive species means fundamental differences in the scale at which the problem needs to be analyzed and addressed. The first global pandemic, the 1918–20 Spanish Flu, was spread anthropogenically in the special circumstances of the World War I—unprecedented movements of men and supplies over the whole world. The scope for similarly virulent human or animal pathogens to spread globally is much greater now, and for such species the scale at which the problem needs to be addressed is clearly global. At the same time, however, many introduced species spread locally at fast enough rates to make them problematic at that scale, but have no implications beyond that. Application of the subsidiarity principle implies that problems of that sort be dealt with at a local scale. Between such polar cases, however, lie problems that occur over a wide range of scales. The challenge in this for the economic, epidemiological, and ecological sciences is to determine the spatial and temporal scale of the problem—including its causes and effects—and to analyze both the problem and the policy and management options accordingly. The homogenization of the world's agricultural systems and their closer integration in world markets for foods, fuels, and fibers, for example, are the key drivers of pest invasions in agroecosystems. This does not mean that national effort to protect agriculture from imported pests and pathogens is inappropriate, but it does mean that it is insufficient. The problem has to be addressed at a global scale.

The third dimension of the global problem of biological invasions we draw attention to here is the least tractable. In this volume, Perrings, Fenichel, and Kinzig (Chapter 4) observe that global invasive species risks depend on the capacity of each country to implement sanitary and phytosanitary measures. Global protection against many invasive species risks is a public good of a very special kind. In the wider literature it is known as a "weakest link public good" (Perrings *et al.* 2002). The level of protection enjoyed by all is fixed by the level of protection offered by the weakest link in the chain. There are two things at issue here. First, the damage done by invasive species—and especially by human, animal and plant pathogens, and agricultural pests—bears especially heavily on people in "poor" countries. This is because a higher proportion of the population in those countries (anywhere up to 90 per cent) depends on agriculture for their livelihood. It is also because those populations are more vulnerable and have fewer resources to either mitigate or accommodate the risks posed by invasive species. Many of the world's poor live on marginal, highly disturbed lands that are often the first to be colonized by invasive agricultural species, or in crowded unsanitary conditions that make them vulnerable to infectious human diseases. The invasive species issue raises an important question of equity.

But even if people living in high and middle income countries do not care about the well-being of those living in low income countries, they still have an interest in building the capacity of poor countries to address the problem. Because global protection is a weakest link public good, the lower the capacity of poor countries to deal with damaging and rapidly-spreading invasive species, the greater the exposure of all their trading partners. It follows that the more closely integrated is the global system, the greater the incentive to high income countries to build capacity in the weakest links in the chain. In the case of human, animal, and plant pathogens, the risk of infection or re-infection can be reduced by direct support of the sanitary and phytosanitary capabilities of low income trading partners.

The implications of this for policy and management are discussed in Chapter 16. The implications for science follow from the fact that the risks posed by the introduction of invasive species in any one country are not independent of the way that other countries deal with the problem. In order to understand the covariance in the risks facing different countries, it is necessary to understand the mechanisms that lie behind the dispersal and re-dispersal of species. The most primitive solutions to risks of this sort have always involved isolation. The introduction of quarantine mechanisms to deal with the threats posed by plague in fourteenth and fifteenth century Europe, for example, were based on biblical injunctions to shun lepers (McNeill 1977). But isolation is not an option for dealing with national risks. While low income countries may currently be less integrated into the world economy than high or middle income countries, they are nevertheless connected to the rest of the world in myriad ways, many of which involve direct or indirect pathways for species. Understanding these pathways and the attendant risks is a precondition for the development of both the science of invasive species and for effective international solutions.

Acknowledgements

Charles Perrings acknowledges support from NSF Grant # 0639252 and DIVERSITAS.

References

Areal, F. J., Touza, J., MacLeod, A., Dehnen-Schmutz, K., Perrings, C., Palmieri, M. G., and Spence, N. J. (2008). Integrating drivers influencing the detection of plant pests carried in the international cut flower trade. *Journal of Environmental Management* **89**(4), 300–307.

Arriaga, L., Castellanos, A. E., Moreno, V. E., and Alarcón, J. (2004). Potential ecological distribution of alien invasive species and risk assessment: a case study of buffel grass in arid regions of Mexico. *Conservation Biology* **18**(6), 1504–14.

Buhs, J. B. (2005). *The Fire Ant Wars*. Chicago University Press, Chicago.

Cassey, P., Blackburn, T. M., Russel, G. J., Jones, K. E., and Lockwood, J. L. (2004). Influences on the transport and establishment of exotic bird species: an analysis of the parrots (Psittaciformes) of the world. *Global Change Biology* **10**, 417–26.

Christen, D. C. and Matlack, G. R. (2009). The habitat and conduit function of roads in the spread of three invasive plant species. *Biological Invasions* **11**, 453–65.

Costello, C., Springborn, M., McAusland, C., and Solow, A. (2007). Unintended biological invasions: Does risk vary by trading partner? *Journal of Environmental Economics and Management* **54**, 262–76.

Dalmazzone, S. (2000). Economic factors affecting vulnerability to biological invasions. In C. Perrings, M. Williamson and S. Dalmazzone (eds) *The Economics of Biological Invasions*, Edward Elgar, Cheltenham, 17–30.

Daszak, P. and Cunningham, A. A. (1999). Extinction by infection. *Trends in Ecology and Evolution* **14**, 279.

Daszak, P., Cunningham, A. A., and Hyatt, A. D. (2000). Emerging infectious diseases of wildlife: threats to biodiversity and human health. *Science* **287**, 443–49.

Diamond, J. (1997). *Guns, Germs and Steel*. W. W. Norton, New York.

Diamond, J. (2005). *Collapse*. Penguin Books, New York.

Diaz, S., Symstad, A. J., Stuart Chapin, F. III., Wardle, D. A. and Huenneke, L. F. (2003). Functional diversity revealed by removal experiments, *Trends in Ecology and Evolution* **18**, 140–46.

Dobson, A. (2005). What links bats to emerging infectious diseases? *Science* **310**, 628–29.

Eisworth, M. E. and Johnson, W. S. (2002). Managing nonindigenous invasive species: insights from dynamic analysis. *Environmental and Resource Economics* **23**, 319–42.

Elton, C. S. (1958). T*he Ecology of Invasions by Animals and Plants*. Methuen, London.

Enserink, M. (1999). Biological invaders sweep In. *Science* **285**, 1834–36.

Evans, E. A. 2003. Economic dimensions of invasive species. *Choices* (June).

Glowka, L., Burhenne-Guilmin, F., and Synge, H. (1994) *A Guide to the Convention on Biological Diversity*. IUCN, Gland, Switzerland.

Harremoës, P., Gee, D., MacGarvin, M., Stirling, A., Keys, J., Wynne, B., and Guedes Vaz S., eds (2001). *Late Lessons from Early Warnings: The Precautionary Principle*, Environment Issue Report No 22, European Environment Agency, Copenhagen.

Harvell, D., Aronson, R., Baron, N., Connell, J., Dobson, A., Ellner, S., Gerber, L., Kuris, Kim, A., McCallum, H., Lafferty, K., McKay, B., Porter, J., Pascual, M., Smith, G., Sutherland, K., Ward, J. (2004). The rising tide of ocean diseases: Unsolved problems and research priorities. *Frontiers in Ecology and the Environment* **2**, 375–82.

Hooper, D. U., Chapin III, F. S., Ewel, J. J., Hector, A., Inchausti, P., Lavorel, S., Lawton, J. H., Lodge, D. M., Loreau, M., Naeem, S., Schmid, B., Setälä, H., Symstad, A. J., Vandermeer, J., and Wardle, D. A. (2005). Effects of biodiversity on ecosystem functioning: a consensus of current knowledge. *Ecological Monographs* **75**(1): 3–35.

Hubalek, Z. (2003). Emerging human infectious diseases: Anthroponoses, zoonoses,and sapronoses. *Emerging Infectious Diseases* **9**, 403–404.

Hughes, T. P., Baird, A. H., Bellwood, D. R., Card, M., Connolly, S. R., Folke, C., Grosberg, R., Hoegh-Guldberg, O., Jackson, J. B. C., Kleypas, J., Lough, J. M., Marshall, P., Nyström, M., Palumbi, S. R., Pandolfi, J. M., Rosen, B., Roughgarden, J. 2003. Climate change, human impacts, and the resilience of coral reefs'. *Science* **301**, 929–33.

Hughes, T. P., Bellwood, D. R., Folke, C., Steneck, R. S. and Wilson, J. (2005). New paradigms for supporting the resilience of marine ecosystems. *Trends in Ecology and Evolution* **20**(7) 380–86.

IPCC (2007). *Climate Change 2007: Synthesis Report. Report of the Intergovernmental Panel on Climate Change*, IPCC, Geneva.

Johnson, C. (2006). *Australia's Mammal Extinctions*. Cambridge University Press, Cambridge.

Jones, K. E, Patel, N. G., Levy, M. A., Storeygard, A., Balk, D., Gittleman, J. L., and Daszak P. (2008). Global trends in emerging infectious diseases. *Nature* **451**(21), 990–94.

Kareiva, P. (1996). Developing a predictive ecology for non-indigenous species and ecological invasions. *Ecology* **77**, 1651–97.

Kareiva, P., Daily, G., Ricketts, T., Tallis, H. and Polasky, S. (2009). *The Theory and Practice of Ecosystem Service Valuation in Conservation*, Oxford University Press, Oxford.

Kilpatrick, A. M., Chmura, A. A., Gibbons, D. W., Fleischer, R. C., Marra, P. P., Daszak, P., (2006). Predicting the global spread of H5N1 avian influenza. *Proceedings of the National Academy of Sciences of the United States of America* **103**, 19368–73.

Kinzig, A. P., Ryan, P., Etienne, M., Elmqvist, T., Allison, H., and Walker, B. H. (2006). Resilience and regime shifts: Assessing cascading effects. *Ecology and Society* **11**(1): article 13 [online].

Lawton, J. (1999). Are there general laws in ecology?, Oikos 84, 177–92.

Le Maitre, D. C., van Wilgen, B. W., Gelderblom, C. M., Bailey, C., Chapman, R. A., and Nel, J. A. (2002). Invasive alien trees and water resources in South Africa: case studies of the costs and benefits of management. *Forest Ecology and Management* **160**, 143–59.

Leung, B., Lodge, D. M., Finnoff, D., Shogren, J. F., Lewis, M. A. and Lamberti, G. (2002). An ounce of prevention or a pound of cure: bioeconomic risk analysis of invasive species. *Proceedings of the Royal Society of London, Biological Sciences* **269**(1508): 2407–413.

Levine, J. M. and D'Antonio, C. M. (2003). Forecasting biological invasions with increasing international trade. *Conservation Biology* **17**, 322–26.

Lewison, R. L., Crowder, L. B., Read, A. J., Freeman, S. A. (2004). Understanding impacts of fisheries bycatch on marine megafauna, *Trends in Ecology and Evolution* **19**, 598–604.

Loreau, M., Mouquet, N., and Gonzalez, A. (2003). Biodiversity as spatial insurance in heterogeneous landscapes. *Proceedings of the National Academy of Sciences* **22**, 12765–70.

Lovell, S. J. and Stone, S. (2005). *The economic impacts of aquatic invasive species: a review of the literature*, Working Paper # 05-02, U. S. Environmental Protection Agency National Center for Environmental Economics, Washington, D. C.

Malthus, T. (1970). *An Essay on the Principle of Population*. Penguin Books, Harmondsworth.

McAusland, C. and Costello, C. (2004). Avoiding invasives: trade related policies for controlling unintentional exotic species introductions. Journal of Environmental Economics and Management 48, 954–977.

McNeely, J. A. (2001). An introduction to human dimensions of invasive alien species. In McNeely, J. A., ed. *The Great Reshuffling. Human Dimensions of Invasive Alien Species*, IUCN, Gland, 5–20.

McNeill, W. H. (1977). *Plagues and People*. Anchor Books, New York.

Millennium Ecosystem Assessment (MA) (2005). *Ecosystems and Human Well-Being: Synthesis*. Island press, Washington D. C.

Myers, R. A. and Worm, B. (2003). Rapid worldwide depletion of predatory fish communities. *Nature* **423**, 280–83.

Naeem, S. and Wright, J. P. (2003). Disentangling biodiversity effects on ecosystem functioning: Deriving solutions to a seemingly insurmountable problem. *Ecology Letters* **6**, 567–79.

National Academy of Sciences (NAS), Committee on the Scientific Basis for Predicting the Invasive Potential of Nonindigenous Plants and Plant Pests in the United States. (2002). *Predicting Invasions of Non-Indigenous Plants and Plant Pests*, National Academy Press, Washington D. C.

Niemelä, J., Koivula, M., and Kotze, J. (2007). The effects of forestry on carabid beetles (Coleoptera: Carabidae) in boreal forests. *Journal of Insect Conservation* **11**, 5–18.

Normile, D. (2008). Rinderpest: driven to extinction. *Science* **319**(5870): 1606–1609.

Oerke, E.-C., Dehne, H.-W., Schönbeck, F. and Weber, A. (1994). *Crop Production and Crop Protection: Estimated Losses in Major Food and Cash Crops*, Elsevier, Amsterdam.

Office of Technology Assessment. U. S. Congress (OTA). (1993). Harmful Non-Indigenous Species in the United States. OTA Publication OTA-F-565. US Government Printing Office, Washington D. C.

Pandolfi, J. M. Bradbury, R. H., Sala, E., Hughes, T. P., Bjorndal, K. A., Cooke, R. G., McArdle, D., McClenachan, L., Newman, M. J. H., Paredes, G., Warner, R. R., Jackson, J. B. C. (2003). Global trajectories of the long-term decline of coral reef ecosystems. *Science* **301**, 955–58.

Pauly, D., Christensen, V., Guenette, S., Pitcher, T. J., Sumaila, U. R., Walters, C. J., Watson, R., Zeller, D. (2002). Towards sustainability in world fisheries. *Nature* **418**, 689–95.

Perrault, A., Bennett, M., Burgiel, S., Delach A., and Muffett, C. (2003). *Invasive Species, Agriculture and Trade: Case Studies from the NAFTA Context*. North American Commission for Environmental Cooperation, Montreal.

Perrings, C., Williamson, M., Barbier, E. B., Delfino, D., Dalmazzone, D., Shogren, J., Simmons, P. and Watkinson, A. (2002). Biological invasion risks and the public good: an economic perspective. *Conservation Ecology* 6(1), 1. [online] URL: http://www.consecol.org/vol6/iss1/art1

Perrings, C. (2007). Pests, pathogens and poverty: biological invasions and agricultural dependence. In A. Kontoleon, U. Pascual and T. Swanson (eds) *Biodiversity Economics: Principles, Methods and Applications*, Cambridge University Press, Cambridge, 133–65.

Perrings, C., Dehnen-Schmutz, K., Touza, J., and Williamson, M. (2005). How to manage biological invasions under globalization. *Trends in Ecology and Evolution* **20**(5), 212–15.

Phillips, B. L., Brown, G. P., Webb, J. K. and Shine, R. (2006). Invasion and the evolution of speed in toads. *Nature* **439**, 803.

Pimentel, D., Lach, L., Zuniga, R. and Morrison, D. (2000). Environmental and economic costs of nonindigenous species in the United States. *Bioscience*, 50(1), 53–56.

Pimentel, D., McNair, S., Janecka, S., Wightman, J., Simmonds, C., O'Connell, C., Wong, E., Russel, L., Zern, C., Aquino T., and Tsomondo, T. (2001). Economic and environmental threats of alien plant, animal and microbe invasions. *Agriculture, Ecosystems and Environment* **84**: 1–20.

Pimentel, D., Zuniga, R., and Morrison, D. (2005). Update on the environmental and economic costs associated with alien-invasive species in the United States. *Ecological Economics* 52, 273–88.

Polasky, S., Costello, C., and McAusland, C. (2004). On trade, land- use and biodiversity. *Journal of Environmental Economics and Management* **48**, 911–25.

Rangi, D. K. (2004). Invasive alien species: agriculture and development. *Proceedings of a global synthesis workshop on biodiversity loss and species extinctions: managing risk in a changing world*. UNEP, Nairobi.

Reich, P. B., Tilman, D., Naeem, S., Ellsworth, D. S., Knops, J., Craine, J., Wedin, D., and Trost, J. (2004). Species and functional group diversity independently influence biomass accumulation and its response to CO2 and N. *Proceedings of the National Academy of Sciences* **101**, 10101–106.

Salisbury, E. (1961). *Weeds and Aliens*. Collins, London.

Scheffer, M., Brock, W. A., and Westley, F. (2000). Socioeconomic mechanisms preventing optimum use of ecosystem services: an interdisciplinary theoretical analysis. *Ecosystems* **3**: 451–471.

Semmens, B. X., Buhle, E. R., Salomon, A. K., Pattengill-Semmens, C. V. (2004). A hotspot of non-native marine fishes: evidence for the aquarium trade as an invasion pathway. *Marine Ecology Progress Series* **266**, 239–44

Simberloff, D. (2009). We can eliminate invasions or live with them. Successful management projects. *Biological Invasions* **11**, 149–57.

Starmer, C. (2000). Development in non-expected utility theory: the hunt for a descriptive theory of choice under risk. *Journal of Economic Literature* **38**, 332–82.

Steneck, R. S., Vavrinec, J., and Leland, A. V. (2004). Accelerating trophic-level dysfunction in kelp forest ecosystems of the Western North Atlantic. *Ecosystems* **7**, 323–32.

Stutzman, S. K. M. Jetter and Klonsky, K. M. (2004). *An Annotated Bibliography on the Economics of Invasive Plants*, University of California, Davis, Agricultural Issues Center.

Tatem, A. J., Hay, S. I., and Rogers, D. J. (2006). Global traffic and disease vector dispersal. *Proceedings of the National Academy of Sciences* **103**(16), 6242–6247.

Turpie, J. K. and Heydenryc, B. J. (2000). Economic consequences of alien infestation of the Cape Floral Kingdom's Fynbos vegetation. In C. Perrings, M. Williamson and S. Dalmazzone, eds, *The Economics of Biological Invasions*, Edward Elgar, Cheltenham, 152–82.

Vila, M. and Pujadas, J. (2001). Land use and socio-economic correlates of plant invasions in European and North African countries, *Biological Conservation* **100**, 397–401.

Walker, B. H., Gunderson, L. H., Kinzig, A. P., Folke, C., Carpenter, S. R., and Schultz, L. (2006). A handful of heuristics and some propositions for understanding resilience in social-ecological systems. *Ecology and Society* **11**(1): 13. [online] URL: http://www.consecol.org/vol11/iss1/art13/

Walker, B. H., Holling, C. S., Carpenter, S. R. and Kinzig, A. P. (2004). Resilience, adaptability, and transformability. *Ecology and Society* **9**(2): 5. [online] URL: http://www.ecologyandsociety.org/vol9/iss2/art5

Williamson, M. (1996). *Biological Invasions*, Chapman and Hall, London.

Williamson, M. (1999). Invasions, *Ecography* **22**, 5–12.

PART I

THE DRIVERS OF BIOLOGICAL INVASIONS

CHAPTER 2

Climate Change and Species' Distributions: An Alien Future?

Chris D. Thomas and Ralf Ohlemüller

2.1 Introduction

Climate change seems likely to alter the balance between currently-native and currently non-native species, most probably leading to a step-change increase in the prevalence of what we currently term non-native, alien or invasive species. As the "old climate" is replaced by a "new climate", we can expect pre-existing local populations and communities of animals and plants to become less well adjusted to local climatic conditions, and a concomitant decline in the resistance of natural communities to invasion.

Some of the invading species are likely to colonize from elsewhere within the same continent, whereas others are expected to originate from further afield. In this chapter, we present analyses that suggest increasing levels of invasion will be observed with increasing climate change, and that invasions of species from far away will increase disproportionately. We also suggest that climate change will require us to reconsider our attitudes to the distinction between native and non-native species, because many species are only expected to survive climate change by establishing in new regions, outside their historical ranges.

2.2 Natives and aliens

Species' distributions are dynamic, not static. Each species originated somewhere, and subsequent range changes have produced its current distribution (Huntley 1991; Williams *et al.* 2004). The vast majority of species contain portions of their current ranges that were colonized since the origination of that species, and where, on some time scale, they could be considered to be non-native. Recent climate change is already changing species' distributions and abundance patterns, and will continue to do so, increasingly blurring the distinction between those species we have traditionally regarded as native and non-native (Parmesan & Yohe 2003; Walther 2003, 2004; Hickling *et al.* 2006). This will require a major shift in attitudes to such species. It is neither practical nor desirable to attempt to weed-out species whenever they turn up in new places.

The current distributions of species emerge from their interactions with their physical and biotic environments, and of the histories of those interactions. Distributions reflect the dynamic consequences of spatial patterns of birth and death, combined with the movement of individuals. If the distributions of physical and biotic (including human) variables change, so will a species' range. Where one considers a species to be an alien is usually an arbitrary decision about time and space. Absences of a species from a particular part of the world may involve evolution in one area and a failure to colonize other regions on a time scale of tens of millions of years, the failure to colonize environments that have become climatically suitable in the last 10,000 years of warm Holocene conditions, and the failure to colonize novel environments created by human activities on time scales of millennia to minutes (Svenning & Skov 2004). Similarly, failed colonization may involve distances of thousands of kilometers around the earth, a few hundred kilometers between lakes or mountains, a few kilometers between naturally patchy habitats, and so on. Over the last million years, species have repeatedly shifted to lower altitudes and latitudes during glacial maxima and to higher elevations and latitudes during interglacial warm periods, including during the current Holocene warm period (Huntley 1991; Pitelka *et al.* 1997; Brewer *et al.* 2002; Williams *et al.* 2004). Large parts of Europe and North America are currently covered by species that traveled at least a thousand kilometers since the end of the last cold period, about 11.5 thousand years ago. This is accepted, and our land management and conservation policies do not distinguish between species in these regions on the basis of the order in which they arrived (except for recent arrivals within the documented historical period). This is part of the natural ebb and flow of species and has been essential to their survival during periods of past climatic variation. Many local endemics ended up surviving in particular mountain ranges or other refugial habitats, rather than having necessarily evolved there (Coope 1973). We attempt to maintain species where they currently are, not where they might have lived many millennia ago.

Because distributions are dynamic, definition of what species are "native" or "non-native" is evidently a construct based on utility to humans interested in the management of biodiversity and of ecosystem goods and services (and those interested in documenting history), rather than a distinction that has any fundamental biological meaning. All species have histories. "From how far away must a species come to classify as non-native?" and "How long must a species be present in a particular location to be classified as native?" are practical or philosophical issues, not questions that have a scientific answer.

Human activities are already part of the recent histories of most species, at least on land and in fresh waters; humans have been part of terrestrial ecosystems for thousands of years. Most terrestrial species have been removed from some areas by human-caused habitat alteration or loss, and human-mediated range expansions are also frequent. Global over-exploitation of the former megafauna, from mammoths to moas, must have had important consequences for the subsequent structure and composition of the vegetation, affecting the range sizes and locations of a wide variety of other species. With fire management, then cultivation, humans altered the distributions of vast numbers of additional species, arguably generating as many expansions as contractions (Huntley & Webb 1988). Species-rich calcareous grasslands in northern and central Europe typically contain a greater diversity of higher plant species than the forests they replaced (Baur *et al.* 2006), and likewise many insects are associated with successional habitats (Thomas 1993). Shifting cultivation in the moist tropics may have greatly extended the range sizes of early successional species, which typically have larger geographic range sizes than do species of late-successional vegetation (Thomas 1991). It is difficult to find anywhere on earth land with no human impact (Willis *et al.* 2004).

For most individual species in most parts of the world, we do not know the detailed history of its distribution, and how this has been affected either directly or indirectly by past human activities. Even if we wanted to, it would not be practical to distinguish between aliens and natives.

Recent range expansions of the majority of species (Pounds *et al.* 1999; Walther *et al.* 2005; Hickling *et al.* 2006; Parmesan 2006; IPCC 2007) to higher latitudes and altitudes reinforce the historical conclusions that species' distributions shift with the climate. Whilst we might not consider species that have expanded a few tens of kilometers to be "alien", the most rapidly-shifting species are achieving about 100 km per decade, and increasing numbers of species are crossing national boundaries (Parmesan *et al.* 1999; Warren *et al.* 2001). Rates of expansion are extremely variable, so community composition will probably take centuries to come to "equilibrium" in the improbable event that the climate and other human impacts were to stabilize in 2100. Biodiversity is on the move and this will continue for the foreseeable future. Distinguishing between species on the basis of how long they have been in a particular location is not sensible.

2.3 Future distributions will be very different

Projections of the distributions of species under future climates are uncertain, but several robust generalizations can be made. First, as climate change proceeds, there will be a decreasing degree of overlap between the current distributions of species and where conditions may be suitable for them in future (Peterson et al. 2002; Thomas et al. 2004; Araújo et al. 2006). Secondly, some species will only have the potential to survive in regions that do not overlap with their current distributions (Midgley et al. 2002; Thomas et al. 2004; Thuiller et al. 2005; Ohlemüller 2006a). This implies that, on average, species will be "alien" (by comparison with 20th century distributions) in an increasing fraction of their future distributions, and that conservation efforts are likely to be needed in such areas. The default "native species good—alien bad" culture that has developed in conservation and alien species control programs is no longer sustainable. Third, the composition of biological communities will be quite different, given the individualistic responses of species to climate change, such that attempts to retain particular species compositions within ecosystems will ultimately be doomed to failure. Virtually every community will contain species that were previously not present, and some of the colonists will become quite abundant (Menéndez et al. 2006). Fourth, the strength of each of these changes will increase with the level of climate change experienced.

The climate envelope (or niche/distribution) models used to develop some of these conclusions are based on matching-up the current distributions of a species with the current (or recent past) climate, and then evaluating where such climatic conditions might exist in future (if anywhere). Amongst a number of assumptions, they presume that species' realized niches will be retained (making some projections pessimistic and others optimistic), that species will manage to spread immediately into new areas that become climatically suitable for them (an optimistic assumption), and that they will die out quickly in regions where the future climate is expected to deteriorate (a pessimistic assumption). In other words, they identify the overall scale of the problem, but do not have sufficient detail in most cases to provide specific prognoses for individual species, and they provide no direct information on expected rates of change. Dynamic colonization and metapopulation models (Collingham et al. 1996; Hill et al. 2001; Thomas et al. 2001) may be used in a more routine way in the near future to estimate likely rates of change, but we are some way off being able to project expansion rates confidently for large numbers of species. Despite these caveats, climate envelope approaches indicate that distributions and community compositions will become increasingly dissimilar with increasing levels of climate change and increasing time. Many species will have to be maintained outside their current distributions, and ecosystem goods and services will be provided by mixtures of currently-native and currently-alien species.

2.4 What are "invasive species"?

The sheer numbers of non-native species establishing has led many authors, especially those working in the policy arena, to shift the emphasis towards the effects rather than origins of species (Ricciardi & Cohen 2007). Species defined as "invasive" are often defined as species that have a net negative effect on human interests (which includes human concern for native biodiversity). This has been a sensible and practical approach, since it focuses control efforts where it is most needed.

The invasive species problem is simply an issue of community composition and assembly. The majority of species that are transported around the world and establish naturalized populations (i.e. show positive net population growth) remain relatively uncommon in the area of establishment (Williamson 1996). Most "invasive species" are recent arrivals that have become abundant, or otherwise had major impacts on the recipient community. When existing community diversity is low and/or lacks specific taxonomic or functional groups (e.g. oceanic islands lacking rodents or snakes), the impact of introduced species can be extremely large, especially if the invasive species have arrived without their own natural enemies (Melbourne et al. 2007; Theoharides & Dukes 2007).

When recipient communities are large and functionally diverse, as in most continent-to-continent introductions, the impact is usually, though not always, much smaller. Rank species-abundance curves inform us that virtually every biological assemblage contains a few high-abundance species, more species of medium-abundance, and many low-abundance species (Tokeshi 1993). Similarly, most species have small geographic range sizes, whilst a few have very large ranges (Gaston 1995) and this generalization applies to species within a particular landscape as well as to distributions at a national or international scale (Cowley *et al.* 1999). These patterns may differ to some extent between native and non-native species, but the similarity of pattern is more remarkable than the difference. In most cases, "invasives" are common (or otherwise high impact) species that happen to have arrived recently: they do not necessarily have any more or less influence on the resulting assemblage than do common "native" species. That we should distinguish between "invasive" and "other" species based on a quantitative measure of impact and duration since arrival may be useful in some contexts, but it is not easy to make an objective distinction.

Invasions are regarded as being most frequent in disturbed habitats. The more dissimilar the new (disturbed) vegetation is from the "climax" vegetation expected for a region, in the absence of humans, the smaller the expected pool of "native" species available to colonize the new vegetation, and the more likely it is that species from other parts of the world will be able to invade. In northern Europe, clearance of forest for crops and pasture facilitated major range expansions of open-habitat "weeds" from the Mediterranean, near-East and other naturally open habitats (sand dunes, rocky outcrops, high-altitude grasslands). Many of these are now prized targets of conservation effort in northern Europe! The non-native open-country species were well suited to the new conditions, whereas the previous forest flora was not.

We argue that the likelihood of invasion depends, in general, on the pool of "native" species (including both the species richness and functional diversity) relative to the pool of species arriving as propagules, and how well suited each group of species is to any particular environment (whether anthropogenic or natural). Where few "native" species are available in the native biota, relative to the continental or international pool of species available to colonize, native communities will be relatively susceptible to invasion. Where the vegetation (e.g. through disturbance) makes conditions less suitable for the native species, the derivative communities will be even more susceptible to invasion because the balance of native and distant species in the pool of potential colonists shifts in favor of those from far away. The question that we now address is whether climate change is likely to alter this balance between currently-native and potential invading species.

2.5 Climatic invasibility

Climate is likely to affect the success of an invasive species during the colonization, establishment and landscape spread stages of a species' invasion (Theoharides & Dukes 2007). Suitable climate conditions form the baseline conditions which determine whether the propagules of a potentially invasive species will be able to grow in a new area. Interactions with other species will then combine with climatic suitability to determine whether the species is able to establish viable populations. Once established, the rate of spread is also likely to increase with climatic suitability, for example by increasing the survival and reproductive success of individuals. Several recent examples have used a climate matching approach to predict likely areas of species invasions of particular species outside their native ranges, hypothesizing that species will establish most readily in regions where the climate resembles that within their native range (Hartley *et al.* 2006; Richardson & Thuiller 2007; Tatem & Hay 2007). We here attempt a generalization of this approach. We assess the extent to which the climate of a particular area is similar to that in other parts of the world (from which future potential invasive species might arise), and whether the likelihood of invasion, so defined, will change with climate warming.

The climatic component of invasibility of a particular location might be expected to depend on the size (and functional diversity) of the native biota, relative to the number of species (and their functional diversity) adapted to similar climates in other parts of the world. We have statistically modeled the climatic invasibility of different parts of the world using the following approach. For each 0.5° grid cell in the world, we have calculated the area of similar climate within 1000 km of that cell, and also the area of similar climate in the rest of the world (>1000 km away) (Fig. 2.1a,b). In these figures, regionally (Fig. 2.1a) and globally (Fig. 2.1b) rare climates are shown as dark grey, and widespread climates are depicted as light grey. We did this for the last "pre-warming climate-normal" period, 1931–60. We defined "similar climates" broadly. For this exercise, two grid cells were regarded as having similar climates if they had average winter and summer precipitation within ±1 mm/day of one another, and winter and summer temperature within ±6 °C. Average values for the winter period were calculated using months DJF (northern hemisphere) and JJA (southern hemisphere), and summer values were calculated using months JJA (northern hemisphere) and DJF (southern hemisphere). We then calculated for each cell the proportion of the world's similar climate that exists within 1000 km. We presumed that if similar climates exist within 1000 km, the majority of well-adapted species will have arrived in the fullness of time (within the last 10,000 years), and will be regarded as part of the native biota; but species from >1000 km are less likely to have arrived. Climatically-similar regions further than 1000 km away represent places where the world's pool of potential invasive species (for that cell) could come from. One minus this proportion (Fig. 2.1c) indicates the relative invasibility of each location: a value close to zero (light grey) indicates that most similar climates are found within 1000 km, so few species are likely to be available worldwide to colonize the area (that have not already done so), whereas a value close to one (dark grey) indicates that most similar climates are a long way away—and hence there is a large potential pool of species in the rest of the world that might be able to invade (given suitable transport and other suitable conditions). The pattern is very similar when thresholds other than 1000 km are used (not shown).

The pattern in Fig. 2.1c is best illustrated by two examples. Oceanic islands have very little land within 1000 km, and so most other places with similar climates in the rest of the world are a long way away; hence, the climatic invasibility index is close to 1. Whilst not so dramatic, Mediterranean climates are also highlighted as having high invasibility indices. This type of climate occurs in scattered locations around the world, so the expected global pool of species adapted to these climates is high relative to the numbers expected within any continent.

Using the 1931–60 period as our reference period, we then calculated the change in the invasibility index between 1931–60 and 2041–50 (Fig. 2.2). To characterize the climate of the later period, we used the simulation outcomes from the HadCM3 climate model, as used in Williams *et al.* (2007), for two greenhouse gas emission scenarios: B1 (moderate) and A2 (severe). For details of calculations, see (Ohlemüller *et al.* 2006b). As the climate changes in future, the balance is tipped in favor of long-distance migrants (larger areas of dark grey than light grey in Fig. 2.2). The amount of area with conditions analogous to pre-warming conditions within 1000 km of a given location tends to decrease, that is, the invasibility index tends to increase. In contrast, the average area that is climatically-similar and >1000 km away, from which invasive species might arrive, tends to stay fairly similar. Thus, the long-distance invasibility index increases in most parts of the world (Fig. 2.2a,b). Effectively, "home advantage" is lost, as the ratio of regional: global potential colonists decreases. Note, however, that every location is likely to show increased levels of invasion from species that currently inhabit slightly different climates within 1000 km, so the light grey areas in Fig. 2.2 should not interpreted as places where the native biota will be resistant to "non-native" species. These are simply the areas where most invaders are expected to come from shorter distances.

2.6 The invasion–climate change interaction

Until now, the greatest impacts of invasive species have mainly affected natural (oceanic islands, lakes) and human-disturbed ecosystems that lack adequate functional and species diversity to resist invasion. In most cases, the global pool of species available to colonize ecosystems of both types is much larger than the regional pool of species. In contrast, relatively undisturbed native vegetation has proven quite resistant to the establishment of

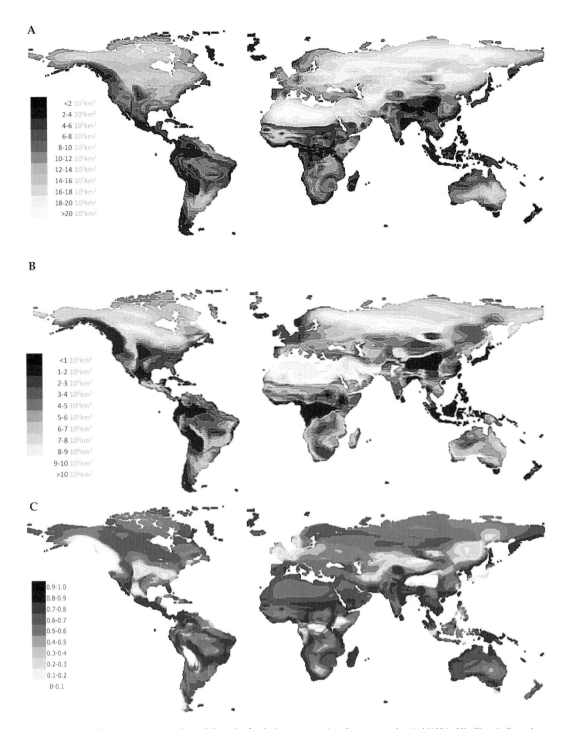

Figure 2.1 Climatically analogous area and invasibility index for the last pre-warming climate-normal period (1931–60). Climatically analogous areas are shown: within a 1000 km search radius (A), and within a global search radius (B). The invasibility index (C) is calculated as 1-(A/B): values close to 0 indicate that a high proportion of the climatically similar area is within 1000 km (i.e. potentially low invasibility of the natural vegetation); values close to 1 indicate that a high proportion of climatically similar area is further away than 1000 km (i.e. potentially high invasibility of the natural vegetation). The climate of each grid cell was quantified using four variables: summer/winter precipitation and temperature (see text for details).

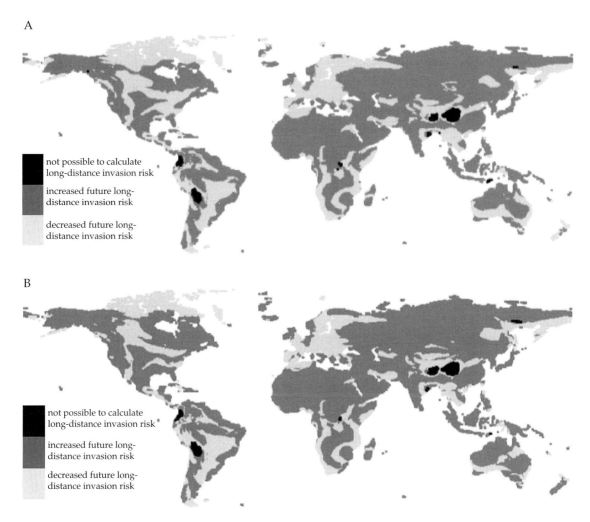

Figure 2.2 Change in Invasibility Index between the pre-warming 1931–60 and 2041–2050 periods using HadCM3 climate model and moderate B1 (A) and more severe A2 (B) greenhouse gas emission scenarios. Positive values (dark grey) indicate that a grid cell has a lower proportion of area with pre-warming analogous conditions within 1000 km in the 2045 period than in the pre-warming period (i.e., increased future long-distance invasion risk); negative values (light grey) indicate that a grid cell has a higher proportion of area with pre-warming analogous conditions within 1000 km in the 2045 period than in the pre-warming period (i.e., decreased future long-distance invasion risk). Changes in Invasibility Index could not be calculated for the black areas which are grid cells for which by 2045 no other grid cell had conditions analogous to those of the reference period.

(common) new species in most parts of the world, presumably because the species (and genotype) composition of the native biota is already usually well-balanced with the recent climate, soils, photoperiods, and so on. However, climate change is expected to reduce some aspects of local adaptation within species, and alter the outcome of interactions between species. Hence, climate change is expected to increase the invasibility of natural vegetation in much of the world (Fig. 2.2).

Climate change is accompanied by unprecedented rates of between-continent dispersal events, associated with ever-increasing levels of international trade and human movement. The combined effects of climate and rapid transport are likely to bring about worldwide changes to the earth's biota that could easily exceed the impact of either climate change or invasion acting on its own. Indeed, many of the species that will become major components of the new globalized biological communities are already present, either as rare naturalized species, or growing in our gardens. Here we provide just two examples to illustrate why we expect that the interaction between climate and invasive species will be such a potent combination.

Our first example is the emerging disease of amphibians, *Batrachochytrium dendrobatidis*, a fungal skin pathogen that has assumed a worldwide distribution in recent decades. This skin pathogen is already responsible for the presumed extinction of approximately two-thirds of harlequin frog species in the New World tropics (Pounds *et al.* 2006). Pathogen outbreaks and frog extinctions coincide with unusually warm years, such that extinctions seem to represent the combined impact of climate and species' invasion. Thus, the majority of currently known species-level extinctions associated with climate change (corresponding to over 1 per cent of the world's amphibian species; Thomas *et al.* 2006) seem to emerge from the interacting effect of climate and an invasive pathogen.

Our second example is the formation of a new international community of broad-leaved evergreen trees in the foothills of the Alps in Europe (Walther *et al.* 2002; Walther *et al.* 2007). As the winters have warmed and hard frosts have become rare, broad-leaved evergreen trees have established a dense understorey beneath the deciduous canopy—the biome is apparently in the process of shifting from deciduous to evergreen forest. The fascinating aspect of this change is that the majority of the broad-leaved tree (and palm) species do not originate from Europe, and have escaped from nearby gardens. This is entirely in line with our previous conclusions. Probably because of the ice-age climatic history of Europe, rather few native broad-leaved evergreen tree species survive in Europe—relative to the number of such species that exist in other parts of the world. As the climate has changed to suit this functional type of woody plant, the new vegetation that is becoming established does not "respect" the continental origins of each species, and the new vegetation has a truly international complexion.

These two examples suffice to illustrate what we believe may be commonplace in future. Species extinctions at both local and global levels will often involve climate-driven changes to the interactions between species, and specifically to the interaction between currently non-native and native species. And major ecosystem changes will commonly involve the formation of novel biotas that draw on species from several continents. The combination of climate change and invasion seems likely to generate virtually unrecognizable biological communities within the next few centuries, if not much sooner, in many parts of the world.

2.7 Love thine enemies—but not all of them!

Faced with potential wholesale re-arrangement of biological communities, how should we proceed? Inevitably, some non-native species will become common in future, although the majority will remain relatively rare. The difficulty is that it is not until the spread is well underway (and perhaps by that time impossible to reverse) that the rate of change and likely eventual abundance of the species becomes clear (Thomas *et al.* 2001; Williamson *et al.* this volume). In the face of climate change, the "invasive species" problem is one that we are now forced to consider for every species that colonizes new areas; an impossible task.

We can no longer presume that the arrival of species from other regions and countries should be regarded as negative—especially for species that are threatened within their existing range. Dispersal will usually be beneficial when species move within biological regions. Some of these species may become abundant or otherwise troublesome for humans, and may require ongoing management in exactly the same way that some existing abundant and/or undesirable members of native biotas require management. In contrast, the movement of species among continents, and from continents to oceanic islands, is expected to carry higher risks. Deliberately translocating species from one centre of endemism to another (e.g. to find a colder mountain) could potentially unleash multiple extinctions within the recipient communities.

Thousands of non-native garden plant species already grow within many countries. It is likely that trans-continental establishment of these plants will accelerate with climate change, and changes such as those described for the deciduous–evergreen transition will be repeated. In some cases, these changes may provide benefits in terms of ecosystem goods and services. If we presume that internationalized biotas will emerge most frequently in locations where the existing continental biota lacks sufficient species/functional types to form an invasion-resistant biological community, then invasion clearly has the potential to deliver more resilient ecosystems in many places (e.g. with increased fixation of carbon and soil/water retaining capacity). We should not presume that new biotas composed predominantly of non-native species are "all bad".

2.8 Conclusion

The "experiment" that humans have been engaged in over the last few centuries, of mixing-up the world's biota, is undergoing a step-change: the likelihood that species will establish in "natural" ecosystems will almost certainly accelerate rapidly as climate change destabilizes the native biota. New biological assemblages are likely to be created and they will recruit species from diverse parts of the world. Given the scale of the expected changes, gardening the planet to maintain the status quo is not an option.

A major concern is that species will not be able to spread across fragmented landscapes, and that species will therefore fail to survive climate change. The interaction between habitat loss/fragmentation and climate change has even been described as a deadly cocktail (Travis 2003). Species that suffer both habitat loss and declines in climatic suitability are undoubtedly a cause for concern. However, we suggest that the interaction between climate change and species' invasions could bring about even more fundamental changes to the world's biota. Introductions in the past have contributed to nearly 40 per cent of species-level extinctions in recent centuries, through predation, disease, competition, and by introgression between related taxa, but the majority of these extinctions have taken place on oceanic islands or in lake ecosystems (Groombridge 1992). In contrast, continental biotas have often been regarded as relatively resistant to invasion, except in disturbed environments. Now that native ecosystems are in the process of change, this resistance can no longer be presumed, and a new spate of continental extinctions and ecosystem changes seem to be emerging.

Acknowledgements

This work was partly funded by the EU FP6 Integrated Project ALARM ("Assessing large-scale risks for biodiversity with tested methods"). Climate data was kindly provided by Jack Williams.

References

Araújo, M.B., Thuiller, W., and Pearson, R.G. (2006). Climate warming and the decline of amphibians and reptiles in Europe. *Journal of Biogeography*, 33, 1712–28.

Baur, B., Cremene, C., Groza, G., *et al.* (2006). Effects of abandonment of subalpine hay meadows on plant and invertebrate diversity in Transylvania, Romania. *Biological Conservation*, 132, 261–73.

Brewer, S., Cheddadi, R., de Beaulieu, J.L., and *Reille, M.* (2002). The spread of deciduous *Quercus* throughout Europe since the last glacial period. *Forest Ecology and Management*, 156, 27–48.

Collingham, Y.C., Hill, M.O., and Huntley, B. (1996). The migration of sessile organisms: A simulation model with measurable parameters. *Journal of Vegetation Science*, 7, 831–46.

Coope, G.R. (1973). Tibetan species of dung beetle from Late Pleistocene deposits in England. *Nature*, 245, 335–36.

Cowley, M.J.R., Thomas, C.D., Thomas, J.A., and Warren, M.S. (1999). Flight areas of British butterflies: assessing species status and decline. *Proceedings of the Royal Society of London Series B-Biological Sciences*, 266, 1587–92.

Gaston, K.J. (1995). *Rarity*. Chapman & Hall, London.

Groombridge, B. (1992). *Global Biodiversity: Status of the Earth's Living Resources*. Chapman and Hall, London.

Hartley, S., Harris, R., and Lester, P.J. (2006). Quantifying uncertainty in the potential distribution of an invasive species: climate and the Argentine ant. *Ecology Letters*, 9, 1068–79.

Hickling, R., Roy, D.B., Hill, J.K., Fox, R., and Thomas, C.D. (2006). The distributions of a wide range of taxonomic groups are expanding polewards. *Global Change Biology*, 12, 450–55.

Hill, J.K., Collingham, Y.C., Thomas, C.D., *et al.* (2001). Impacts of landscape structure on butterfly range expansion. *Ecology Letters*, 4, 313–21.

Huntley, B. (1991). European postglacial forests—compositional changes in response to climate change. *Journal of Vegetation Science*, 1, 507–18.

Huntley, B.J. and Webb, T. (1998). *Vegetation History*. Kluwer Academic Publishers, Dordrecht.

IPCC (2007). *Climate Change 2007: The Physical Science Basis. Contribution of Working Group I to the Fourth Assessment Report of the Intergovernmental Panel on Climate Change*. In S. Solomon, D. Qin, M. Manning, Z. Chen, M. Marquis, K.B. Averyt, M. Tignor, and H.L. Miller eds. pp. 996. Cambridge University Press, Cambridge, UK and New York, NY, USA.

Melbourne, B.A., Cornell, H.V., Davies, K.F., *et al.* (2007). Invasion in a heterogeneous world: resistance, coexistence or hostile takeover? *Ecology Letters*, 10, 77–94.

Menéndez, R., Megias, A.G., Hill, J.K., *et al.* (2006). Species richness changes lag behind climate change. *Proceedings of the Royal Society B-Biological Sciences*, 273, 1465–70.

Midgley, G.F., Hannah, L., Millar, D., Rutherford, M.C., and Powrie, L.W. (2002). Assessing the vulnerability of species richness to anthropogenic climate change in a biodiversity hotspot. *Global Ecology and Biogeography*, 11, 445–51.

Ohlemüller, R., Gritti, E.S., Sykes, M.T., and Thomas, C.D. (2006a). Quantifying components of risk for European woody species under climate change. *Global Change Biology*, 12, 1788–99.

Ohlemüller, R., Gritti, E.S., Sykes, M.T., and Thomas, C.D. (2006b). Towards European climate risk surfaces: the extent and distribution of analogous and non-analogous climates 1931–2100. *Global Ecology and Biogeography*, 15, 395–405.

Parmesan, C. (2006). Ecological and evolutionary responses to recent climate change. *Annual Review of Ecology, Evolution and Systematics*, 37, 637–69.

Parmesan, C., Ryrholm, N., Stefanescu, C., *et al.* (1999). Poleward shifts in geographical ranges of butterfly species associated with regional warming. *Nature*, 399, 579–83.

Parmesan, C. and Yohe, G. (2003). A globally coherent fingerprint of climate change impacts across natural systems. *Nature*, 421, 37–42.

Peterson, A.T., Ortega-Huerta, M.A., Bartley, J., *et al.* (2002). Future projections for Mexican faunas under global climate change scenarios. *Nature*, 416, 626–29.

Pitelka, L.F., Gardner, R.H., Ash J., Berry, S., *et al.* (1997). Plant migration and climate change. *American Scientist*, 85, 464–73.

Pounds, J.A., Bustamante, M.R., Coloma, L.A., *et al.* (2006). Widespread amphibian extinctions from epidemic disease driven by global warming. *Nature*, 439, 161–167.

Pounds, J.A., Fogden, M.P.L., and Campbell, J.H. (1999). Biological response to climate change on a tropical mountain. *Nature*, 398, 611–15.

Ricciardi, A. and Cohen, J. (2007). The invasiveness of an introduced species does not predict its impact. *Biological Invasions*, 9, 309–15.

Richardson, D.M. and Thuiller, W. (2007). Home away from home—objective mapping of high-risk source areas for plant introductions. *Diversity and Distributions*, 13, 299–312.

Svenning, J.C. and Skov, F. (2004). Limited filling of the potential range in European tree species. *Ecology Letters*, 7, 565–73.

Tatem, A.J. and Hay, S.I. (2007). Climatic similarity and biological exchange in the worldwide airline transportation network. *Proceedings of the Royal Society B-Biological Sciences*, 274, 1489–96.

Theoharides, K.A. and Dukes, J.S. (2007). Plant invasion across space and time: factors affecting nonindigenous species success during four stages of invasion. *New Phytologist*, 176, 256–73.

Thomas, C.D. (1991). Habitat use and geographic ranges of butterflies from the wet lowlands of Costa-Rica. *Biological Conservation*, 55, 269–81.

Thomas, C.D., Bodsworth, E.J., Wilson, R.J., *et al.* (2001). Ecological and evolutionary processes at expanding range margins. *Nature*, 411, 577–81.

Thomas, C.D., Cameron, A., Green, R.E., *et al.* (2004). Extinction risk from climate change. *Nature*, 427, 145–148.

Thomas, C.D., Franco, A.M.A., and Hill, J.K. (2006). Range retractions and extinction in the face of climate warming. *Trends in Ecology & Evolution*, 21, 415–16.

Thomas, J.A. (1993). Holocene climate changes and warm man-made refugia may explain why a 6th of British butterflies possess unnatural early-successional habitats. *Ecography*, 16, 278–84.

Thuiller, W., Lavorel, S., Araújo, M.B., Sykes, M.T., and Prentice, I.C. (2005). Climate change threats to plant diversity in Europe. *Proceedings of the National Academy of Sciences of the United States Of America*, 102, 8245–50.

Tokeshi, M. (1993). Species abundance patterns and community structure. *Advances in Ecological Research*, 24, 112–86.

Travis, J.M.J. (2003). Climate change and habitat destruction: a deadly anthropogenic cocktail. *Proceedings of the*

Royal Society of London Series B-Biological Sciences, 270, 467–73.

Walther, G.R. (2003). Are there indigenous palms in Switzerland? *Botanica Helvetica*, 113, 159–80.

Walther, G.R. (2004). Plants in a warmer world. *Perspectives in Plant Ecology Evolution and Systematics*, 6, 169–85.

Walther, G.R., Beissner, S., and Burga, C.A. (2005). Trends in the upward shift of alpine plants. *Journal of Vegetation Science*, 16, 541–48.

Walther, G.R., Gritti, E.S., Berger, S., Hickler, T., Tang, Z.Y., and Sykes, M.T. (2007). Palms tracking climate change. *Global Ecology and Biogeography*, 16, 801–809.

Walther, G.R., Post, E., Convey, P., *et al.* (2002). Ecological responses to recent climate change. *Nature*, 416, 389–95.

Warren, M.S., Hill, J.K., Thomas, J.A., *et al.* (2001). Rapid responses of British butterflies to opposing forces of climate and habitat change. *Nature*, 414, 65–69.

Williams, J.W., Jackson, S.T., and Kutzbach, J.E. (2007). Projected distributions of novel and disappearing climates by 2100 AD. *Proceedings of the National Academy of Sciences of the United States of America*, 104, 5738–42.

Williams, J.W., Shuman, B.N., Webb, T., Bartlein, P.J., and Leduc, P.L. (2004). Late-quaternary vegetation dynamics in North America: Scaling from taxa to biomes. *Ecological Monographs*, 74, 309–34.

Williamson, M. (1996). *Biological Invasions*, Chapman and Hall, London.

Willis, K.J., Gillson, L., and Brncic, T.M. (2004). How "virgin" is virgin rainforest? *Science*, 304, 402–403.

CHAPTER 3

Climate and Invasive Species: The Limits to Climate Information

Mark New and Carol McSweeney

3.1 Introduction

Climate is a key explanatory variable for the distribution and diversity of species in space and time (Whittaker *et al.* 2001; Willis *et al.* 2007; Woodward 1987). It follows that climate should be a determining factor in both establishment potential and range of invasive alien species (IAS)—and climate has indeed been used as a variable (often along with other factors) to predict potential and observed ranges for IAS (Chen *et al.* 2007; Ficetola *et al.* 2007; Peterson *et al.* 2003; Robertson *et al.* 2004; Zhu *et al.* 2007).

These approaches to forecasting the range of (IAS) typically use an empirical "niche" or "envelope" based approach similar to that used to predict the range of native species under climate change (Araujo and New 2007; Pearson and Dawson 2003)—the environmental niche of a potential or detected IAS is defined using its native range and then used to project potential ranges in new regions. In some cases, the approach has been shown to work once invasion has been detected (e.g. Zhu *et al.* 2007), but in other cases there is evidence that the realized niche can change in the new environment. Reasons for this might include that the fundamental niche is not well observed in the source region, that the realized niche in the invaded region changes because of a "different biotic environment in the invaded region" (Broennimann *et al.* 2007), or a combination of these two factors. Furthermore, predicting the rate of expansion into a potential range is rarely possible as time referenced information on the spatial expansion of the IAS is required (Welk 2004).

Predicting establishment is even more difficult, and requires understanding of the many factors that affect the ability of a species to establish (Facon *et al.* 2006; Moles *et al.* 2008). Most invasions have occurred where environmental conditions have recently changed, for example through disturbance or altered resource availability, or conceivably in the future, through climate change. Other factors that may be important are release from natural predators or pathogens and availability of an unfilled ecological niche; the latter is potentially very important on islands where biodiversity is often impoverished because isolation presents barriers to species establishment.

Limitations to the use of environmental envelope models to predict the impact of climate change on species distributions have been widely explored (e.g. Botkin *et al.* 2007; e.g., Davis *et al.* 1998; Hampe 2004; Lawton 2000; Pearson and Dawson 2003) and include the need to represent dispersal, interspecies competition/interactions, and phenotypic plasticity. However, even more basically, bioclimatic models that are inadequately conceived may fail to predict invasive ranges because the incorrect climate variables are used. For example, Sutherst and Bourne (2008) show for an invasive tick species in Africa that a regression model conditioned on southern hemisphere climate data and distributions cannot predict its potential range in other areas (e.g. North Africa) because the seasonality of the climate drivers is different, and show that a model that links climate predictors to the seasonal phenology of the species is more appropriate. While there is clearly a need for process-based models that include ecological interactions and processes (e.g. Sutherst *et al.* 2007; Woodward and Beerling 1997), they require detailed ecological knowledge that is difficult to come by, difficult to model, and difficult to validate. In the meantime, envelope models remain the most widely used precisely because they bypass the need for representation of ecological processes (Pearson and Dawson 2004); for example the BIO-CLIM model has been used to assess more than 800 species (Hughes *et al.* 1996). A current area of development in envelope models aims to incorporate some mechanistic components, such as generalized population dynamics and dispersion (e.g. Keith *et al.* 2008).

Clearly, assessing whether an alien species poses a risk, in terms of whether it will become invasive and, if so, defining its potential invasive range, in a static climate is in itself a complex business [see refs in this volume]. A changing climate adds a further layer of difficulty through a number of new factors. At the most basic level, a changing climate can firstly expand the locations at which invasion/establishment might occur and secondly expand the potential range of any invading species. Climate change may also have secondary effects, for example through altering disturbance regimes such as fire or forcing changes in land use. There might also be an effect on the potential sources for IAS: changing patterns of production and export may create new sources, and new species may also expand into existing source areas. Finally, including climate change in an assessment of IAS risk requires information on how climatic variables of relevance to the case at hand will change.

Regardless of whether a statistical, process-based, or mixed process–statistical model is used to project species distribution changes, climatic data are needed to drive the projections. There is evidence that envelope models can be sensitive to different climate inputs during calibration (Peterson and Nakazawa 2008) and that projections of (non-invasive) species distributions in a changed climate are highly sensitive to the choice of driving climate data from global climate models (e.g. Thomas *et al.* 2004). The purpose of this chapter is therefore to describe the current state-of-the-science in observational climate dataset development and in climate change modeling and assess the ability of this science to support projections that are useful for assessing the effects of climate change on IAS risk.

3.2 Climate observations

Most models that aim to predict ranges for IAS are dependent on observed climate data for calibration and cross-validation. Typically, "gridded" climate datasets, derived from interpolation of an irregular network of meteorological station data (Hijmans *et al.* 2005; New *et al.* 1999, 2002) have been used as they provide complete spatial coverage. Indeed, most envelope type models rely on spatial autocorrelation of climate in these gridded datasets. Alternative sources of data include those derived from satellite remote sensing measurements (e.g. Huffman *et al.* 1997; Xie *et al.* 1996) or numerical weather prediction model re-analyses such as NCEP (Kalnay *et al.* 1996) and ERA40 (Uppala *et al.* 2005).

A common misconception is that the observed data are "truth" and that a poor fit of species data to climate information is due to inadequate IAS data or incorrectly designed IAS predictive modeling. A poor fit may simply be because the invasive species may be insensitive to climate with other barriers constraining its range. However, a poor fit to climate data may also be due to inaccuracies in the climate data—all the climate sources described above contain errors and/or uncertainties. For example, the $0.10°$ latitude/longitude climatology of New *et al.* (2002) reports mean generalized cross-validation errors (GCVEs) in interpolated temperature and precipitation. For precipitation, mean GCVEs of 10–20 per cent were estimated for areas with dense stations networks, such as North America and Western Europe, while GCVEs of 20–60 per cent were estimated for other regions with less dense station networks. For mean, maximum and minimum temperatures GCVEs vary from $0.5°C$ to $1.8°C$, depending on station density. While these mean error estimates, averaged over all stations in a given domain, provide an indication of gridded data accuracy it should be noted that true estimation error varies spatially as a function of interpolation scheme, station network, and the spatial heterogeneity of the climate variable of interest (for example, precipitation is generally more spatially heterogeneous and has larger interpolation errors). A number of the interpolation methods used in commonly available gridded data are unable to provide spatial error estimates; these include most distance based area averaging and polygon averaging. Geostatistical approaches such as thin-plate splines and kriging can provide spatially variable error surfaces; while the global climate datasets most commonly used in species modeling (Hijmans et al. 2005; New et al. 1999) are derived using thin-plate splines (Hutchinson and Gessler 1994), the associated error fields have not been incorporated into the publically available versions of these datasets. An exception is the gridded dataset of daily temperature and precipitation for Europe (Haylock et al. 2008) which includes a separate standard error field for each daily grid.

However, even if geostatistical error estimates are incorporated, they tend to underestimate the true interpolation error, as they provide an estimate of the goodness of fit of the interpolated surface to the available data. In many cases, the station network is not dense enough to capture finer spatial variability in the climate element of interest, but the interpolation produces a good fit to the sparser network, resulting in relatively small error estimates (Haylock et al. 2008). Thus a dataset such as WORLDCLIM (Hijmans et al. 2005) with a nominal 1 km resolution cannot be expected to capture the effects of local meteorology on mean climate at these scales unless the input station data are of similar spatial resolution; station network densities that can support such a high resolution of spatial interpolation are extremely rare.

While most envelope models are trained on long term average data, there is likely to be a need for time-varying climate data as species models become more mechanistic. For example, species invasions can be "triggered" by anomalous climate conditions that create a temporary change in the niche structure (Sutherst and Bourne 2008), or a species range may be more sensitive to climatic extremes such as frost occurrence, which have to be derived from daily data. If monthly (and also daily) data are required, additional accuracy issues arise, as the station network is almost invariably less dense than those used to construct climatologies, and also varies over time. This is illustrated for tropical Africa and South America in Fig. 3.1, which

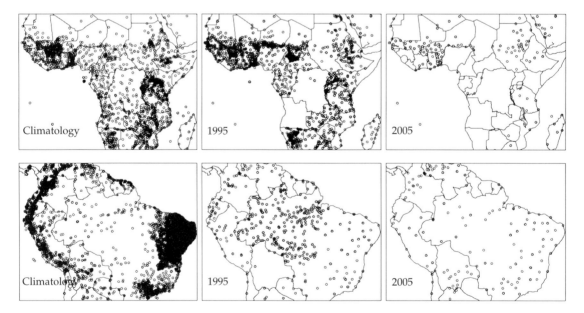

Figure 3.1 The station network that underlies the Climatic Research Unit gridded precipitation dataset (New et al. 1999; New et al. 2000), over example domains in South America and Africa. The left hand panel shows the station network for the 1961–90 long-term climatology, while the other panels show stations available for interpolation of at least one monthly anomaly field in each of two selected years, 1995 and 2005.

shows the changes in precipitation station network density between climatological means, and monthly data for the mid-1990s and mid-2000s. Monthly climate fields (especially precipitation) are less coherent than long term climatological fields, so for a given level of accuracy, more station data would be needed for monthly grids than climatology grids. One should therefore expect monthly grids to be less accurate than climatology grids.

The magnitude of errors one might expect from the existing station networks is illustrated through use of high resolution satellite data, in this case the 0.25° latitude/longitude TRMM (Tropical Rainfall Monitoring Mission) satellite precipitation dataset over tropical Africa. We use the spatially complete dataset as "reality"—assuming that while the satellite dataset may be biased in the absolute amounts of precipitation (New et al. 2001) it captures the spatial variability in monthly and long term mean rainfall. We select those satellite pixels that are co-located with actual meteorological stations in the region as our "station network", and then interpolate the "station network" to estimate the full 0.25° fields, on a monthly and long-term climatologic basis.

Results for interpolation of an example monthly climatology and an individual month are shown in Fig. 3.2, for tropical Africa, during July which is the middle of the West African rainy season. For the interpolation of long term climatologic data, errors away from the "virtual stations" can be quite large, with only 40 per cent of pixels falling within 10 per cent of the true values, 40 per cent of pixels having errors of between 10 and 50 per cent, and 20 per cent having even larger relative errors. In absolute terms, over 25 per cent of the grid points in the analysis domain have absolute errors greater than 10 mm. When interpolation for an individual month is considered (Fig. 3.2, right) the accuracy of the interpolated field deteriorates. Now only 20 per cent of interpolated pixels are within 10 per cent of the "true" values, 60 per cent of pixels have errors between 10–50 per cent and 20 per cent of pixels have errors greater than 50 per cent. Absolute errors also increase, with 62 per cent of pixels having errors greater than 10 mm. Of further note is that at both monthly and climatology scales, interpolation is biased towards positive errors (overestimation), most likely because station networks themselves tend to be preferentially located in wetter areas.

There has not been any assessment of the extent to which uncertainty in climate inputs to IAS distribution models may affect model accuracy, though Peterson and Nakazawa (2008) have shown that different datasets describing nominally the same variables can produce different results. It is likely that where the IAS distribution data itself is poor, climate data will not be *the* major source of error. However, as new climate datasets with better uncertainty estimates become available, these uncertainties should be incorporated in the training/fitting of IAS distribution models.

3.3 Climate models

Global climate models (GCMs) are generally considered to be the only approach to providing physically-based quantitative data on how the climate system might respond to future changes in radiative forcing of the atmosphere—either human or natural in origin (Meehl et al. 2007). GCMs have evolved from atmospheric numerical weather prediction models; those used in the most recent IPCC assessment (IPCC 2007a) represent the major components of the climate system of relevance to climate change over timescales of 25 to 100 years: atmosphere and radiation processes, ocean, land surface and topography, sea, and land ice. A few models now also incorporate biogeochemical processes, such as the terrestrial and ocean carbon cycle and atmospheric chemistry (Friedlingstein et al. 2006), but most currently available data are from GCMs without built-in biogeochemical processes.

GCMs simulate most of the processes important for climate, either through physically-based fluid-dynamic equations or through parameterization—for processes that cannot be resolved at the scales of numerical space-time discretization in the GCM or where there is inadequate scientific understanding for process simulation. Examples of parameterizations include cloud formation, cloud microphysics, precipitation processes, momentum and heat exchanges between the atmosphere and the ocean, ocean eddies/convection, and sea-ice dynamics.

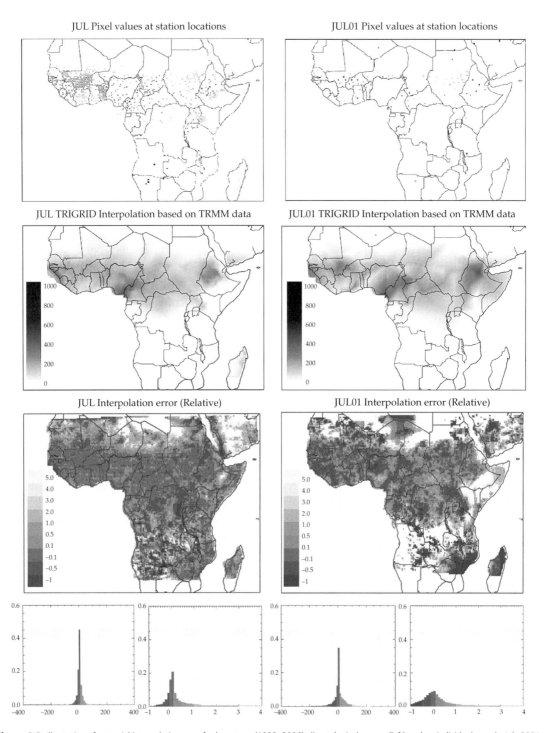

Figure 3.2 Illustration of potential interpolation error for long-term (1998–2006) climatological means (left) and an individual month, July 2001 (right). The top panel shows the TRMM Satellite Rainfall data, sampled only at grid points (pixels) at which station data are available. The second row shows precipitation interpolated from the sampled grid points, using a simple triangulation gridding scheme. The third panel shows the interpolation "error" at each grid point, defined as the relative difference between the interpolated values and the measured TRMM values (interpolated divided by TRMM). The bottom panel shows the frequency distribution of errors of different magnitude, with errors expressed as mm/day and relative terms (as for third panel) in the left and right of each pair of histograms.

Despite their physical basis and parameterization of key processes important for climate change all GCMs struggle to simulate aspects of contemporary climate patterns and processes. Similarly, projections of climate change from different GCMs vary widely; at the most basic level, GCMs used by the IPCC show a range of end-century global temperature changes of up to 2 °C, for any one greenhouse gas concentration scenario. These arise from differences among models in a number of factors (Meehl *et al.* 2007), including radiative processes simulated by the models, rates of heat uptake by the oceans, and numerous interacting feedbacks such as clouds, water vapour, and snow and sea ice. The upper limit to these ranges is particularly difficult to constrain and recent experiments with simpler models, as well as large-ensembles of GCMs that explore parameter uncertainties, suggest that global temperature changes due to a doubled CO_2 concentrations of up to 10 °C cannot be ruled out (e.g. Andronova and Schlesinger 2001; Forest *et al.* 2001; Knutti *et al.* 2003; Lopez *et al.* 2006; Murphy *et al.* 2004).

At regional and more local scales, GCM climate change projections often diverge even more, due to a combination of global and local responses to the global forcing (e.g. changes in atmospheric and ocean circulation; representation of topography and the land-surface in a particular model). This is particularly the case for thermally-indirect variables such precipitation and climatic extremes; the latter often related to specific, intense, circulation features. The lower resolution limit for data from a GCM is, of course, the grid resolution, but for many variables the "skill resolution" at which predictions are meteorologically plausible is much larger.

3.4 Emissions and other forcing of climate

In addition to uncertainties inherent to climate models, future anthropogenic and emissions are impossible to predict precisely, as they depend both on the future "unknowable" socio-economic futures and stochastic or semi-stochastic variations in natural forcing, primarily volcanism, and solar activity.

In the face of this emissions uncertainty, the climate change community makes use of emissions scenarios, which estimate future emissions as a function of "storylines" of future population growth, socio-economic wealth, and energy demand and sources—the so-called IPCC SRES (Special Report on Emissions Scenarios) scenarios. The SRES emissions vary widely, as do more recent scenarios for stabilization or emissions-reductions policy. An important point to note is that since the SRES scenarios were published in the 1990s, real-world emissions have tracked and are now exceeding the most extreme of the scenarios (Anderson and Bows 2008; Rahmstorf *et al.* 2007). Thus GCM projections driven by top of the range emissions scenarios are probably more realistic than lower emissions scenarios, at least for the next few decades.

An often unrecognized additional uncertainty in estimating forcing of climate relates to biogeochemical cycling of GHGs and aerosols. The majority of model runs in the IPCC (IPCC 2007a) report do not have integrated carbon and aerosol cycles and make use of concentrations estimated by off-line biogeochemical cycling models. These offline concentration estimates provide a single concentration trajectory associated with each emissions scenario, whereas uncertainties in carbon cycling and additional uncertain feedbacks between biogeochemical cycling and global warming mean that any emissions scenario can be associated with a wide range of future concentrations. Recent work has suggested that current uncertainties in concentrations arising from *one* emissions scenario are as large as the range between the single concentration trajectories associated with each emissions scenario in the 2007 IPCC GCM simulations (Friedlingstein *et al.* 2006).

3.5 Approaches to the use of uncertain climate information

In the absence of exact predictions of climate change from climate models a number of approaches have been used to assess the potential consequences of climate change on a system or feature of interest.

3.5.1 Scenarios

Scenarios acknowledge the limitations in models and forcing information by taking one or a few combinations of models and forcing to derive one or a few trajectories through forcing—climate-response space. The emphasis is therefore on an exploration of possible impacts in the future, rather than a quantitative assessment of risk; typically all scenarios are considered equally likely. This has been the approach in the nearly all of the published literature, including that reviewed in recent IPCC reports (IPCC 2001, 2007b).

3.5.2 Probabilistic projections

Probabilistic projections attempt, using a range of approaches, to address climate change uncertainty in quantitative manner. The emphasis therefore is on the likelihood of climate change and/or system response across climate change space. Climate model data underlying probabilistic projections come from two main sources: "ensembles of opportunity" and perturbed-physics ensembles.

Ensembles of opportunity (EOEs) make use of existing outputs from climate modeling centers around the world and therefore do not have a pre-defined experimental design; however, the underlying rationale is that a large sample of models will necessarily span climate prediction uncertainty. In reality, there is an informal experimental design, in that the IPCC has encouraged modeling centers to use specific emissions scenarios and specific time periods for their simulations. For the most commonly used scenarios and common climate variables, such as temperature and precipitation, as many as 21 different climate models are available, some as small (up to 4 members) initial-condition ensembles.

Perturbed-physics ensembles (PPEs) are Monte-Carlo type experiments where several hundred to many thousands of climate model simulations are performed in order to sample uncertainty in the representation of key physical processes in the model. To date the main examples of this approach are the UK Met Office QUMP and the Oxford University led climate*prediction*.net projects (Murphy *et al*. 2004; Stainforth *et al*. 2005).

Both EOEs and PPEs have limitations. To date, PPEs have been restricted to a single model, the UK Met Office third generation coupled model, and so only sample the climate response across physics parameters in a single model type. Conversely, EOEs sample a single physics configuration in each model type. The ideal is clearly a multi-model PPE, but this has yet to be achieved.

Deriving a probability distribution for climate change involves estimating the characteristics of the "population" of climate model responses from the available model data (the sample). A number of different approaches have been adopted, most of which are Bayesian in philosophy (Fowler *et al*. 2007; Knutti *et al*. 2003; Lopez *et al*. 2006; Murphy *et al*. 2007; New and Hulme 2000; Rougier 2007; Stott and Kettleborough 2002) Nearly all make assumptions about the prior distribution of model parameters and/or variables and most use observational constraints to weight the posterior distribution. Observational constraints are used to assess model "goodness"; constraints that have been used include a model's ability to simulate the present day patterns of key variables such as temperature and precipitation (e.g. Lopez *et al*. 2006), multivariate indices of present day climate (Murphy *et al*. 2004) and ability to simulate the observed transient climate change response (Stott and Kettleborough 2002).

3.5.3 A middle ground?

A key characteristic—or criticism (e.g. Dessai and Hulme 2004; Hall 2007)—of any probability distribution of climate change is that the distribution is highly conditional upon, for example, experimental design, sample size, model type(s), observational constraints, and the statistical model used. Thus the probabilities are not a statement of the likelihood of future climate changes, but a "numerical summary of [the] state of knowledge about a proposition" (Rougier 2007: 248), in this case climate change. There is therefore real danger for the misuse of probabilities to support quantitative risk-based decisions on climate change, only for the probabilities to change as new data, observations, and methods are employed. Indeed, some authors suggest that we should avoid using probabilities and assess impacts within a "bounding box" or range of climate that we cannot rule out in the face of available data (Smith 2002; Stainforth et al. 2007).

A middle ground between probability distributions and bounding boxes is to use "qualitative" likelihoods that provide information on the relative likelihood of climate change, but prevent the use of probabilities in a quantitative manner. This approach recognizes that post-processing of model data can add value by providing some information on the likelihood of changes across a bounding box, but also that the exact shape of a distribution might change as new data/methods are employed.

This is illustrated in Fig. 3.3, which shows an analysis of results over North America from an early PPE experiment of climate*prediction*.net. The experiment, described in detail in Stainforth et al. (2005), comprises some 2400 simulations with the HadSM3 climate model. In this experiment, each perturbed-physics model version is run for 15 years at preindustrial CO_2 concentrations and then a further 15 years after an instantaneous doubling of CO_2. As models with a high climate sensitivity have not equilibrated to doubled CO_2 after 15 years, the spatial pattern of climate change at equilibrium is estimated using the scaling approach of New et al. (2007). The top panel shows the type of result one might achieve using a probabilistic analysis, where the likelihood or frequency of a given index—in this case August mean temperature exceeding 25 °C—is displayed. While the analysis here is in fact based on the empirical frequencies across the ensemble, similar results can be achieved through post-processing to estimate conditional probabilities. The map shows both the spread of model predictions—shown by the boundary of the non-zero probabilities—and the probability of the threshold being reached at any particular location. As might be expected, the highest probabilities are closer to the present day distribution of the threshold, and lowest probabilities further to the north, where fewer model realizations have a large enough temperature response to cross the threshold.

The lower panel in Fig. 3.3 presents the same underlying model data, but presented in an alternative form. Recognizing that the distribution would change if a different set of models and/or post-processing is used, we present the information in a way that does not assign probabilities anywhere within the predicted range. Each model run is assigned a skill score, based on the joint spatial correlation between present day simulated and observed temperature and precipitation. This score is similar to the "Climate Prediction Index" used by Murphy et al. (2004), but uses only two variables. Based on this score, one model is considered relatively "better" than another if the overall correlation is higher. Then, for each location where at least one model run produces a climate change that exceeds the 25 °C threshold, the skill score of the best performing model is mapped. This provides a map showing the range of locations at which one cannot rule out a threshold being exceeded (the bounding box) *and* a measure of how good the best performing model is. This method therefore makes no assumptions about the probability, which is partly dependent on the underlying sample. The advantage of this approach is that it provides more information than just defining a bounding box—it enables a user of the information to make use of qualitative information on relative model performance across the distribution, but doesn't permit them to assign probabilities.

3.6 Discussion

A large weight of empirical evidence has shown that climate is a strong predictor of actual and potential distributions of species, both native and alien invasive. This relationship forms the basis for climate-envelope species distribution models that are used to predict potential ranges of IAS in the present day, and also both IAS and native species under climate change. These correlative models have many well-documented limitations, but in the absence of more detailed data and improved methods to calibrate process-based models more rapidly, we are hampered in our ability to develop process-based models at the species level. Thus, correlative models will remain a tool for projecting potential ranges of IAS under climate change, perhaps with enhancement to account for dispersal and other dynamics. Studies of the prediction of future ranges of native species show that the use of ensembles of species distribution models provides a better

38 THE DRIVERS OF BIOLOGICAL INVASIONS

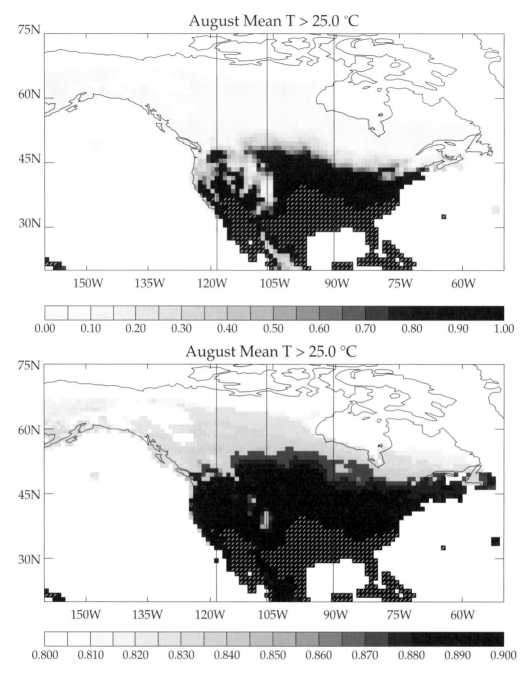

Figure 3.3 Different approaches to the use of a large perturbed physics ensemble to provide climate change likelihoods. Each map shows the likelihood of mean temperature in August exceeding 25 °C in a 2 × CO_2 climate in a 2500 member ensemble of the climateprediction.net experiment, but with the likelihoods expressed in different ways. Top: a simple accounting of the spatially-varying frequency of model runs that exceed the 25 °C threshold at 2 × CO_2. For reference, white dashes indicate areas of the domain in the present day where August mean temperature exceeds 25 °C. Bottom: spatial pattern of occurrence of mean August temperature exceeding 25 °C, colour-coded by the performance index of the "best" model among all those that exceed the threshold at each location; the performance index is a composite correlation index for each model compared to observed climate, and the "best" model is defined as the one with the highest index.

representation of the predictive uncertainty than use of a single model (Araujo and New 2007). Similarly, ensemble modeling of IAS is necessary if predictive uncertainty is to be quantified.

An important component of quantifying the predictive uncertainty of species distribution models is the assessment of the role of uncertainties in the climate data used to calibrate the model—as we have shown these can be quite large in areas with poor meteorological station networks and in areas with complex terrain. While most climate datasets do not come with associated uncertainty estimates, this is likely to change in the next few years; as these become available, their inclusion in the model calibration procedure will enable quantification of the uncertainty in this part of the prediction process.

Model projections of future climate result in a very wide spread of possible changes. These arise from well documented uncertainties in: future greenhouse gas emissions; biogeochemical cycling of GHGs and aerosols; representation of key climate processes in GCMs. Thus, use of GCM projections to drive IAS models requires careful analysis of the climate model uncertainty, in combination with the IAS model uncertainties. A number of approaches to assessing GCM uncertainty through analysis of inter-model variance in ensembles have been developed, all with advantages and limitations. However, all methods rely on the assumption that GCMs project in some realistic (but limited) way onto the real climate system. This is partly provable for the present climate, but cannot be proved for the future, and GCM uncertainties may not reflect the true prediction uncertainty.

References

Anderson, K. and Bows, A. (2008). Reframing the climate change challenge in light of post-2000 emission trends. *Philosophical Transactions of the Royal Society A-Mathematical Physical and Engineering Sciences*, **366**(1882), 3863–82.

Andronova N. G. and Schlesinger, M. E. (2001). Objective estimation of the probability density function for climate sensitivity. *Journal of Geophysical Research-Atmospheres*, 106 (D19), 22605–11.

Araujo, M. B. and New, M. (2007). Ensemble forecasting of species distributions. *Trends in Ecology & Evolution*, **22**(1), 42–47.

Botkin, D. B., *et al.* (2007). Forecasting the effects of global warming on biodiversity. *Bioscience*, **57**(3), 227–36.

Broennimann, O., Treier, U. A., Muller-Scharer, H., Thuiller, W., Peterson, A. T., and Guisan, A. (2007). Evidence of climatic niche shift during biological invasion. *Ecology Letters*, **10**(8), 701–09.

Chen, H., Chen, L. J., and Albright, T. P. (2007). Predicting the potential distribution of invasive exotic species using GIS and information-theoretic approaches: A case of ragweed (Ambrosia artemisiifolia L.) distribution in China. *Chinese Science Bulletin*, **52**(9), 1223–30.

Davis, A. J., Jenkinson, L. S., Lawton, J. H., Shorrocks, B., and Wood, S. (1998). Making mistakes when predicting shifts in species range in response to global warming. *Nature*, **391**(6669), 783–86.

Dessai, S. and Hulme, M. (2004). Does climate adaptation policy need probabilities? *Climate Policy*. **4**(2), 107–28.

Facon, B., Genton, B. J., Shykoff, J., Jarne, P., Estoup, A., and David, P. (2006). A general eco-evolutionary framework for understanding bioinvasions. *Trends in Ecology & Evolution*, **21**(3), 130–35.

Ficetola, G. F., Thuiller, W., and Miaud, C. (2007). Prediction and validation of the potential global distribution of a problematic alien invasive species - the American bullfrog. *Diversity and Distributions*, **13**(4), 476–85.

Forest, C. E., Allen, M. R., Sokolov, A. P., and Stone, P. H. (2001). Constraining climate model properties using optimal fingerprint detection methods. *Climate Dynamics*, **18**(3–4), 277–95.

Fowler, H. J., Blenkinsop, S., and Tebaldi, C. (2007). Linking climate change modelling to impacts studies: recent advances in downscaling techniques for hydrological modelling. *International Journal of Climatology*, **27**(12), 1547–78.

Friedlingstein, P., *et al.* (2006). 'Climate-carbon cycle feedback analysis: Results from the (CMIP)-M-4 model intercomparison'. *Journal of Climate*, **19**(14), 3337–53.

Hall, J. (2007). Probabilistic climate scenarios may misrepresent uncertainty and lead to bad adaptation decisions. *Hydrological Processes*, **21**(8), 1127–29.

Hampe, A. (2004). Bioclimate envelope models: what they detect and what they hide. *Global Ecology and Biogeography*, **13**(5), 469–71.

Haylock, M., Hofstra, N., Klein Tank, A. M. G., New, M., Klok, E. J., and Jones, P. D. (2008). A European daily high-resolution gridded dataset of surface temperature and precipitation for 1950–2006. *Journal of Geophysical Research - Atmospheres*, in press.

Hijmans, R. J., Cameron, S. E., Parra, J. L., Jones, P. G., and Jarvis, A. (2005). Very high resolution interpolated climate surfaces for global land areas. *International Journal of Climatology*, **25**(15), 1965–78.

Huffman, G. J., et al. (1997). The Global Precipitation Climatology Project (GPCP) Combined Precipitation Dataset. *Bulletin of the American Meteorological Society*, **78**, 5–20.

Hughes, L., Cawsey, E. M., and Westoby, M. (1996). Climatic range sizes of Eucalyptus species in relation to future climate change. *Global Ecology and Biogeography Letters*, **5**(1), 23–29.

Hutchinson, M. F. and Gessler, P. E. (1994). Splines - more than just a smooth interpolator. *Geoderma*, **62**, 45–67.

IPCC (2001). *Climate Change 2001: Impacts, Adaptation, and Vulnerability. Contribution of Working Group II to the third assessment report of the Intergovernmental Panel on Climate Change*, eds J. J. McCarthy, O. F. Canziani, N. A. Leary, D. J. Dokken, and K. S. White Cambridge University Press, Cambridge, UK, 1032.

IPCC (2007a). *Climate Change 2007: The Physical Science Basis. Contribution of Working Group I to the Fourth Assessment Report of the Intergovernmental Panel on Climate Change*, eds S. Solomon, et al., Cambridge University Press Cambridge, UK, 996.

IPCC (2007b). *Climate Change 2007: Impacts, Adaptation and Vulnerability. Contribution of Working Group II to the Fourth Assessment Report of the Intergovernmental Panel on Climate Change*, eds M. L. Parry, O. F. Canziani, J. P. Palutikof, P. J. v. d. Linden, and C. E. Hanson, Cambridge University Press, Cambridge, UK, 976.

Kalnay, E., et al. (1996). The NCEP/NCAR 40-year reanalysis project. *Bulletin of the American Meteorological Society*, **77**, 437–71.

Keith, D. A., et al. (2008). 'Predicting extinction risks under climate change: coupling stochastic population models with dynamic bioclimatic habitat models'. *Biology Letters*, **4**(5), 560–63.

Knutti, R., Stocker, T. F., Joos, F., and Plattner, G. K. (2003). Probabilistic climate change projections using neural networks. *Climate Dynamics*, **21**(3–4), 257–72.

Lawton, J. H. (2000). Concluding remarks: a review of some open questions, in M. J. Hutchings, E. John, and A. J. A. Stewart (eds.). *Ecological Consequences of Heterogeneity*, Cambridge University Press, Cambridge, UK, 401–24.

Lopez, A., Tebaldi, C., New, M., Stainforth, D., Allen, M., and Kettleborough, J. (2006). Two approaches to quantifying uncertainty in global temperature changes. *Journal of Climate*, **19**(19), 4785–96.

Meehl, G. A., et al. (2007). Global Climate Projections, in S. Solomon, et al. (eds.), *Climate Change 2007: The Physical Science Basis. Contribution of Working Group I to the Fourth Assessment Report of the Intergovernmental Panel on Climate Change*, Cambridge University Press, Cambridge), 747–846.

Moles, A. T., Gruber, M. A. M., and Bonser, S. P. (2008). A new framework for predicting invasive plant species. *Journal of Ecology*, **96**(1), 13–17.

Murphy, J. M., et al. (2004). Quantification of modelling uncertainties in a large ensemble of climate change simulations. *Nature*, **430**, 768–72.

Murphy, J. M., Booth, B. B. B., Collins, M., Harris, G. R., Sexton, D. M. H., and Webb, M. J. (2007). A methodology for probabilistic predictions of regional climate change from perturbed physics ensembles. *Philosophical Transactions of the Royal Society A: Mathematical, Physical and Engineering Sciences*, **365**(1857), 1993–2028.

New, M., Hulme, M., and Jones, P. (1999). Representing twentieth-century space-time climate variability. Part I: Development of a 1961–90 mean monthly terrestrial climatology. *Journal of Climate*, **12**, 829–56.

New, M. and Hulme, M. (2000). Representing uncertainty in climate change scenarios: a monte-carlo approach. *Integrated Assessment*, **1**, 203–13.

New, M., Hulme, M., and Jones, P. (2000). Representing twentieth-century space-time climate variability. Part II: Development of 1901–1996 monthly grids of terrestrial surface climate. *Journal of Climate*. **13**(13), 2217–38.

New, M., Todd, M., Hulme, M., and Jones, P. (2001). Precipitation measurements and trends in the twentieth century. *International Journal of Climatology*, **21**(15), 1889–922.

New, M., Lopez, A., Dessai, S., and Wilby, R. (2007). Challenges in using probabilistic climate change information for impact assessments: an example from the water sector. *Philosophical Transactions of the Royal Society A: Mathematical, Physical and Engineering Sciences*, **365**(1857), 2117–31.

New, M. G., Lister, D., Hulme, M., and Makin, I. (2002). A high-resolution data set of surface climate for terrestrial land areas. *Climate Research*, **21**, 1–25.

Pearson, R. G. and Dawson, T. P. (2003). Predicting the impacts of climate change on the distribution of species: are bioclimate envelope models useful?'. *Global Ecology and Biogeography*, **12**(5), 361–71.

—— (2004). Bioclimate envelope models: what they detect and what they hide - response to Hampe (2004). *Global Ecology and Biogeography*, **13**(5), 471–73.

Peterson, A. T., Papes, M., and Kluza, D. A. (2003). Predicting the potential invasive distributions of four alien plant species in North America. *Weed Science*, **51**(6), 863–68.

Peterson, A. T. and Nakazawa, Y. (2008). Environmental data sets matter in ecological niche modelling: an example with Solenopsis invicta and Solenopsis richteri. *Global Ecology and Biogeography*, **17**(1), 135–44.

Rahmstorf, S., *et al.* (2007). Recent climate observations compared to projections. *Science*, **316**(5825), 709–09.

Robertson, M. P., Villet, M. H., and Palmer, A. R. (2004). A fuzzy classification technique for predicting species' distributions: applications using invasive alien plants and indigenous insects. *Diversity and Distributions*, **10**(5–6), 461–74.

Rougier, J. (2007). Probabilistic inference for future climate using an ensemble of climate model evaluations. *Climatic Change*, **81**(3), 247–64.

Smith, L. A. (2002). What might we learn from climate forecasts?. *PNAS*, **99**(90001), 2487–92.

Stainforth, D. A., *et al.* (2005). Uncertainty in predictions of the climate response to rising levels of greenhouse gases. *Nature*, **433**(7024), 403–06.

Stainforth, D. A., Allen, M. R., Tredger, E. R., and Smith, L. A. (2007). Confidence, uncertainty and decision-support relevance in climate predictions. *Philosophical Transactions of the Royal Society A: Mathematical, Physical and Engineering Sciences*, **365**(1857), 2145–61.

Stott, P. A. and Kettleborough, J. A. (2002). Origins and estimates of uncertainty in predictions of twenty-first century temperature rise. *Nature*, **416**(6882), 723–26.

Sutherst, R. and Bourne, A. (2008). Modelling non-equilibrium distributions of invasive species: a tale of two modelling paradigms. *Biological Invasions*, published online, August 2008.

Sutherst, R. W., Maywald, G. F., and Bourne, A. S. (2007). Including species interactions in risk assessments for global change. *Global Change Biology*, **13**(9), 1843–59.

Thomas, C. D., *et al.* (2004). Extinction risk from climate change. *Nature*, **427**(6970), 145–48.

Uppala, S. M., *et al.* (2005). The ERA-40 re-analysis. *Quarterly Journal of the Royal Meteorological Society*, **131**(612), 2961–3012.

Welk, E. (2004). Constraints in range predictions of invasive plant species due to non-equilibrium distribution patterns: Purple loosestrife (Lythrum salicaria) in North America. *Ecological Modelling*, **179**(4), 551–67.

Whittaker, R. J., Willis, K. J., and Field, R. (2001). Scale and species richness: towards a general, hierarchical theory of species diversity. *Journal of Biogeography*, **28**(4), 453–70.

Willis, K. J., Kleczkowski, A., New, M., and Whittaker, R. J. (2007). Testing the impact of climate variability on European plant diversity: 320 000 years of water-energy dynamics and its long-term influence on plant taxonomic richness. *Ecology Letters*, **10**, 673–79.

Woodward, F. I. (1987). *Climate and Plant Distribution*, Cambridge University Press, Cambridge, UK.

Woodward, F. I. and Beerling, D. J. (1997). The dynamics of vegetation change: health warnings for equilibrium 'dodo' models. *Global Ecology and Biogeography Letters*, **6**(6), 413–18.

Xie, P. P., Rudolf, B., Schneider, U., and Arkin, P. A. (1996). Gauge-based monthly analysis of global land precipitation from 1971 to 1994. *Journal of Geophysical Research-Atmospheres*, **101**, 19023–34.

Zhu, L., Sun, O. J., Sang, W. G., Li, Z. Y., and Ma, K. P. (2007). Predicting the spatial distribution of an invasive plant species (Eupatorium adenophorum) in China. *Landscape Ecology*, **22**(8), 1143–54.

CHAPTER 4

Globalization and Invasive Alien Species: Trade, Pests, and Pathogens

Charles Perrings, Eli Fenichel, and Ann Kinzig

4.1 Identifying the problem

Globalization—especially the widening and deepening of international trade—has had a number of environmental consequences, of which the most significant may well be the redistribution of pest species in general, and pathogens in particular. Particular attention is now being given to emergent epizootic diseases in humans (Jones *et al.* 2008; Hubalek 2003), although these are merely the most publicized examples of a widespread and long-standing process of species redistribution (Williamson 1996; Daszak *et al.* 2000; McNeely 2001; Perrings *et al.* 2002; Perrings *et al.* 2005). The process has been accelerated by the growth of world trade. The opening of new markets or trade routes has resulted in the introduction of new species either as the object of trade or as the unintended consequence of trade. The growth in the volume of trade along existing routes has increased the frequency with which new introductions are repeated, and hence the probability that an introduced species will establish and spread (e.g. Enserink 1999; Cassey *et al.* 2004; Semmens *et al.* 2004). Other things being equal, the more open economies are, the more vulnerable they are to species introductions (Dalmazzone 2000; Vilà and Pujadas 2001). Indeed, the volume and direction of trade turn out to be good empirical predictors of invasions (Levine and d'Antonio 2003; Costello *et al.* 2007).

The costs of invasive species include both direct and indirect effects. Direct effects comprise losses associated with pests and pathogens in valued local species, whether native or exotic. While there are no very good estimates of these costs, it is clear that they disproportionately affect countries that are heavily dependent on the exploitation of natural resources. Pimentel *et al.* (2001) concluded that at the close of 19th century invasive species cause annual damage equal to 53 per cent of agricultural GDP in the USA, 31 per cent in the UK and 48 per cent in Australia, but 96 per cent, 78 per cent, and 112 per cent of agricultural GDP in South Africa, India, and Brazil respectively. The indirect effects have not been evaluated at all. Introduced pathogens, predators, or competitors have, for example, been implicated in the loss of native species over a wide range of ecosystems (Daszak and Cunningham 1999; Williamson 1996). This in turn affects the capacity of ecosystems to deliver the services that underpin much economic activity (Millennium Ecosystem Assessment 2005), and to absorb anthropogenic and environmental stresses and shocks without losing resilience (Kinzig *et al.* 2006; Walker *et al.* 2004; Walker *et al.* 2006). Maintenance of functional diversity, in particular, supports the provision of ecosystem services over a range of environmental conditions (e.g. Kinzig *et al.* 2002; Loreau *et al.* 2003; Naeem and Wright 2003; Reich *et al.* 2004; Hooper *et al.* 2005).

The risk posed by an introduced species depends on a number of factors: some of which are within the control of regulatory authorities and some of which are not. It is a product of the probability that imports will introduce new species and the expected damage done by those species. Among the risk factors taken as exogenous by the regulatory authorities are the invasiveness of the species itself (whether it is a generalist or a specialist, its plasticity, and so on), the bioclimatic similarity between the trading partners, the volume and composition of trade, and the vulnerability of the host (economic and ecological) system. The sanitary and phytosanitary (SPS) measures taken by the exporter may or may not be taken as exogenous. These together determine both the likelihood that a unit of trade will introduce species of different types and the expected damage if a particular species (or class of species) is introduced. The risk factors taken as endogenous by the regulatory authorities comprise the sanitary and phytosanitary efforts of the host country and the degree to which introduced species that have already established are controlled (Williamson 1999).

To the extent that damage costs are taken into account, they are reflected in SPS measures either at the point of export, the point of import, or en route. In general, this implies inspection and interception programs (IIP). One reason for this is that IIP at the port of entry is one of a very few actions admitted under the General Agreement on Tariffs and Trade (GATT). Signatories to the GATT may take precautionary defensive action against trading partners under the Sanitary and Phytosanitary Agreement that include temporary trade bans, but the default strategy is IIP at the port of entry. A number of studies have considered a wider range of policy responses to the trade-related risks posed by biological invasions including the use of tariffs on imports to reduce the damage costs from accidental introductions, but these are not currently consistent with the GATT (Perrings *et al.* 2002; Kohn and Capen 2002; Costello and McAusland 2003; McAusland and Costello 2004).

The questions we address in this chapter concern the nature of the SPS response to the invasive species risks of international trade, and the effect this has on the dispersion of species. More particularly, we ask what determines current country-level decisions to mitigate the risks of species introductions or to adapt to the consequences of establishment and spread, and how this is reflected in the dispersion data. We are, in addition, especially interested in the relationship between country-level responses and the global interest. We ask what the relationship is between the mitigation and adaptation decisions taken by individual countries and the risks confronting others. We focus on invasive pests—specifically animal diseases notified to the World Animal Health Organization (OIE). We are especially interested in the role of inspection and interception in modifying the relation between the level of imports of risk materials and disease outbreaks. We explore the relationship between these factors in the context of a national model of optimal inspection and interception effort for individual species, where the external effects of inspection and interception efforts are not taken into account. We then estimate a model of aggregate species introductions as a function of the factors driving optimal effort, using data on reported animal disease outbreaks.

The set of pathogens that countries are required to report to the OIE comprise a set of both animal diseases and zoonoses. Until recently countries were required to report a sub-set of especially damaging pathogens, the "List A pathogens", separately. There is an interesting asymmetry in the results for both the set of all pathogens taken together and the List A pathogens. We find that vulnerability to trade-induced biological invasions as measured by the total number of disease outbreaks notified to the OIE is increasing in the volume of trade in risk materials and the value at risk. However, for List A species, we find that the number of notified outbreaks falls with both imports and value at risk. This indicates that importing countries focus inspection and interception effort on species with known and severe damage costs. However, it also indicates that action by exporting countries reduces the probability that exports will be contaminated—which indicates that at least for some potentially invasive species the external benefits of national inspection and interception are taken into account. At the same time, the data show a very clear divide between low and high income countries in this respect. The countries with the greatest dependence on agriculture for both output and employment are also the countries with the highest incidence of damaging animal pathogens.

4.2 Species dispersal and inspection and interception effort

The existing literature includes both theoretical studies of optimal inspection policy (McAusland and Costello 2004; Springborn, Costello and Ferrier 2008) and empirical studies of the effectiveness of existing inspection regimes (e.g. Areal *et al.* 2008). While we are interested in the choice of inspection and interception effort, our primary concern is with the consequences of this choice for the pattern of trade-related invasive species. Hence we model the impacts of optimal inspection and interception effort on the introduction of potentially invasive species in a particular country. There are two potential sources of introduced disease growth—resident and imported host populations. In the first case disease growth may depend on host density. In the second it will be density-independent (Norberg *et al.* 2001). Density-dependent factors only come into play if the pathogen establishes and spreads (McCallum *et al.* 2001). For such established pathogens we follow Anderson and May's (1979) compartmental framework for susceptible, x, and infected hosts, y, where x and y are densities. In the naïve system $y \approx 0$, with $y + x = N$. Other health classes may be added to the basic model including immune and latent stages. Assuming that susceptible hosts (flaura or fauna) are part of a country's resource base, R, and that R also comprises other forms of natural as well as produced capital the equation of motion for hosts infected with the ith of n potentially invasive pathogens in an importing country in our generic model takes the form:

$$\dot{y}^i = f^i\left(y^i, R(x), u^i\right) + \left(p^i - q^i\right)M;$$
$$i = 1, \ldots, n \qquad (4.1)$$

u^i is *in situ* control of the species, which we hold constant here; f^i is the density-dependent growth of the infected population in the importing country; and $(p^i - q^i)M$ the density independent growth of the infected population through imports. In most cases we would expect this to be an increasing function of imports M, p^iM being the probability that M units of imports will introduce an individual of species i (taken to be exogenous by the importing country). We would also expect it to be a decreasing function of SPS effort per unit of imports, q^iM being the reduction in the probability that M units of imports will introduce an individual of species i due to SPS effort for that species. Since SPS measures cannot directly affect the probability of introduction, $(p^i - q^i) \geq 0$.

SPS measures confer two benefits. The first is a benefit to the country itself through the damage avoided in that country. The second is a benefit to other countries through the reduction in the probability that exports to those countries will be contaminated. This makes SPS an "impure public good": it confers a non-exclusive indirect benefit on others whilst yielding a direct benefit to the provider. The problem for the relevant agency is to choose the level of SPS for all potentially invasive pathogens, **q**, so as to maximize the expected present value of net benefits, E(W), from two sources: (a) exploitation of a domestic resource base, R, and (b) imports, M. M and R are related by the fact that the domestic resource base may be adversely affected by the invasive pathogens that accompany imports. Managers must balance the costs of *preventing* pathogen emergence and the costs of pathogen emergence. In the generic model, we assume M is given (managers cannot change import volumes) and that R is affected by regulating the spread of pathogens (this assumption is relaxed in more detailed models). Suppressing time arguments and the expectation operator, the objective function takes the form:

$$\underset{q^i}{Max}\, W = \int_{t=0}^{\infty} e^{\delta t} W\left(R\left(y^i, x\right), q^i, M\right)dt,$$
$$i = 1, \ldots, n, \qquad (4.2)$$

where δ is the discount rate, a proxy for the opportunity cost or growth potential of capital. We assume that $R_{y^i} \leq 0$ (the resource base is potentially damaged by invasive pathogens), and that the partial derivatives of the function carry the following signs $W_R > 0$, $W_{RR} < 0$, $W_M > 0$, $W_{MM} < 0$, $W_{q^i} < 0$, $W_{q^iq^i} < 0$. Welfare is increasing in both R and M. It is decreasing in q^i (since SPS effort requires resources), but is increasing in the effects of q on R and y (since $R_{y^i} \leq 0$ and $\dot{y}_{q^i} \leq 0$). The marginal welfare cost of the ith pathogen is indicated by its shadow value, λ^i. The first order necessary conditions for maximizing [4.2] subject to [4.1] yield the following conditions on (a) the rate of change in the shadow value of pathogens:

$$\dot{\lambda}^i = \frac{W_{q^i}}{M}\left(\delta - f_{y^i}^i\right) - W_R R_{y^i}; i = 1\ldots n \quad (4.3)$$

and (b) on SPS effort:

$$\dot{q}^i = \frac{W_{q^i}}{W_{q^i q^i}}\left[\delta - f_{y^i}^i - \frac{W_R R_{y^i}}{W_{q^i}}M\right]; i = 1\ldots n. \quad (4.4)$$

These two expressions together enable us to characterize optimal SPS effort for given levels of M. The time rate of change in SPS effort is increasing in the potential marginal damage averted by SPS effort, $W_R R_{s^i}$, but is decreasing in the relative marginal growth rate of the pathogen (if established), $\delta - f_{y^i}$, and the marginal cost of SPS effort, W_{q^i}. This implies the following "golden rule" equation or rate-of-return condition:

$$\delta = \frac{\dot{\lambda}^i}{\lambda^i} + \frac{W_R R_{y^i}}{\lambda^i} + f_{y^i}^i; i = 1\ldots n. \quad (4.5)$$

The left hand side of this expression is the rate of return society could have earn by investing resources in activities other than SPS effort, and is therefore the opportunity cost of investment in SPS effort. The right hand side is the return that society gains from investing in SPS effort. In the steady state, the first term on the right hand side will be zero. The second term represents the averted loses associated with screening and is positive ($\lambda_i < 0$, $R_{y^i} < 0$, $W_R > 0$). The final term is the marginal growth rate of the pathogen if established. If there exists a steady state optimum level of SPS effort—that is $\dot{q}^i = 0; \dot{y}^i = 0, i = 1\ldots n$ – this term will be strictly less than the opportunity cost of capital. This basic structure enables us to characterize optimal SPS effort along an optimal trajectory and from this to derive a set of testable hypotheses about the way that SPS impacts disease risks. The following propositions follow directly.

Proposition 4.1. A steady state level of SPS effort will exist if and only if $\delta = f_{y^i}^i + W_R R_{y^i} M/W_{q^i}$. Since $W_R R_{y^i} M/W_{q^i} > 0, \dot{q}^i = 0$ implies that $f_{y^i}^i < \delta$: i.e. that the ith pathogen present in the country is 'slow growing' relative to the economy. This follows from the fact that if $f_{y^i}^i > \delta$ then $\delta - f_{y^i}^i - W_R R_{y^i} M/W_{q^i} \neq 0$. For pathogens in the country that are "fast growing", there is no positive steady state level of SPS effort. A corollary is that if the density-dependent growth rate of the ith pathogen is strictly zero (because that pathogen is unable to reproduce in the particular bioclimatic conditions of the country or it is excluded by domestic eradication efforts), then the optimal steady state level of SPS will be such that $W_R R_{y^i} M/W_{q^i} = \delta$. That is, SPS effort will increase up to the point where the marginal welfare gains of SPS at import level M (the marginal reduction in welfare losses from introductions at that level of imports) is equal to the opportunity cost of capital. This implies the following hypothesis: *If an introduced pathogen has established and spread, there will exist a positive optimal steady state (sustained) level of import risk mitigation only for "slow growing" pathogens. If a potentially dangerous pathogen is not locally established, then the optimal steady state level of inspection and interception will be chosen such that the marginal damage avoided at current import levels will be equal to the opportunity cost of resources committed to SPS.*

Now consider the conditions that hold away from the steady state: i.e. $\dot{q}^i \neq 0$, $\dot{s}^i \neq 0$, $i = 1, \ldots, n$. We are interested in factors that induce an increase or decrease SPS effort along an optimal trajectory, and hence in the abundance of potentially invasive pathogens that are introduced along with imports.

Proposition 4.2. Away from the steady state ($\dot{q}^i \neq 0$, $\dot{y}^i \neq 0$, $i = 1, \ldots, n$) optimal SPS effort for the ith pathogen will be decreasing over time if that species is established and is 'fast-growing' relative to the economy. Since $W_{q^i}/W_{q^i q^i} > 0$ and $W_R R_{y^i} M/W_{q^i} > 0$, $f_{y^i}^i > \delta \Rightarrow \dot{q}^i < 0$. The consequences for the growth in abundance of potentially invasive species follow from [4.1]. Specifically, since growth of the ith potentially invasive pathogen along an optimal path is

$$\dot{y}^i = f^i(y^i) + \left(p^i - \left(\int \dot{q}^i dt - A\right)\right)M;$$

$$i = 1, \ldots, n \quad (4.6)$$

(where A is a constant of integration—the baseline level of SPS effort), abundance of the ith potentially invasive pathogen is increasing in its own rate of growth and imports, and is decreasing in the factors that increase optimal SPS effort: the level of imports, M, and the value at risk, $W_R R_{y^i}$. Since optimal SPS efforts rise with the potential cost of introduced pathogens in terms of damage to the resource base, but fall with the growth rate of established pathogens, the decision-maker will balance the harm potentially done by established pathogens with the cost of reducing pathogen abundance through SPS measures. This implies the following hypothesis: *Optimal SPS effort will be decreasing over time if the established population of those pathogens is "fast-growing" relative to the economy, and will be increasing if it is "slow-growing". If a pathogen is not controllable through the regulation of imports (because it is already established in the country) it will not be optimal to commit resources to SPS, while SPS effort will be greatest for species that are not yet established, but that are potentially highly damaging.*

Where a country selects q^i so as to maximize an index of well-being in that country without taking into account the impact this level of SPS effort has on other countries, the additional risk to those other countries is said to be an externality of trade. The level of trade is fixed in the generic model at M_k; but this level of import implies a corresponding level of export risk that is carried by k's trading partners in the form of an import risk $p_j^i \left(q_k^i \right) M_j$. Mechanisms do exist to encourage countries to internalize these costs directly—including the threat of defensive actions admitted both under Article XX of the GATT and the Sanitary and Phytosanitary Agreement. These mechanisms are designed to ensure feedback between export risks to other countries and SPS effort within a country. Specifically, for countries k and j they ensure that q_k^i is selected to satisfy:

$$\lambda_k^i = \left[\partial W_k / \partial q_k^i + \alpha \left(\partial W_j \left(p_j^i \left(q_k^i \right) M_j \right) / \partial q_k^i \right) \right] / M_k;$$
$$1 = 1, \ldots, n \quad (4.7)$$

in which $\partial W_k / \partial q_k^i$ the partial derivative of W_j with respect to q_k^i is the marginal welfare cost to country j of SPS effort in country k, and α is the share of that cost levied on country k. Our third key hypothesis follows: *The share of the external cost of trade that is internalized by exporting countries is an increasing function of the expected damages associated with infected exports. Where mechanisms to internalize the external costs of disease exist, the probability that exports will be infected is lower the more harmful the disease.*

4.3 Data and methods

The evidence across a range of species is that for most invasive plants, mammals, birds, and reptiles the invasion process is relatively slow. The lag between introduction and establishment for many plants, for example, can be upwards of a hundred years (Williamson 1996). While some work has been done on the linkages between trade patterns and invasions of such species (Dalmazzone 2000; Vilà and Pujadas 2001; Levine and d'Antonio 2003; Dehnen-Schmutz *et al*. 2007) obtaining a well-defined relationship is complicated by the existence of such long lags. To address this problem we focus on species whose dynamics are such that there is minimal lag between introduction and establishment. While some insect pests fall into this category, the best known examples are pathogens. By concentrating on these species, we are able to explore the relationship between changes in the trade regime and the introduction of alien species.

Within the set of introduced pathogens, we further specialize to a sub-set of species for which monitoring data are readily available: animal diseases notified to the World Animal Health Organization (OIE). Because emergent zoonotic diseases like SARS and HPN51 (avian) flu were genuinely novel there were no screens in place for them at the time they were discovered. However, many other pathogens are familiar enough that they are routinely monitored. Indeed, most inspection and interception services focus their attention on a set of known pests. Familiar examples are foot and mouth disease (FMD), bovine spongiform encephalopathy (BSE), or rabbit hemmorhagic disease (RCD). While we do not have detailed data on inspection and interception efforts, we do have mandatory reports on the presence of a range of infectious diseases. The relation between abundance, intrinsic growth rates, and the welfare cost of both the inspection and interception regime and the damage to the resource base accordingly enables us to evaluate trade-induced changes in the potentially invasive pests of this kind using observations on imports, together with proxies for both the damage associated with those species and their relative rates of growth.

Although we do not have direct observations on growth rates for introduced diseases, f_{si}^i, the OIE until recently categorized species according to both their rate of spread and potential damage. We evaluate the data reported to the OIE in each of two categories. The first category, List A species, comprises transmissible diseases that have the potential for very serious and rapid spread, that have serious socioeconomic implications, and have potentially major negative effects on public health. The second category, List B species, comprises transmissible diseases that are important from a socioeconomic perspective, and that particularly affect countries which are significant in the international trade of animals and animal products (see Table 4.1).

List A species are accordingly differentiated from List B species by two things: their rate of growth

Table 4.1 List A and List B diseases reported to the OIE

OIE code	List A Diseases	OIE code	List B Diseases
	All species	B106	Bovine cysticercosis
A010	Foot and mouth disease	B107	Dermatophilosis
A020	Vesicular stomatitis	B108	Enzootic bovine leukosis
A030	Swine vesicular disease	B109	Haemorrhagic septicaemia
A040	Rinderpest	B110	Infectious bovine rhinotracheitis
A050	Peste des petits ruminants	B111	Theileriosis
A060	Contagious bovine pleuropneumonia	B112	Trichomonosis
A070	Lumpy skin disease	B113	Trypanosomosis (*tsetse-transmitted*)
A080	Rift Valley fever	B114	Malignant catarrhal fever
A090	Bluetongue	B115	Bovine spongiform encephalopathy
A100	Sheep pox and goat pox		**Sheep and goats**
A110	African horse sickness	B151	Ovine epididymitis (*Brucella ovis*)
A120	African swine fever	B152	Caprine and ovine brucellosis (excluding *B. ovis*)
A130	Classical swine fever	B153	Caprine arthritis/encephalitis
A150	Highly pathogenic avian influenza	B154	Contagious agalactia
A160	Newcastle disease	B155	Contagious caprine pleuropneumonia
	List B Diseases	B156	Enzootic abortion of ewes (*ovine chlamydiosis*)
	Multiple species	B157	Ovine pulmonary adenomatosis
B051	Anthrax	B158	Nairobi sheep disease
B052	Aujeszky's disease	B159	Salmonellosis (*S. abortusovis*)
B053	Echinococcosis/hydatidosis	B160	Scrapie
B055	Heartwater	B161	Maedi-visna
B056	Leptospirosis		**Equidae**
B057	Q fever	B201	Contagious equine metritis
B058	Rabies	B202	Dourine
B059	Paratuberculosis	B203	Epizootic lymphangitis
B060	New world screwworm (*Cochliomyia hominivorax*)	B204	Equine encephalomyelitis (*Eastern and Western*)
B061	Old world screwworm (*Chrysomya bezziana*)		**Equine species**
B062	Trichinellosis	B205	Equine infectious anaemia
	Cattle	B206	Equine influenza
B101	Bovine anaplasmosis	B207	Equine piroplasmosis
B102	Bovine babesiosis	B208	Equine rhinopneumonitis
B103	Bovine brucellosis	B209	Glanders
B104	Bovine genital campylobacteriosis	B210	Horse pox
B105	Bovine tuberculosis	B211	Equine viral arteritis

(*cont.*)

Table 4.1 (Continued)

OIE code	List B Diseases	OIE code	List B Diseases
B212	Japanese encephalitis	B352	Tularemia
B213	Horse mange	B353	Rabbit haemorrhagic disease
B215	Surra (*Trypanosoma evansi*)		**Fish**
B216	Venezuelan equine encephalomyelitis	B401	Viral haemorrhagic septicaemia
	Swine	B404	Spring viraemia of carp
B251	Atrophic rhinitis of swine	B405	Infectious haematopoietic necrosis
B252	Porcine cysticercosis	B413	Epizootic haematopoietic necrosis
B253	Porcine brucellosis	B415	Oncorhynchus masou virus disease
B254	Transmissible gastroenteritis		**Molluscs**
B256	Enterovirus encephalomyelitis	B431	Bonamiosis (*B. exitiosus, B. ostreae, Mikrocytos roughleyi*)
B257	Porcine reproductive and respiratory syndrome	B432	MSX disease (*Haplosporidium nelsoni*)
	Birds	B433	Perkinsosis (*Perkinsus marinus, P. olseni/atlanticus*)
B301	Avian infectious bronchitis	B434	Marteiliosis (*Marteilia refringens, M. sydneyi*)
B302	Avian infectious laryngotracheitis	B436	Mikrocytosis (*Mikrocytos mackini*)
B303	Avian tuberculosis		**Crustaceans**
B304	Duck virus hepatitis	B445	Taura syndrome
B305	Duck virus enteritis	B446	White spot disease
B306	Fowl cholera	B447	Yellowhead disease
B307	Fowl pox		**Bees**
B308	Fowl typhoid	B451	Acariosis of bees
B309	Infectious bursal disease (*Gumboro disease*)	B452	American foulbrood
B310	Marek's disease	B453	European foulbrood
B311	Avian mycoplasmosis (*M. gallisepticum*)	B454	Nosemosis of bees
B312	Avian chlamydiosis	B455	Varroosis
B313	Pullorum disease		**Other**
	Lagomorphs	B501	Leishmaniosis
B351	Myxomatosis		

Source: Terrestrial Animal Health Code (2007), http://www.oie.int/eng/normes/Mcode/en_sommaire.htm. Articles 1.1.2.1. to 1.1.2.10.

and the damage they are capable of doing to the resource base. We do not have direct measures of either intrinsic growth rates or damage potential for individual invasive pathogens. However, we are able to use the OIE lists to aggregate species that have broadly similar growth and damage potential. Data on each list are available for a number of countries with some consistency from 1996.

We estimate two models of the form:

$$\sum_{j=1}^{k} \hat{y}_t^j = \beta_0 + \beta_1 \sum_{j=1}^{k} W_R R_{y^j} + \beta_2 M_t + \varepsilon_t; \quad j = 1\ldots m; \, k = \{m, n\} \qquad (4.8)$$

where $\hat{y}_t^j = 1$ if $y_t^j > 0$ and $\hat{y}_t^j = 0$ otherwise and $m < n$ is number of List A species. There is no explicit expression for $\sum_i f^i(y_t^j)$ in this equation as there is in [4.1]. It is, however, implicit in the distinction between List A and other species and is revealed in differences in the constant in the estimated equations. M is measured by import volumes of food and live animals—commodity group "0" of the Standard International Trade Classification Revision 1. This commodity group includes all of the risk-materials associated with animal diseases. Value at risk is approximated by per capita agricultural GDP derived from the World Bank (2007) http://go.worldbank.org/3JU2HA60D0; trade volumes derive from the UN Comtrade Database, http://comtrade.un.org/db/; pathogens derive from the World Animal Health Organization (OIE) http://www.oie.int/eng/maladies/en_alpha.htm;

population from UNSO. There is some potential to exploit the increasingly fine disaggregation of traded commodities to identify the risk materials with greatest predictive power for particular classes of pathogens. The disaggregated data in this case form a panel of 369 observations in 41 groups (countries). Since the sample comprises countries of highly variable size and income levels, it is characterized by heteroskedasticity in errors across panels (plus autocorrelation within panels).

The set of countries analyzed are indicated in Table 4.2, along with averaged or aggregated data for the period 1996–2004. The disaggregated data form a panel of 369 observations in 41 groups with which we can test the impact of the main determinants of inspection and interception effort on changes in the abundance of the species in the OIE A and Combined Lists.

Note that the dataset comprises countries of highly variable size and income levels, implying the existence heteroskedasticity in errors across panels (along with autocorrelation within panels). We adopt two strategies to address this in estimating [4.8] (a) a generalized least squares regression which accommodates both AR(1) autocorrelation within panels as well as heteroskedasticity across panels and (b) a Prais–Winsten regression with corrected standard errors (PCSEs) for correlated panels. The results of both procedures are reported in the next section, and are shown to be consistent across methods. Given the count nature of the data on disease outbreaks and the fact that they are over-dispersed we also estimate a negative binomial model, taking the number of trading partners as our measure of exposure. While the fit is not as good, the results offer additional and useful insights into the gap between national and global interests.

4.4 Empirical results

Tables 4.3 and 4.4 report the estimated coefficients in each of the first two procedures adopted. Table 4.3 refers to the Combined List model, and Table 4.4 to the List A model. Both models assume that the panels are heteroskedastic and that there is panel-specific autocorrelation. The Combined List model fit is good in both estimation procedures. The Prais–Winsten regression has an R-square value of 0.7441, a Wald Chi-square value of 66.71 and an associated p-value of 0.0000. The FGLS model has a log-likelihood of -980.2182, a Wald Chi-square value 94.97 with an associated p-value of 0.0000.

Table 4.4 refers to the List A model. Though not as good as the Combined List model, it still fits the data. The Prais–Winsten regression has an R-square value of 0.5107, a Wald Chi-square value of 175.35 and an associated p-value of 0.0000. The FGLS model has a log-likelihood of -392.0437, a Wald Chi-square value 255.07 with an associated p-value of 0.0000. Note that the greater Chi-square value in the List A model relative to the Combined List model results from the high frequency of low numbers of disease outbreaks in the List A data. In both models the generalized least squares approach yields slightly higher z values for the estimated coefficients.

The results may be interpreted in terms of the unobserved inspection and interception efforts. We have assumed that inspection and interception effort is an increasing function of (a) the growth rate of the introduced pathogen, (b) value at risk, and (c) the volume of imported risk materials. However, the form of the function reflects expectations of the potential damage associated with particular species.

First, consider the difference between the constant term in the Combined List and the List A models. If this captures the impact of differences in the inspection and interception response to the characteristics of the species in each of the two the two lists, i.e. to differences in $\sum_i f^i(s_t^i)$, we would expect the value of the constant term to be lower in List A than in the Combined List model—as is the case. That is, we would expect concern over the effects of List A species to lead to inspection and interception regimes that reduce the abundance of introduced species on that list relative to all species.

Second, consider the estimated coefficients on imports and the potential cost of diseases in the Combined List and the List A models. The change of sign on both coefficients indicates a fundamental difference both in mean responses to List A diseases relative to the set of all notifiable diseases and, more importantly, in the range of responses

Table 4.2 Reported diseases, average potential damage costs, and imports of risk materials 1996–2004

	Lists A + B Diseases (Total)	List A Diseases (Total)	GDP Per capita in constant (2000) USD (Average)	% Share of GDP from agriculture (Average)	Population '000 (Average)	SITC Rev1 imports of group 0 commodities (Average)
	1	2	3	4	5	6
Algeria	117	24	1827	10.67	30757	2393
Argentina	339	14	7488	7.09	37061	2647
Australia	331	15	20846	3.62	19271	2493
Austria	170	11	23601	2.14	8145	2250
Brazil	290	33	3440	8.49	175436	2372
Burkina Faso	146	43	235	32.89	12124	2384
Cameroon	173	50	678	40.64	16069	2848
Canada	339	2	22592	2.43	30871	2828
Chile	248	6	4924	5.90	15492	3578
China	90	40	979	15.47	1273212	1125
Denmark	235	3	29320	2.58	5342	1493
Egypt	139	5	1456	16.71	67186	1433
France	488	7	22030	2.90	59458	1155
Germany	402	11	22660	1.22	82357	1184
India	296	54	458	24.10	1054792	1042
Indonesia	196	28	842	16.59	213154	472
Ireland	268	2	24386	3.72	3849	698
Israel	275	35	17576	3.02	6140	800
Italy	249	45	18889	2.86	57858	2231
Japan	243	10	36613	1.77	127079	2241
Kenya	293	44	420	30.69	31717	2156
Mexico	322	41	5682	4.68	99830	2302
Morocco	152	8	1240	16.14	28972	2445
Netherlands	300	6	23506	2.63	15965	2338
New Zealand	230	8	13814	7.42	3887	2647
Niger	78	42	158	39.81	11354	3219
Pakistan	224	40	531	25.18	145393	3933
Peru	217	40	2078	8.37	25811	3324
Philippines	189	35	996	16.58	77110	3285
Russia	348	29	1811	6.18	146935	3267
Senegal	191	63	424	18.71	10488	3499
Singapore	36	3	22011	0.14	4027	3475
Spain	313	7	14093	4.34	40793	3603
Sudan	134	22	377	42.64	33653	3980
Sweden	211	4	26605	2.03	8900	5038
Switzerland	215	4	33459	1.58	7286	5923
Thailand	165	18	2085	9.67	60828	6180
Turkey	81	25	2915	14.94	68583	6165
Uganda	254	48	242	37.43	25174	5628
United Kingdom	333	4	24147	1.18	59049	3949
USA	507	23	33817	1.32	286389	4099

Sources:
1, 2: World Organization for Animal Health (OIE), http://www.oie.int/eng/maladies/en_alpha.htm.
3, 4: The World Bank. 2007. 2007 World Development Indicators Online. Washington, DC: The World Bank. http://go.worldbank.org/3JU2HA60D0.
5: Population Division of the Department of Economic and Social Affairs of the United Nations Secretariat. 2007. World Population Prospects: The 2006 Revision. Dataset on CD-ROM. New York: United Nations. http://www.un.org/esa/population/ordering.htm.
6: UN Comtrade Database, http://comtrade.un.org/db/.

Table 4.3 List A + B estimated coefficients on the economic determinants of species introductions

List A + B	Prais–Winston regression, heteroskedastic panels corrected standard errors			FGLS heteroskedastic panels, panel-specific AR(1)correlation		
	coeff	z	P > |z|	coeff	z	P > |z|
M_t	4.45-07	6.71	0.000	4.48e-07	8.26	0.000
C_t	.0067475	4.18	0.000	.0058489	4.61	0.000
Constant	19.59435	18.55	0.000	20.17623	29.15	0.000

Table 4.4 List A estimated coefficients on the economic determinants of species introductions

List A diseases	Prais–Winston regression, heteroskedastic panels corrected standard errors			FGLS heteroskedastic panels, panel-specific AR(1)correlation		
	coefficient	z	P > |z|	coefficient	z	P > |z|
M_t	−2.63e-08	−2.99	0.003	−2.64e-08	−3.75	0.000
C_t	−.0038699	−13.12	0.000	−.0036982	−15.26	0.000
Constant	4.008636	19.82	0.000	3.869829	26.02	0.000

observed between countries with different propensities to import risk materials and different potential losses from invasive pathogens.

In addition to these, given the count nature of the data we estimated two random-effects negative binomial regression models for the same groups of species. In both cases, the constant is suppressed. We assume that exposure increases in the number of trading partners. The results from the combined list model are reported in Table 4.5, and from the List A model in Table 4.6. The fit is weaker in both cases but is still significant at commonly used significance levels (for the Combined List model the Wald chi-square is 13.95, and p-value of 0.0009; and for the List A model the Wald chi-square is 8.01, and p-value of 0.0182).

While the sign on value at risk and import variables, C_t and M_t, is consistent with the linear models in both cases, the coefficient on imports is not significant in either. The implication of this is that the most important driver of change is the value at risk from invasive species. Countries with the most to lose (in terms of agricultural GDP per capita) are proportionately most responsive to changes in trade related disease risks.

The implication of the change in the sign of the coefficient on import volumes is interesting. These coefficients are not significant in the negative bionomial models, but are highly significant in the FGLS and P-W models. Aggregate abundance of potentially invasive species can decrease with imports only if the probability of species introduction

Table 4.5 Combined list estimated coefficients (negative binomial model)

Combined list	Coef.	Std. Err.	z	P > |z|
C_t	.0005442	.000149	3.65	0.000
M_t	9.18e-06	.0000146	0.63	0.530

Table 4.6 List A estimated coefficients (negative binomial model)

List A Diseases	Coef.	Std. Err.	z	P > |z|
C_t	−.0010857	.0004619	−2.35	0.019
M_t	−.0001018	.0000674	−1.51	0.131

falls as imports rise. Since this is unrelated to inspection and interception at the port of entry, it implies increasing inspection and interception efforts undertaken by trading partners, or enhanced sanitary and phytosanitary measures at the point of export (we are unable to say which).

4.5 Discussion

While increasing trade increases the probability that new species (including new pest species) will be introduced, countries have the power to reduce the associated risks through inspection and interception regimes. We consider the implications of this for the dispersal of one class of potentially invasive species: animal pathogens reported to the OIE. We find that when the set of all notifiable pathogens is analyzed as a whole, inspection and interception efforts have some moderating effect on the risks of potentially invasive pathogens, but that the aggregate abundance of invasive species grows with both the volume of imports of risk materials and the value at risk. Since the marginal damage cost of potentially invasive species is assessed as the marginal impact of imported species on the domestic resource base, $\sum_i W_R R_{Si}$, and ignores the marginal impact of exported species on the foreign resource base, $\sum_i W_R^E R_{Si}^E$, national inspection and interception effort will be below the globally optimal level wherever $\sum_i W_R^E R_{Si}^E < 0$. In other words, sanitary and phytosanitary measures are an impure public good. Reduction in the local abundance of pest species confers benefits on the individual country, but as it also reduces the probability of transmission of species in exports it confers benefits on other countries. However, these benefits are not typically taken into account by individual countries.

When the set of only List A species is analyzed, however, increasing trade volumes are associated with reduced risks. This is partly because of the increase in inspection and abundance efforts, but since the probability of introductions declines with imports for this class of pathogens independently of inspection and interception efforts, sanitary and phytosanitary actions in exporting countries must be having an effect. Notice that our measure of value at risk, per capita agricultural GDP, is also a measure of the capacity of a country to invest in sanitary and phytosanitary measures. Consider, for example, the data for List A species illustrated in Fig. 4.1.

Our proxy for the potential damage done by these diseases is agricultural GDP per capita (per 1000 people in this figure). Average agricultural GDP per capita in the set of countries with the most frequently reported outbreaks of List A diseases for the period 1996–2004—from Senegal to Algeria—was US$ 220. Average agricultural GDP per capita in the remaining set of countries was US$ 542. At the same time the share of the population engaged in agriculture in the first set of countries is 47 per cent compared to 13 per cent for the second set. A much greater proportion of the labor force is adversely affected by List A pathogens in low income countries with ineffective inspection and interception regimes than in the higher income countries. Although this dimension of the invasive species problem is well recognized (Perrings 2007), and although the resource constraints in low income countries are one of the best predictors of inspection and interception efforts, the equity implications are worth addressing directly. The Convention on Biological Diversity refers to the equitable sharing of the benefits of biodiversity conservation as an important principle. This is an example of the inequitable sharing of the costs of biodiversity loss.

Two final points: the relationship between the growth in abundance of potentially invasive species and changes in patterns of trade is more complex than the stylized problem analyzed in this chapter. The risks posed by changes in trade routes, trade volumes, and trade goods are sensitive to species traits, the bioclimatic similarity of trade partners, the role of intermediate importers as well as conditions along (trade) pathways. None of these are modeled here. Moreover, given that we focus only on the default strategy of inspection and interception, we ignore other activities that have the potential to contain the spread of established species. Nevertheless, the approach enables us to identify several factors that are both critical to the management of invasive species risks at the national level, and that explain the global pattern of species redistribution through trade.

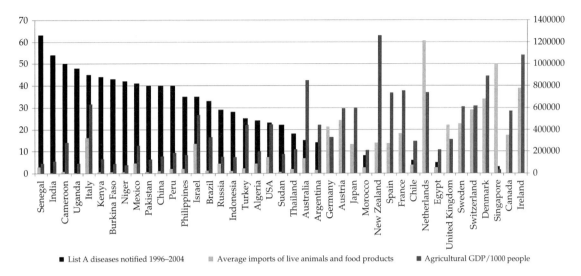

Figure 4.1 List A diseases reported to the OIE 1996–2004 by selected countries, ranked by the number of outbreaks of diseases. Also reported is a proxy for the volume of risk materials (average imports of live animals and food products) and value at risk (agricultural GDP per 1000 people). The data show the negative relation between outbreaks and both the volume of risk materials and the value at risk reported in Tables 3–6.

Sources: World Organization for Animal Health (OIE), http://www.oie.int/eng/maladies/en_alpha.htm; The World Bank. 2007. 2007 World Development Indicators Online. Washington, DC: The World Bank. http://go.worldbank.org/3JU2HA60D0; Population Division of the Department of Economic and Social Affairs of the United Nations Secretariat. 2007. World Population Prospects: The 2006 Revision. Dataset on CD-ROM. New York: United Nations. http://www.un.org/esa/population/ordering.htm; UN Comtrade Database, http://comtrade.un.org/db/.

Since species introductions have the potential to impose damage on host societies, and since the mitigation of invasive species risks is costly, the best predictors of invasive species success are a combination of trade patterns, and the economic factors that determine inspection and interception regimes. Trade patterns offer information on the set of species being sampled through imports: i.e. the likelihood that trade along a given route will result in the introduction of particular species. The economic factors that determine inspection and interception regimes offer information on the effort committed to modifying invasive species risks: i.e. the change in likelihood that trade along a given route will result in the introduction of particular species. The latter is affected by the direct costs of inspection and interception, the risk of damage if the species is not intercepted, and the resources available to mitigate that risk. Differences in these things between countries ensure that the inspection and interception regime for species with the same traits may be expected to differ from one country to another. In turn, differences in inspection and interception regimes may be expected to be reflected in dispersion rates, and this is in fact precisely what we see in the data on notifiable animal diseases.

Lastly, we have explicitly considered only the risks associated with imports and the inspection and interception response. An important aspect of the problem is the dependence of import risks on both the impact of regional trade agreements on the pattern of trade, and especially on the pattern of exports and re-exports between bioclimatically similar countries. While the frequency of introductions per unit of imports may not differ between bioclimatically similar and dissimilar trading partners, the likelihood of establishment and hence the damage associated with those introductions will differ. Extensions of this work will address the relationship between invasive species risks and the pattern of trade between bioclimatically similar trading partners. We observe that one of the main sources of world trade growth in the last decade has been the expansion of exports from China to the USA and Europe—regions that are bioclimatically similar. Understanding both how climate change may affect the distribution of bioclimatically similar zones and how the growth and pattern of international trade may affect the dispersion of species between those zones is a challenge yet to be met.

Acknowledgements

Helpful comments on an earlier draft were received from Buz Brock. We acknowledge support from NSF Grant # 0639252 and DIVERSITAS.

References

Anderson, R. and R. May. (1979). Population biology of infectious diseases: Part I. *Nature* **280**, 361–67.

Areal, F. J., Touza, J., MacLeod, A., *et al* (2008). Integrating drivers influencing the detection of plant pests carried in the international cut flower trade. *Journal of Environmental Management*, **89**, 300–307.

Cassey, P., Blackburn, T. M., Russel, G. J., Jones, K. E., and Lockwood, J. L. (2004). Influences on the transport and establishment of exotic bird species: an analysis of the parrots (Psittaciformes) of the world. *Global Change Biology* **10**, 417–26.

Costello, C. and McAusland, C. (2003). Protectionism, Trade and Measures of Damage from Exotic Species Introduction. *American Journal of Agricultural Economic*, **85**(4), 964–975.

Dalmazzone, S. (2000). Economic Factors affecting vulnerability to biological invasions, in C. Perrings, M. Williamson and S. Dalmazzone, eds. *The Economics of Biological Invasions*, Edward Elgar, Cheltenham, pp. 17–30.

Daszak, P., Cunningham, A. A., and Hyatt, A. D. (2000). Emerging infectious diseases of wildlife: threats to biodiversity and human health. *Science*, **287**, 443–49.

Daszak, P., and Cunningham, A. A. (1999). Extinction by infection. *Trends in Ecology & Evolution*, **14**, 279.

Dehnen-Schmutz, K., Touza, J., Perrings, C. and Williamson, M. (2007). A century of the ornamental plant trade and its impact on invasion success. *Diversity and Distributions*, **13**, 527–34.

Enserink, M. (1999). Biological invaders sweep in. *Science*, **285**, 1834–36.

Hooper, D. U., Chapin, III, Ewel, F. S., *et al.* (2005). Effects of biodiversity on ecosystem functioning: a consensus of current knowledge. *Ecological Monographs*, **75**(1), 3–35.

Hubalek, Z. (2003). Emerging human infectious diseases: Anthroponoses, zoonoses,and sapronoses. *Emerging Infectious Diseases*, **9**, 403–404.

Jones, K. E., Patel, N. G., Levy, M. A., *et al*. (2008). Global trends in emerging infectious diseases. *Nature* **451**(21), 990–994.

Kinzig, A. P., Pacala, S. and Tilman, D. (2002). *Functional Consequences of Biodiversity: Empirical Progress and Theoretical Extensions*. Princeton University Press, Princeton, N. J.

Kinzig, A. P., Ryan, P., Etienne, M., Elmqvist, T., Allison, H., and Walker, B. H. (2006). Resilience and regime shifts: Assessing cascading effects. *Ecology and Society*, **11**(1), 20.

Kohn, R. E. and Capen, D. (2002). Optimal volume of environmentally damaging trade. *Scottish Journal of Political Economy*, **49**(1), 22–38.

Levine, J. M. and D'Antonio, C. M. (2003). Forecasting biological invasions with increasing international trade. *Conservation Biology*, **17**, 322–26.

Loreau, M., Mouquet, N., and Gonzalez, A. (2003). Biodiversity as spatial insurance in heterogeneous landscapes. *PNAS*, **22**, 12765–70.

McAusland, C. and Costello, C. (2004). Avoiding invasives: trade related policies for controlling unintentional exotic species introductions. *Journal of Environmental Economics and Management*, **48**, 954–77.

McNeely, J. A. (2001). An introduction to human dimensions of invasive alien species. In J. A. McNeely, ed. *The Great Reshuffling: Human Dimensions of Invasive Alien Species*. IUCN: Gland.

Millennium Ecosystem Assessment (MA). (2005). *Ecosystems and Human Well-Being: Synthesis*. Island Press, Washington D. C.

Naeem, S. and Wright, J. P. (2003). Disentangling biodiversity effects on ecosystem functioning: Deriving solutions to a seemingly insurmountable problem. *Ecology Letters*, **6**, 567–79.

Norberg, J., Swaney, D. P., Dushoff, J., Lin, J., Casagrandi, R., and Levin, S. A., *et al*. (2001). Phenotypic diversity and ecosystem functioning in changing environments: a theoretical framework. *Proceedings of the National Academy of Sciences*, **98**(20), 11376–81.

Perrings, C. (2007). Pests, pathogens and poverty: biological invasions and agricultural dependence. In A. Kontoleon, U. Pascual and T. Swanson, eds. *Biodiversity Economics: Principles, Methods and Applications*, Cambridge University Press: Cambridge, pp. 133–165.

Perrings, C., Dehnen-Schmutz, K., Touza, J. and Williamson, M. (2005). How to manage biological invasions under globalization. *Trends in Ecology and Evolution*, **20**(5), 212–215.

Perrings, C., Williamson, M., Barbier, E. B., *et al.* (2002). Biological invasion risks and the public good: an economic perspective. *Conservation Ecology*, **6**(1), 1.

Pimentel, D., McNair, S., Janecka, S., *et al.* (2001). Economic and environmental threats of alien plant, animal and microbe invasions, Agriculture. *Ecosystems and Environment*, **84**, 1–20.

Reich, P. B., Tilman, D., Naeem, S., *et al.* (2004). Species and functional group diversity independently influence biomass accumulation and its response to CO_2 and N. *Proceedings of the National Academy of Sciences of the United States of America*, **101**, 10101–106.

Semmens, B. X., Buhle, E. R., Salomon, A. K., Pattengill-Semmens, C. V. (2004). A hotspot of non-native marine fishes: evidence for the aquarium trade as an invasion pathway. *Marine Ecology Progress Series* **266**, 239–44.

Vila, M. and Pujadas, J. (2001). Land use and socio-economic correlates of plant invasions in European and North African countries. *Biological Conservation*, **100**, 397–401.

Walker, B. H., Holling, C. S., Carpenter, S. R., and Kinzig, A. P. (2004). Resilience, adaptability, and transformability. *Ecology and Society*, **9**(2), 5.

Walker, B. H., Gunderson, L. H., Kinzig, A. P., Folke, C., Carpenter, S. R. and Schultz, L. (2006). A handful of heuristics and some propositions for understanding resilience in social-ecological systems. *Ecology and Society*, **11**(1), 13.

Williamson, M. (1996). *Biological Invasions*. Chapman and Hall, London.

Williamson, M. (1999). Invasions. *Ecography*, **22**, 5–12.

CHAPTER 5

Variation in the Rate and Pattern of Spread in Introduced Species and its Implications

Mark Williamson

5.1 Introduction

Biological invasions are well known to go through a series of stages: introduction, establishment, spread, full range. Managers and policy makers tend to concentrate on the first and last of these, attempting to prevent the introduction of new potentially harmful species and attempting to control or eradicate those that have completed their spread and are as widespread as they are going to be. Nevertheless the intermediate stages are also potential targets and the likely course of spread needs more attention. It is important in a static environment but even more so in environments that are likely to alter in relation to climate change (see Thomas & Ohlemüller, Chapter 2 this volume).

Although spread is the easiest of the stages of invasion to study quantitatively (Williamson 1996), there is still a lot to learn about its drivers, rates and patterns, and about the relative importance of biological, environmental, social, and economic factors. Here I consider, in particular, variation in rates and patterns between plant species in Europe and consider what the conclusions from that say about policy for and management of introduced species.

The classical model of spread, the Fisher–KPP–Skellam model, uses a reaction-diffusion equation and predicts that an invading species will spread out at a constant speed (Hastings *et al.* 2005; Williamson *et al.* 2005) after a short initial slower phase (Sherratt 1998). The speed is a simple function of the intrinsic rate of natural increase and the diffusion coefficient. Although these are two basic parameters of the dynamics of any population, they have been determined accurately in very few cases (Williamson 1996). The tail of the dispersal function in that model is normal, with the shape $\exp(-x^2)$. More recent models suggest more varied patterns, particularly with thicker tailed dispersal functions, that is with more long distance dispersal, giving what is sometimes called a popcorn pattern with scattered colonies appearing. Even so, these models do not give clear predictions of how the rate of spread will vary in one species in different conditions or vary between species in the same place. That is, they concentrate on homogeneous environments and consider the species to have constant characters, not evolving. None of that is adequate for the spread patterns likely to be seen under climate change. This chapter is particularly concerned about the recorded variation between species but that needs the context of variation within species and across habitats.

5.2 Variation in spread rates within species

A number of studies showing that the rate of spread can vary within one species in different places are listed in Table 5.1. I will discuss features of interest in each in turn.

The American populations of *Drosophila subobscura* are "a grand experiment in evolution" (Ayala

Table 5.1 Some examples of varying rates of spread within species, showing the effects of varying habitats and of evolution. Rates in kilometers per year

Drosophila subobscura in the Americas, first decade		
	north–south	>850
	east–west	ca. 85
Africanized bee *Apis mellifera* in the Americas		
	north–west	185
	south–west	85
Rabbits *Oryctolagus cuniculus*		
Britain, medieval		1–2
New Zealand		<16
Australia	east & south	15
	Nullarbor Plain	100
	Murray-Darling	125
	Northern Territories	270–390
Grey squirrel *Sciurus carolinensis*		
England	East Anglia	7.7
	north of River Tees	0
Ireland	up to 1996	0–13.4
Himalayan balsam *Impatiens glandulifera*		
Czech Republic, 4 river tributaries		0, 0.0083, 0.0204, 0.0310
grid mapping records		3.66
Britain, nearest hectads		1920–40 1.9, 1940–60 3–5
cumulative hectad records		3.86
vice-county records		initial 2.6, maximum 38
Cane toads *Chaunus* [previously *Bufo*] *marinus* in Australia		
	1945–64	10
	1965–74	20
	1980–84	25
	2001–05	55

Sources: *Drosophila subobscura* Ayala *et al.* (1989); Africanized bee Williamson (1996); rabbits Flux (1994), Gibb and Williams (1994), Myers *et al.* (1994); squirrels Williamson and Brown (1986), Okubo *et al.* (1989), Yalden (1999), Teangana *et al.* (2000); *Impatiens glandulifera* (Usher 1986), Perrins *et al.* (1993), Williamson *et al.* (2003), Williamson *et al.* (2005), Malíková and Prach (in press); cane toad Phillips *et al.* (2006)

et al. 1989). This small fly, wing span only 2 mm, is often the commonest drosophilid in traps in Europe, where it ranges from Norway to Algeria and from the Azores to the southern edge of the Caspian Sea. The American populations were probably started by 4 to 12 flies from near Barcelona in Spain getting to Chile in the late 1970s, and from there a couple of years later by 100–150 flies to the Pacific northwest of the USA (Pascual *et al.* 2007). The "experiment" is one of evolution after dispersal. The fly is notable for the convergent evolution to a cline in wing size and shape which mimics, in both North and South America but in different ways, the cline in Europe. These clines developed after the north–south spread was complete. In any case, it is not known if these wing changes affect dispersal—though they well might. Evolution during dispersal, the spread phase, is discussed below. In Europe *Drosophila subobscura* is found in temperate, Mediterranean and, to some extent, in steppe-desert climates. Its north–south spread in the Americas into the first two was very fast in its first decade there, but there was scarcely any spread east–west into steppe (Table 5.1) though there has been since then. A clear case of more rapid spread into more favorable environments.

The Africanized honey-bee in South America, with many genes from tropical Africa, shows the same effect, with spread being faster into favorable environments. There was a rapid spread northwest into the tropics from its release point near Sao Paulo, Brazil, but a slower spread southwest towards temperate areas (Table 5.1).

The rabbit *Oryctolagus cuniculus*, originally from Spain, shows the contrary effect: rapid spread across the pastures and unfavorable deserts of Australia—the more unfavorable the faster—and much slower spreads in temperate Britain and New Zealand (Table 5.1). To show that this is not a contrast of mammals with insects, the grey squirrel *Sciurus carolinensis* again shows the slowing effect of unfavorable environment in Britain and Ireland (Table 5.1) where a barrier such as the tree-poor country north of the River Tees can hold up the spread for decades, though in that case it has now been overcome.

A different effect again is shown by the Himalayan balsam *Impatiens glandulifera*. This plant has explosive seed capsules that throw the seed a couple of meters. Spreading away from rivers in the Czech Republic its spread rate is not much more than that, but at the landscape scale the spread is in kilometers per year, not meters, showing reasonably unambiguously the importance of human transport of various kinds. In controlling this species, now commonly regarded as a pest since it has become widespread, natural dispersion is of little account.

The final point about these within-species variations is that spread rates can evolve; an important consideration for management strategies. In Australia the cane toad *Bufo* (now *Chaunus*) *marinus*, a notorious alien, spread at around 10 kilometers per year in the sugar cane areas of Queensland, but the rate accelerated markedly in the Northern Territories where the toads move faster and have evolved longer hind legs (Phillips *et al.* 2006); the genetics of these changes has yet to be studied but they are, of course, related to the environment. The cane toad is the only alien species in which such evolution has been documented. None of the 50 plants species showing well-defined patterns of spread in the Czech Republic, discussed below, show any signs of increasing spread rate. Although such evolution should be subject to strong natural selection (Phillips *et al.* 2008), it may nevertheless be fairly rare.

Similar evolution has been found in a few cases of native species increasing their range. The best case is the Western Bluebird *Sialia mexicana* reoccupying territory it had lost in western Montana, USA which had become occupied by Mountain Bluebirds *S. currucoides* (Duckworth and Badyaev 2007). Both species are hole-nesters, and the holes in lowland western Montana in old trees had been lost to logging, leading to the loss of all bluebirds in the valleys. Later, nest boxes were put up in large numbers, initially colonized by *currucoides* from higher in the mountains. In advancing from the west and recapturing lost ground, the first *mexicana* males are aggressive but poor fathers. Those characters have been shown to be heritable and are at an advantage while the displacement takes place but not later (Duckworth and Badyaev 2007; Duckworth 2008).

The other cases of native species changing as they spread are less definitely genetic. Hill *et al.* (1999) showed change in the flight morphology of the speckled wood butterfly *Parage aegeria* in Britain and Madeira, very likely a genetic change. In females, but not males, the thorax was larger and the abdomen smaller in colonists, indicating a shift of resources from reproduction to flight. There was also a shift in wing aspect in Madeira. Hill *et al.* give references to similar older papers. Simmons and Thomas (2004) showed an increase in the frequency of macropterous, and so dispersing, morphs in two bush crickets, *Conocephalus discolor* and *Metrioptera roeselii*, in spreading populations in England, but their results only suggested a genetic difference without proving one. There are also cases were there are morphological differences towards the edge of a range but these have neither been shown to be selected for in advancing populations nor their genetic basis established. One example is seed size in lodgepole pine *Pinus contorta* in western North America (Cwynar and MacDonald 1987; Cousens *et al.* 2008), another is flower size, seed (anthocarp) wing size, herkogamy (spatial separation of the anthers and stigma) and self-incompatibility in the pink sand Verbena *Abronia umbellata* on the western shores of North America (Darling *et al.* 2008). In both cases the relationship of phenotype to spread and the genetics of the phenotype are unclear.

5.3 Variation in pattern and rate of spread between species

All those examples are of variation in the linear rate of spread within species. They serve as a background for the main theme of this chapter, the variation in both speed and pattern between species. So far I have quoted rates in distance per time, kilometers per year, which would be appropriate for populations spreading in the manner predicted by the Fisher–KPP–Skellam equations, where there is a circular wave of advance moving at a constant linear speed. This pattern comes from random, Gaussian, dispersal with many patterns of reproduction. It has been realized for almost a century that the spread of forests after the last glaciation, indeed after all the glaciations of the Pleistocene, was much faster than could be explained by random dispersal, a phenomenon sometimes called "Reid's paradox". For a range of tree species in North America and Europe in the Holocene, the rates of advance are mostly between 100 m and 1000 m per year, which implies at least ten times that (1–10 km) per generation (Williamson 1996: Fig 4.11). Acorns never fall off oak trees (*Quercus* spp.) and then roll or are blown to several kilometers away, they must be carried, though other lighter seeds, of birch say (*Betula* spp.), might be blown that far by strong storms.

The difference between the observable rates of spread by random dispersal and the observed rates of spread across landscapes are brought out in Fig. 5.1. The rates plotted there are those culled from the literature by Pyšek and Hulme (2005). The distinction between rates found by studying local populations, the left hand hump, and those found by studying spread at landscape scales, the right hand hump, is striking; the latter is two and a half orders of magnitude, three hundred times, faster. The explanation of this from current models is that there is a thick dispersal tail (Williamson 2002), which still leaves the question

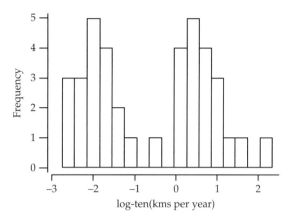

Figure 5.1 The distribution of reported spread rates in the literature, plotted from Table 5.1 in Pyšek and Hulme (2005). Note the logarithmic scale of the abscissa.

of what processes lead to such a tail. Such models, though, lead to accelerating speeds of spread and no obvious wave front, more a set of scattered new colonies.

One way of distinguishing what might be called ripple spread, a constant-speed wave of advance, from what might be called jump spread, with new colonies appearing at increasing distances, is to plot functions of the area occupied against time. A line straight on a square root plot is consistent with ripple spread, one straight on stronger transformations such as the logarithm or logit, with jump spread. However, maps suggest something more complicated, as can be seen in the example in Fig. 5.2. This is a map of the distribution of an invading ladybird, up to 2006, using hectads (10 km × 10 km grid squares). Counting hectads in earlier years leads to an apparently straight line on a square root plot and an estimated rate of spread of about 70 km per year. That rate seems to be the rate of spread of the bulk of the invasion shown by the solid mass of hectad records. But ahead of that mass there is a scattering of records far more than 70 km (a line of seven hectads) from the main mass. Some of those records are transients, some are not, and those are not distinguished when counting cumulative records. But it is clear that a simple ripple spread is not the whole story. In other cases, the square root plot is clearly not straight. Perrins *et al.* (1993) (and Williamson 1996) showed logistic spread in Britain for three alien

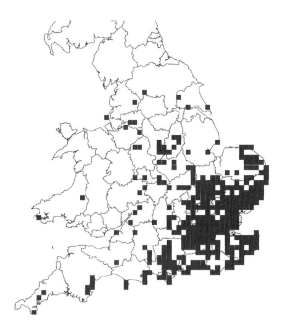

Figure 5.2 The distribution of the invading and undesirable harlequin ladybird, *Harmonia oxyridis*, in Britain in 2006. The recording units are hectads (10 km × 10 km squares). From Harlequin ladybird survey (2007).

plant species, balsams in the genus *Impatiens*. The apparent acceleration from 2.6 to 38 kilometers per year for *I. glandulifera* is in Table 5.1; *I. capensis* and *I. parviflora* showed similar changes in rate, 1.4 and 13 km, 1.6 and 24 km respectively.

Understanding the variation in rate and pattern between species requires sets of good long-term data and such sets are scarce. Williamson *et al.* (2003) and Williamson *et al.* (2005) examined different sets of grid records of alien plants in Europe. The first only used two time points and so was unable to distinguish differences in pattern between species. But it did show, for all 118 species with reliable records, that the modal rate of spread was zero, which means that some were shrinking in range between 1958 and 1988, and found average rates over those thirty years of up to nine kilometers a year. That is consistent with the data for *Impatiens glandulifera*, the only plant in Table 5.1, except for the maximal rate found by Perrins *et al.* (1993). There may be a technical explanation for the difference in maximal rates as Williamson *et al.* (2003) used hectads which are only about 5 per cent of the area of vice-counties used by Perrins *et al.* Where there is jump spread, using larger reference areas will produce larger apparent rates of spread; the logistic curves found by Perrins *et al.* are consistent with jump spread.

Williamson *et al.* (2005) used a marvelous Czech data set with annual records of range over, in some cases, a couple of centuries for selected alien plant species, supplemented by a handful of such species in Britain and Ireland for which there are several records of range. Some details of those data are in Table 5.2 and a summary of the patterns found in the Czech data in Table 5.3. For patterns, the important result is that many species gave straight square root plots, and slightly fewer gave straight logarithmic plots. Slower rates, lags, before and after the main expansion phase were also common. There is good evidence for both ripple type spread and jump type spread, for variable rates, for lags, and for slowing of spread (bends) as the range is filled.

For rates, when the square root plot is straight, a figure in kilometers per year is appropriate. When

Table 5.2 Background information for the study (Williamson *et al.* 2005) of spread rates in three countries. The data go back from the present to 200 years ago

Country	Area k.km^2	Human Population density, #/km^2	Spp. used	Mapping unit	Size km^2	Number
Czech Rep.	79	130	63	map. pole 10′×6′	133	679
Britain	229	235	7	hectad 10km × 10km	100	229
Ireland	84	60	4	hectad 10km × 10km	100	1007

Table 5.3 Summary of the shapes of increase in area with time of invading alien plants in the Czech Republic, from Williamson et al. (2005). The numbers are the number of species showing that spread type

Expansion of the headings:
log	some part of the data straight on a logarithmic plot
log → sqrt	two successive sections, sometimes with an intermediate period, found to be straight on transformation, the first on a logarithmic plot, the second on a square root one
sqrt/log	some of the data statistically straight on both a logarithmic and a square root plot, for an overlapping period but not necessarily for the same period of years
sqrt	some part of the data straight on a square root plot
Σ	sum of the number of species to the left of this column
arith	the data are only straight on an arithmetic plot (these species are casuals)
none	the data are not straight on any plot
ALL	the sum of all the entries below
bend	the data decrease in rate of spread after the straight section
lag	the data show a slow increase before the log or sqrt straight section, this increase sometimes straight arithmetically, indicating a casual period

	log	log → sqrt	sqrt/log	sqrt	Σ	arith	none
ALL	12	6	12	20	50	10	3
bend & lag	1	3	2	3	9		
bend	9	0	5	8	22		
lag	0	1	4	8	13		
neither	2	2	1	1	6		

the logarithmic plot is straight, the distance per year accelerates and it is better to consider doubling time, the number of years needed for the area occupied to double. Histograms for the all the species-country records of the appropriate shape are given in Fig. 5.3. The medians are 1.67 kilometers per year and a doubling time of 10.02 years. From those species which changed from jump type to ripple type, usually suddenly (the log → sqrt species in Table 5.3) jump spread leads to faster invasions in the area occupied as well as in the furthest point reached than ripple spread. That conclusion seems to apply generally including to sqrt/log species in Table 5.3 which are straight in part on both plots. While the spread of the fastest species, *Sisymbrium volgense* Russian mustard, with a doubling time of 2.04 years over 9 years or 4.39 km yr^{-1} over 17, is spectacular, even the slowest spreader, *Oxybaphus* (or *Mirabilis*) *nyctagineus* Four o'clock with a doubling time of 39.02 years or 0.31 km yr^{-1}, both over 65 years, is steadily expanding its range and needs to be watched by managers.

Lag time, though, is a greater difficulty for managers because it is variable, unpredictable, and liable to end at any time. Species in lag are sometime called sleeper weeds. Note that for both ripple and jump spread types, lag cannot usually be assessed from arithmetic plots of range, even though many authors have claimed to have done so. For the Czech set, where the lag has necessarily been completed or the species would not be in the dataset, the distribution of lag times is given in Fig. 5.4. Almost half the species, 22 out of 50, show no lag but the lag time in the majority varied from 7 to 154 years with a median of 41 (longer than the career span of most managers).

Although there is a great deal of biological information in databases on this set of alien species, no character was found that related either to the pattern or rate of spread. Socio-economic factors can be postulated, but there is no information of that sort at the appropriate grain size.

5.4 Time to complete spreading

Another way to think about the rate of spread is to ask how long it will take, on average, for an alien plant to cease spreading in a particular country:

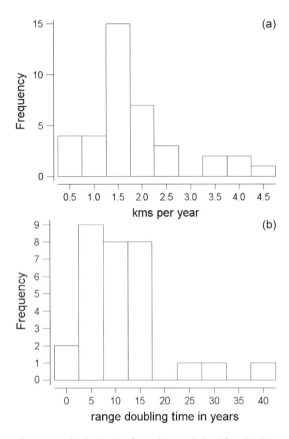

Figure 5.3 The distribution of spread rates calculated from Czech data (see Tables 5.2 and 5.3) for the 19th and 20th centuries. Upper histogram [3a]: spread in kilometers per year for those species with straight lines on square root plots. Lower histogram [3b]: spread as doubling time for those species with straight lines on logarithmic plots. From Table 5.1 in Williamson *et al.* (2005).

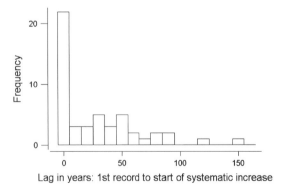

Figure 5.4 The distribution of apparent lag times in the Czech data (Tables 5.2 and 5.3) from Table 5.1 of Williamson *et al.* (2005). Note that 22 out of 50 species show no lag.

how long will it take to fill its range. Williamson *et al.* (2009) have studied the effect of residence time, the time since an alien was first recorded in the wild, on the range size of aliens in Ireland, Britain, Germany, and the Czech Republic. With the marked variation in spread rates between species noted above, it is perhaps not surprising that residence time only accounts for around 10 per cent of the variance in range. Nevertheless the relationship of range to the date of the first record can be used to estimate the time to range completion.

First, though, consider the box and whisker plots of the logits of range size in Fig. 5.5. Logits simplify comparisons between countries as they relate to the proportion of the total area in each country. The main point here is that, in all countries at the time of the censuses in the 1990s, native plants were more widespread than naturalized (established) neophytes, and both much more widespread than casual neophytes. That holds across countries too, except that German natives are less widespread than Czech naturalized neophytes. Neophytes are plants that have escaped since 1500 AD, archaeophytes those that escaped before. Archaeophytes are in fact a touch more widespread than natives, except in Ireland. As archaeophytes are recognizable to a large extent only because they are arable weeds, that result probably just reflects the commonness of arable land in all countries except Ireland.

That naturalized neophytes are less widespread than natives reflects, to some extent, the fact that many have not yet had time to fill their alien ranges. Of the 30 British aliens graphed in Williamson *et al.* (2003), only 6 were not spreading in 1958–88. The relationship of range to residence time produces an estimate of how long it will take, on average, for the mean range of naturalized neophytes to become the same as the mean range of natives. There are two possible relationships that can be used for this. The first is an ordinary least squares regression (ols), the second a reduced major axis (RMA). The first, ols, is appropriate for predicting range at a particular time and assumes that all the error is in the range and that the residence time is known. However, as both range and the date of the first record are subject to various uncertainties, RMA may be more satisfactory for estimating the time

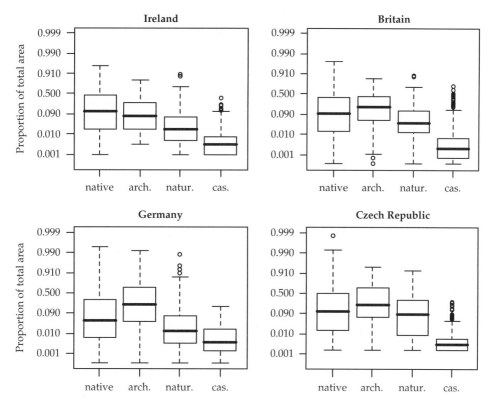

Figure 5.5 Box and whisker plots of the logits of ranges of natives, archaeophytes, naturalized neophytes, and casual neophytes in Ireland, Britain, Germany, and the Czech Republic. From Williamson et al. (2009).

to reach the maximum range as it assumes equal error in both axes. Certainly RMAs are more consistent than regressions across the four counties. The results of using both regressions and RMAs to predict the time needed for naturalized neophytes to match natives in range size is given in Fig. 5.6. The rather flat regressions of Ireland and Britain predict a period of, very roughly, 300 years, but all the RMAs and the other two regressions indicate only half that time, around 150 years. Both estimates indicate that most alien plants will still be spreading in all the countries, only the oldest introductions or the fastest spreaders will have completed their spread.

5.5 Implications and conclusions

The variation in spread rate both within and between species is considerable. Managers' attention is inevitably on those that spread fast. The

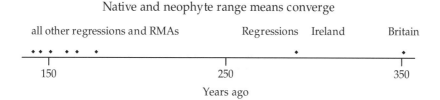

Figure 5.6 The time predicted from ols (ordinary least squares) regressions and from RMAs (reduced major axes) of logits against the time for the mean range of established neophytes to match the mean range of natives. Data from Williamson et al. (2009).

message from this set of studies is that those that spread slowly need to be watched too. Just because an alien species has a small range now does not mean that it will not eventually be widespread. The literature is full of instances of plants that were initially thought benign, even beneficial, that came to be regarded as pests when they became widespread.

Lags in spread are common and measured in decades. The time to become widespread after the lag period is finished is often considerable, measured in centuries even in state sized areas. Both these effects raise problems for managers. Often, resources seem only sufficient to tackle species that are already pests, but it might nevertheless be a good long-term strategy to devote some effort to studying those established aliens which appear benign to see if they are likely to become pests. Action against such species might be very worthwhile.

Acknowledgements

Calvin Dytham and Chris Thomas have helped with information and discussion and I thank both of them.

References

Ayala, F.J., Serra, L., and Prevosti, A. (1989). A grand experiment in evolution: the *Drosophila subobscura* colonization of the Americas. *Genome*, **31**, 246–55.

Cousens, R., Dytham, C., and Law, R. (2008). *Dispersal in Plants*. Oxford University Press, Oxford.

Cwynar, L.C. and MacDonald, G.M. (1987). Geographical variation in lodgepole pine in relation to population history. *American Naturalist*, **129**, 463–69.

Darling, E., Samis, K.E. and Eckert, C.G. (2008). Increased seed dispersal potential towards geographic range limits in a Pacific coast dune plant. *New Phytologist*, **178**, 424–435.

Duckworth, R.A. (2008). Adaptive dispersal strategies and the dynamics of range expansion. *American Naturalist*, **172**, S4–S17.

Duckworth, R.A. and Badyaev, A.V. (2007). Coupling of dispersal and aggression facilitates the rapid range expansion of a passerine bird. *Proceedings of the National Academy of Sciences of the United States of America*, **104**, 15017–22.

Flux, J.E.C. (1994). World distribution. In H.V. Thompson and C.M. King, eds. *The European Rabbit*, Oxford University Press, Oxford, pp. 8–21.

Gibb, J.A. and Williams, J.M. (1994). The rabbit in New Zealand. In H.V. Thompson and C.M. King, eds. *The European Rabbit*, Oxford University Press, Oxford, pp. 158–204.

Harlequin ladybird survey (2007). http://www.harlequin-survey.org

Hastings, A., Cuddington, K., Davies, K.F. *et al.* (2005). The spatial spread of invasions: new developments in theory and evidence. *Ecology Letters*, **8**, 91–101.

Hill, J.K., Thomas, C.D., and Blakeley, D.S. (1999). Evolution of flight morphology in a butterfly that has recently expanded its geographic range. *Oecologia*, **121**, 165–170.

Malíková, L. and Prach, K. (in press) Present occurrence of *Impatiens glandulifera* along four rivers invaded at different times, *Ecohydrology and Hydrobiology*.

Myers, K., Parer, I., Wood, D. and Cooke, B.D. (1994). The rabbit in Australia. In H.V. Thompson and C.M. King, eds. *The European Rabbit*, Oxford University Press, Oxford, pp. 108–57.

Okubo, A., Maini, P.K., Williamson, M., and Murray, J.D. (1989). On the spatial spread of the grey squirrel in Britain. *Proceedings of the Royal Society of London B*, **238**, 113–25.

Pascual, M., Chapuis, M.P., Mestres, F., *et al.* (2007). Introduction history of *Drosophila subobscura* in the New World: a microsatellite-based survey using ABC methods. *Molecular Ecology*, **16**, 3069–83.

Perrins. J., Fitter, A., and Williamson, M. (1993). Population biology and rates of invasion of three introduced *Impatiens* species in the British Isles. *Journal of Biogeography*, **20**, 33–44.

Phillips, B.L., Brown, G.P., Webb, J.K. and Shine, R. (2006). Invasion and the evolution of speed in toads. *Nature*, **439**, 803.

Phillips, B.L., Brown, G.P., Travis, J.M.J., and Shine, R. (2008). Reid's paradox revisited: the evolution of dispersal kernels during range expansion. *American Naturalist*, **172**, S34–S48.

Pyšek, P. and Hulme, P.E. (2005). Spatio-temporal dynamics of plant invasions: Linking pattern to process. *Écoscience*, **12**, 302–15.

Sherratt, J.A. (1998). On the transition from initial data to travelling waves in the Fisher-KPP equation. *Dynamics and Stability of Systems*, **13**, 167–74.

Simmons, A.D. and Thomas, C.D. (2004). Changes in dispersal during species' range expansions. *American Naturalist*, **164**, 378–95.

Teangana, D.Ó., Reilly, S., Montgomery, W.J. and Rochford, J. (2000). Distribution and status of the red squirrel (*Sciurus vulgaris*) and grey squirrel (*Sciurus carolinensis*) in Ireland. *Mammal Review*, **30**, 45–56.

Usher, M.B. (1986). Invasibility and wildlife conservation: invasive species on nature reserves. *Philosophical Transactions of the Royal Society of London B*, **314**, 695–710.

Williamson, M. (1996). *Biological Invasions*. Chapman & Hall, London.

Williamson, M. (2002). Overview and synthesis: The tale of the tail. In J.M. Bullock, R.E. Kenwood and R.S. Hails, eds. *Dispersal Ecology*, Blackwell Science, Oxford, pp. 431–443.

Williamson, M. and Brown, K.C. (1986). The analysis and modelling of British invasions. *Philosophical Transactions of the Royal Society of London B*, **314**, 505–22.

Williamson, M., Preston, C. and Telfer, M. (2003). On the rates of spread of alien plants in Britain. In Child, L., Brock, J.H., Brundu, G., Prach, K., Pyšek, P., Wade, P.M. and Williamson, M., eds. *Plant Invasions: Ecological Threats and Management Solutions*, Backhuys, Leiden, pp. 63–74.

Williamson, M., Pyšek, P., Jarošík, V. and Prach, K. (2005). On the rates and patterns of spread of alien plants in the Czech Republic, Britain, and Ireland. *Écoscience*, **12**, 424–33.

Williamson, M., Dehnen-Schmutz, K., Kühn, I. *et al.* (2009). The distribution of range sizes of native and alien plants in four European countries and the effect of residence time. *Diversity and Distributions* **15**, 158–66.

Yalden, D. (1999). *The History of British Mammals*. T & AD Poyser, London.

CHAPTER 6

Habitats and Land Use as Determinants of Plant Invasions in the Temperate Zone of Europe

Petr Pyšek, Milan Chytrý, and Vojtěch Jarošík

6.1 Introduction

In the current literature on plant invasions, solid information on macroecological patterns has accumulated, which contributes to our understanding of the invasion process. For example, it has been firmly established that temperate mainland regions are more invaded than tropical mainland, islands are more invaded than corresponding areas of mainland, and that the proportion of naturalized alien species to native species in temperate zone decreases with latitude and altitude (see Pyšek and Richardson 2006 for a review). It is symptomatic that these generalizations are based on the numbers of alien species, mostly naturalized, in individual regions or states, and ignore that the emergent patterns on the coarse scale can be an outcome of patterns and processes occurring on finer scales, namely of differences in the level of invasion among different habitats.

Because large-scale quantitative information on the distribution of species, not only alien but also native, in habitats is scarce, the extent to which individual habitats are invaded was sometimes estimated from expert assignments of alien plant species represented in a regional flora to their respective habitats (Crawley 1987; Rejmánek et al. 2005; Walter et al. 2005). This approach essentially defines alien species pools of individual habitats (Zobel 1992; Sádlo et al. 2007), but has a disadvantage that sizes of the regional species pools of aliens cannot be simply scaled down to individual sites so that their level of invasion can be quantified.

In the present chapter, we review studies dealing with invasions in habitats and, by using an extensive dataset from the Czech Republic, we demonstrate differences between the level of invasion (defined as the actual representation of alien species, Chytrý et al. 2005; Richardson and Pyšek 2006) and invasibility (defined as the inherent vulnerability of habitats to invasion, Lonsdale 1999). We classify individual habitats according to these two measures and explore the relative importance of factors which determine both the level of invasion and invasibility. In analyses, two traditional groups of European aliens are considered, distinguished on the basis of residence time (Pyšek et al. 2004): archaeophytes (arrived before 1500 AD, mainly from the Middle East and Mediterranean), and neophytes (arrived after 1500 AD, mainly from North America and Asia). Finally, the potential for application of the knowledge of invasion in individual habitats is discussed and possibilities of using it in risk assessment outlined.

6.2 Overview of studies on the level of invasion in habitats

Until recently our knowledge of which habitats are invaded, and how much, was based on anecdotal evidence rather than rigorous testing. A pioneering study documenting quantitatively invasions in various habitats was published by Kowarik (1995) for the city of Berlin, and papers comparing several habitats over large areas, using large numbers of vegetation plot records, started to appear only recently. Quantitative data on the representation of alien species in various habitats are still surprisingly scarce; only a few such assessments are available from Europe (Chytrý et al. 2005, 2008a, b; Maskell et al. 2006; Vilà et al. 2007) and the United States (Stohlgren et al. 1999, 2006; Spyreas et al. 2004). Most surveys conducted at a landscape scale, are performed within habitat types (Gilbert and Lechowicz 2005) or biased towards particular habitats, which makes the range of habitats considered limited (DeFerrari and Naiman 1994; Planty-Tabacchi et al. 1996; Sobrino et al. 2002; Brown and Peet 2003; Campos et al. 2004; Pyšek et al. 2005). Such studies therefore cannot provide broader insights into the differences in the level of invasion among habitats.

6.2.1 Representation of alien species in habitats

Despite some differences in plot size and habitat classification systems, European studies that evaluated invasions in individual habitats, based on thousands of vegetation plots, yielded consistent results in terms of the representation of alien species in the most invaded habitats. Chytrý et al. (2005), working on the scale of units to hundreds m^2 and using 32 habitat types in the Czech Republic, found that the 6 most invaded habitats harbored on average 4.4–9.6 per cent of neophytes (2.3 per cent on average across all vegetation types). For Catalonia, Vilà et al. (2007), using 34 habitat types and plot size of ca 20–90 m^2, also found a low mean number of neophytes per plot (less than 2.0 per cent pooled across habitats, and less than 9.0 per cent in the most invaded habitats). The highest proportions of neophytes of the total species numbers per plot were reported from the UK, with maxima of 10.0–24.8 per cent in the 3 most invaded of 19 habitat types (Chytrý et al. 2008b). There are also regional surveys, where percentages of alien species higher than 10 per cent are reported, but this is because these studies deliberately focused on highly invaded habitats such as riverine (Sobrino et al. 2002; Planty-Tabacchi et al. 1996) and coastal (Campos et al. 2004).

Outside Europe, the most extensive studies on the occurrence of alien and native species were done on 37 natural vegetation types in 7 states in the central United States (Stohlgren et al. 1999; Stohlgren 2007) in plots ranging in size from 1 to 1000 m^2. In several of these vegetation types, the proportion of alien species considerably exceeded 10 per cent (Stohlgren et al. 2006). The figures given by these authors can be compared with data from the Czech Republic, Catalonia, and Britain, because "exotic" and "alien" species in the US studies (Stohlgren et al. 1999, 2006) correspond to neophytes in the European studies (Chytrý et al. 2005, 2008b; Vilà et al. 2007). It is striking that the proportion of aliens in US natural vegetation types is often much higher than in vegetation of human-made habitats in Europe.

6.2.2 Which habitats are most invaded?

A study summarizing levels of invasion based on vegetation plot data from three European regions, and using standardized classification of habitats (Chytrý et al. 2008b), showed that habitats generally associated with human- and water-induced disturbances, high fertility, and high propagule pressure, exhibit the highest levels of invasions. Pooled across regions, arable land, coastal sediments, and ruderal habitats—including trampled areas—harbor the highest proportions of neophyte species. This is in accordance with regional and habitat-specific studies (DeFerrari and Naiman 1994; Planty-Tabacchi et al. 1996; Sobrino et al. 2002; Campos et al. 2004; Chytrý et al. 2005; Vilà et al. 2007; Simonová and Lososová 2008). The highest level of invasion in Europe was found in coniferous woodlands (with 24.8 per cent of neophytes), arable land (14.3 per cent) and coastal sediments (10.0 per cent) in Britain (Chytrý et al. 2008b). Two subtypes of annual anthropogenic vegetation in the Czech Republic also contained on average 17–22 per cent of neophytes (Simonová and Lososová 2008). However, coniferous woodlands were highly invaded only in Britain, where most of them are plantations of exotic conifers, while native and even planted coniferous woodlands in the other countries had very low proportions of alien species. In a study of the British countryside, Maskell *et al.* (2006) also investigated temporal trends in the representation of alien and native species and concluded that changes such as eutrophication, nitrogen deposition, and increased fertility in infertile habitats currently benefit native species more often than aliens.

The figures reported for archaeophytes are much higher, reaching 55.5 per cent, 35.5 per cent, and 21.8 per cent on arable land, ruderal vegetation, and trampled habitats in the Czech Republic, and 16.2 per cent on arable land in Britain. British habitats generally contain less archaeophytes than Czech habitats, which reflects the differences in the total pools of archaeophytes in the two countries (Sádlo *et al.* 2007; Chytrý *et al.* 2008b).

6.2.3 Importance of scale

When assessing the role of alien species in vegetation, the effect of scale must be taken into account (Chytrý *et al.* 2005; Stohlgren *et al.* 2006; Stohlgren 2007). The proportional representation of alien species has been traditionally assessed in larger areas, such as countries (Essl and Rabitsch 2002; Pyšek *et al.* 2002b), counties (Stohlgren *et al.* 2005), bioregions (Dark 2004), cities (Pyšek 1998), or grid squares of floristic mapping (Deutschewitz *et al.* 2003; Kühn *et al.* 2003; Pino *et al.* 2005), but much less so in vegetation plots ranging in size from units to hundreds of square meters. In general, the shift to a finer scale strongly affects the representation of alien species. For example, neophytes make up 26.8 per cent of the flora of the Czech Republic (Pyšek *et al.* 2002b) and 25.2 per cent of the flora of an average Central European city (Pyšek 1998), but only 2.3 per cent of the species found in an average vegetation plot (Chytrý *et al.* 2005). In the same vein, the 2 per cent of neophytes per vegetation plot recorded by Vilà *et al.* (2007) in Catalonia are much less than the mean of 6.9 per cent reported from 10 km grid cells sampled in the same region (Pino *et al.* 2005). The multi-scale plot design adopted by Stohlgren *et al.* (1999, 2006), who used four nested plot sizes of 1, 10, 100, and 1000 m^2, allows for a rigorous assessment of the effect of scale on the richness of alien species. For example, in the most invaded habitat, irrigated shortgrass prairie, the mean number of aliens per plot increased from 2.9, 5.2, 6.9, to 10.0 from the smallest to largest plots, and the pattern was consistent across habitats. In addition, not only numbers of alien species but also their proportions depend on the scale of observation (Stohlgren *et al.* 2006).

The above examples point to the importance of choosing an appropriate scale for studying the level of invasion in habitats. A test of commonly used rangeland quadrat sampling methods revealed that small quadrats failed to capture about half of the native and alien plant species occurring in several prairie and grassland habitats (Stohlgren *et al.* 1998). The plot size ranging from 10s to 400 m^2, common in phytosociological studies and used in the comparative assessment of European habitats (Chytrý *et al.* 2008b) and in the analysis of factors determining their level of invasion (Chytrý *et al.* 2008a), seems optimal for this purpose. Plots of this size capture a reasonably high proportion of species present in the habitat, and the scale allows for a fine assessment of the effects of environmental variables, because detailed information on environmental settings can be obtained in each sampling site—something that cannot be achieved at larger scales where average values for, for example, regions or states, embody a great variation among sites. To get an insight into factors determining habitat invasibility, it is necessary to obtain good information on habitat variables, because invasions of alien species are determined by a multitude of factors that are likely to co-vary with spatial and temporal scales such as climate, vegetation structure, micro- and macro-disturbances, resource availability, species pools and propagule pressure, and associated ecosystem processes (Stohlgren *et al.* 2006).

6.2.4 Native–alien relationship

Strongly associated with the issue of scale is the relationship between the numbers of native and alien plant species. Within a vegetation type, native species richness and cover may vary considerably (Stohlgren *et al.* 2006). With increasing spatial scale invasions of alien species and their coexistence with native plants are likely to increase (Stohlgren *et al.* 1999, 2006; Knight and Reich 2005). At subcontinental scales there is convincing evidence of strong positive relationships between native and alien plant species richness (Stohlgren *et al.* 2003; Richardson and Pyšek 2006). These results complement regional landscape-scale observational studies (e.g. Stohlgren *et al.* 1999; Brown and Peet 2003; Keeley *et al.* 2003), but do not provide an understanding of the factors that determine the patterns of alien species diversity and the mechanism for the observed patterns across scales (Levine *et al.* 2003, 2004).

Generally, positive relationships between native and alien species richness strengthen at scales of more than 1 m^2, becoming more significant at scales of more than 100 m^2. This is consistent with null models of community invasibility, which show that the relationship between native and alien species richness tends to be negative at small spatial scales, but more positive at larger scales (Fridley *et al.* 2004; Herben *et al.* 2004). It has been proposed that similar factors to those associated with the increase in native species with area, such as habitat heterogeneity, extensions of environmental gradients, and increased probabilities of encountering disturbed habitats are also responsible for increasing establishment of alien species (Stohlgren *et al.* 1999, 2001, 2002, 2003). The positive relationship can be explained by the similarity of native and alien species in the abundance of propagules entering a community (Levine 2000) or by both groups of species occurring more frequently in resource-rich and moderately disturbed sites (Davis *et al.* 2000, Vilà *et al.* 2007). Stohlgren *et al.* (2006) suggested that in the invasion of any area greater than 1 m^2 native species outnumber newly arriving invaders, but their biotic resistance (Elton 1958; see Richardson and Pyšek 2006 for a review) becomes overwhelmed by biotic acceptance, where co-existence is a stronger force than competitive exclusion, resulting in the broad-scale establishment of many alien species. The Theory of Biotic Acceptance (Stohlgren *et al.* 2006) suggests that where environmental heterogeneity, environmental gradients, disturbance, and species turnover increase with spatial scale, natural ecosystems tend to accommodate alien species despite the presence and abundance of native species.

Recent European studies arrived at similar conclusions on the prevailing positive relationship between native and alien species richness at the scale of units to hundreds of square meters (Chytrý *et al.* 2005; Maskell *et al.* 2006). In addition, a similar positive relationship exists between archaeophytes and native species, and even more strongly between neophytes and archaeophytes. Neophytes are found commonly in habitats also occupied by archaeophytes, and archaeophytes can thus serve as predictors of the neophyte invasion risk (Chytrý *et al.* 2005, 2008b). However, the relationships between the numbers of neophytes and native species were mostly positive only if individual plots were compared separately within habitats; the correlations calculated with habitat mean values were non-significant (Chytrý *et al.* 2005). Vilà *et al.* (2007) found both high and low numbers of alien species at intermediate values of native species richness and low values at both extremes of native species richness.

6.3 Theoretical background of community invasibility

If we want to obtain a deeper insight into community (or habitat) invasibility, we face a major limitation. The majority of data reported in the literature are species numbers—how many alien species are present in a given habitat (ecosystem, region), or what is the proportion of aliens to all species. For this measure the term *level of invasion* was suggested (Hierro *et al.* 2005; Chytrý *et al.* 2005; Richardson and Pyšek 2006). However, the level of invasion cannot be used to infer whether or not, and to what extent certain habitat is prone to invasion. Therefore it is necessary to distinguish this measure from *invasibility*, that is, habitat susceptibility to invasion.

As pointed out by Williamson (1996), looking for real differences in invasibility requires looking at the residuals from the regression between invasion success and propagule pressure. From this it follows that successful invasion of a habitat requires dispersal, establishment, and survival, with the number of species determined by the balance between extinction and immigration

(Lonsdale 1999). The number of alien species existing in the habitat, A, is given by the product of the number of alien species introduced to the habitat, I, and their survival rate S, which differs in individual habitats based on their properties:

$$A = I \times S$$

The survival rate S, which is the measure of habitat invasibility, is determined by several components:

$$S = S_v \times S_h \times S_c \times S_m$$

These components represent losses, or constraints to invasion, due to competition with species already present in the habitat (in natural and semi-natural habitats majority of them are native), S_v, the effects of herbivores and pathogens, S_h, chance events, including extreme climatic events, S_c, and maladaptation, S_m. To invade, the species must survive the effect of all these factors; hence the overall survival rate is a product of all the factors listed above (Lonsdale 1999). To compare the invasibility of two habitats, we need to compare their S values rather than their A values. A habitat is more prone to invasions (i.e. more invasible) if the survival rate of alien species, introduced by means of propagule pressure, is higher than in another habitat with lower S.

6.4 Separating the level of invasion from invasibility

Lonsdale's model (1999) implies that a certain fraction of the variation in alien species richness among sites can be attributed to propagule pressure, defined as the rate of influx of alien propagules into the target site. To answer the question why some habitats are more invaded than others, one must separate the effects of habitat properties from those of propagule pressure and from other potentially confounding factors, such as climate (Chytrý *et al.* 2008a). Up to now, surprisingly little was known about the importance of habitat properties, relative to that of propagule pressure and other factors, mostly due to methodological reasons. Seed addition experiments, which suggest that increased propagule pressure strongly contributes to the level of invasion (e.g. Tilman 1997; Kennedy *et al.* 2002) are usually confined to a single habitat and single site, and do not explain differences between broader ranges of habitats. Observational studies have not provided significant insights either, as they are mostly restricted to a few habitats, a single or a few species, use limited numbers of replicates, or do not attempt to separate the effects of habitat properties from those of propagule pressure.

However, recent compilations of large databases of vegetation survey plots (e.g. Font and Ninot 1995; Hennekens and Schaminée 2001; Chytrý and Rafajová 2003; Firbank *et al.* 2003), which include thousands of records of species composition from all the major habitats of a country or a large region, provide an excellent opportunity to rigorously compare the levels of invasion between habitats. Continental Europe is a region with a strong phytosociological tradition—classification of vegetation into units is based on field data collated in a standard way. Over several decades phytosociologists collected huge amount of data which can now, profiting from the development in computing facilities, be used to study macroecological questions, if some limitations such as preferential sampling or varying plot size are taken into account (Chytrý 2001; Knollová *et al.* 2005; Roleček *et al.* 2007; Haveman and Janssen 2008).

6.4.1 Comparison of the levels of habitat invasions with their invasibility

Vegetation plots accumulated over the past 30 years of phytosociological vegetation surveys were used in a study of invasions in habitats of the Czech Republic (Chytrý *et al.* 2005, 2008a, b; see Fig. 6.1 for the outline of methods). Due to its geographical position and variety of habitats (Sádlo *et al.* 2007), the results can be considered as representative of a wider temperate region of Central Europe. Regression tree analysis (see Chytrý *et al.* 2008a for details) identified habitats with the highest and lowest levels of invasion in the Czech Republic. The lowest proportion of neophytes (on average 0.3 per cent) is found in natural and semi-natural habitats at altitudes above 465 m above sea level (a.s.l), while the highest proportion of neophytes (20.3 per cent) occurs in human-made

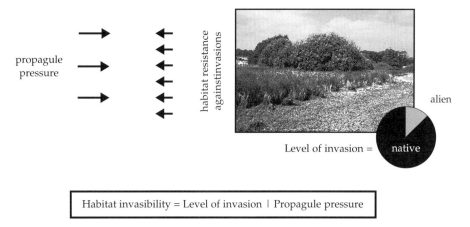

Figure 6.1 A conceptual model for studying the relationship between the level of invasion (an actual number or proportion of alien species present in a habitat) and the invasibility (inherent vulnerability of a habitat to invasion, when the effect of propagule pressure is held constant). To translate Londsdale's equation (see text) into the terminology used in this paper, the Level of Invasion (number or proportion of alien species we observe in a habitat) is a product of Propagule Pressure (number of species introduced there) and Habitat Invasibility (survival rate of invading species). The level of invasion was expressed as proportional numbers of aliens in 20,468 vegetation plots collected by phytosociologists on a regional scale of 78,000 km^2 (Czech Republic) in the last three decades, and stored in the national phytosociological database (Chytrý and Rafajová 2003). To study habitat invasibility, the effect of habitat (classified into 32 types using the standard classification of European habitats, EUNIS; Davies & Moss 2003, available at http://eunis.eea.europa.eu/habitats.jsp) on the level of invasion was tested, holding the effect of propagule pressure and climate, expressed by mean annual temperature and precipitation in the given site, constant. To account for the effect of propagule pressure, the following proxies were measured within 500 m circles around each plot: human population density; proportion of the area that is residential, industrial or agricultural; the distance of a site from the nearest river; and the altitudinal floristic region from the national classification, which reflects the history of human colonization. See Chytrý et al. (2008a) for details.

habitats, disturbed woodlands, or cliffs and walls at altitudes below 365 m a.s.l. that are surrounded by urban and industrial land and have open vegetation cover, less than 23 per cent (Chytrý et al. 2008a). Central European habitats can be divided into three groups based on the level of invasion, increasing from low levels in alpine and subalpine habitats, bogs and coniferous woodland, through intermediate in most grasslands and broad-leaved woodlands, to high in human-made habitats, including arable land, and deciduous plantations. In general, the pattern found for archaeophytes and neophytes is similar, although some habitats are more invaded by one group of aliens or the other (Table 6.1).

Then, between-habitat comparisons of invasibility were made using statistical models in which habitat was the predictor variable and the residuals from the regression of the level of invasion on the confounding variables (i.e. measures of propagule pressure and climate) were the response variables (see Chytrý et al. 2008a for details). Habitats were ranked by (i) the actual proportions of aliens in habitats, that is the level of invasion, and (ii) the residuals after subtracting the effect of confounding variables, that is invasibility. This comparison indicates that there is an overall correspondence between the two measures of invasion—the most invaded habitats are also highly invasible, but some habitats differ markedly if their levels of invasion and invasibility are compared (Fig. 6.2). This especially concerns some moderately invaded habitats whose invasibility is actually low (Table 6.1), but the resistance to invasion is overcome by a high propagule pressure. Consistent with the theoretical models (Alpert et al. 2000; Davis et al. 2000; Shea and Chesson 2002), the pattern we found can be interpreted in terms of disturbance regime and resource availability (Table 6.1):

(i) Most invasible habitats are strongly and/or frequently disturbed. In arable land, disturbance completely removes the above-ground biomass at least once a year. Vegetation of ruderal and trampled sites is also strongly and frequently

Table 6.1 Central European habitats grouped according to their level of invasion and invasibility and characteristics of disturbance regime and nutrient availability. Where the group (archaeophytes, neophytes) is not specified following the habitat name, the pattern holds for all alien plants. See Fig. 2 for quantitative comparison of the level of invasion and invasibility in individual habitats

Habitat	Disturbances	Nutrient availability
High level of invasion, very high invasibility		
Arable land	Frequent, strong	High, frequent and strong pulses due to external input
High level of invasion, high invasibility		
Annual ruderal vegetation	Frequent, strong	High, pulses due to external input
Trampled areas	Frequent, strong	High, pulses due to external input
Perennial ruderal vegetation	Frequent	High, pulses due to external input
Deciduous plantations (neophytes)	Strong at the initial establishment	High, pulses at the initial establishment
Intermediate level of invasion, intermediate invasibility		
Coniferous plantations (archaeophytes)	Strong at the initial establishment	Pulses at the initial establishment
Deciduous plantations (archaeophytes)	Strong at the initial establishment	High, pulses at the initial establishment
Disturbed woodlands	Strong	High, pulses after disturbance
Screes	Frequent	Low, occasional pulses
Riverine scrub	Common	High, frequent pulses
Running waters (neophytes)	Common	Medium, rare pulses
Sedge-reed beds	Occasional	High, occasional pulses
Standing waters (neophytes)	Occasional	High, occasional pulses
Moist tall-herb grasslands	Occasional	High, rare pulses
Temperate scrub	Occasional	Medium to high, rare pulses
Base-rich fens (neophytes)	Occasional	Low, occasional pulses
Cliffs and walls	Occasional	Low, occasional pulses
Wet grasslands (neophytes)	Infrequent, predictable	High, low pulses
Fen scrub	Rare	Low, rare pulses
Mixed woodlands (neophytes)	Rare	High, rare pulses
Intermediate level of invasion, low invasibility		
Saline marshes	Occasional	Medium, rare pulses
Dry grasslands	Occasional	Low, rare pulses
Woodland fringes	Occasional	Low, rare pulses
Mesic grasslands	Infrequent, predictable	High, occasional pulses
Wet grasslands (archaeophytes)	Infrequent, predictable	High, low pulses
Saline grasslands	Infrequent, predictable	Medium, rare pulses
Base-rich fens (archaeophytes)	Infrequent, predictable	Low, occasional pulses
Deciduous woodlands	Rare	High, rare pulses
Low to zero level of invasion, probably low invasibility		
Coniferous plantations (neophytes)	Strong at the initial establishment	Pulses at the initial establishment
Running waters (archaeophytes)	Common	Medium, rare pulses
Standing waters (archaeophytes)	Occasional	High, occasional pulses
Subalpine scrub	Occasional	High, occasional pulses
Subalpine tall forbs	Occasional	High, occasional pulses
Temperate heaths	Occasional	High, occasional pulses
Mixed woodlands (archaeophytes)	Rare	High, rare pulses
Alpine grasslands	Rare	Low, rare pulses
Bogs	Rare	Low, rare pulses
Coniferous woodlands	Rare	Low, rare pulses
Poor fens	Rare	Low, rare pulses

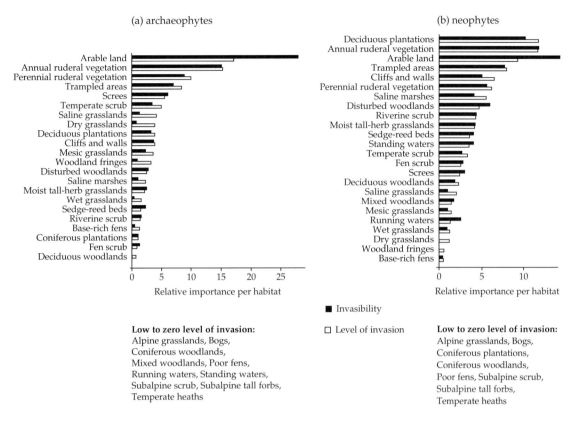

Figure 6.2 Comparison of the level of invasion and invasibility of Czech habitats by two groups of alien plants, archaeophytes and neophytes. The level of invasion is defined as the mean proportion of archaeophytes or neophytes to all species encountered in the vegetation survey plots belonging to particular habitats. Invasibility is defined as the same measure keeping propagule pressure and climate constant across the plots; it was quantified by using residuals of the linear model that subtracted the effects of propagule pressure and climate from the relation between the level of invasion and invasibility. To make both measures comparable, they were normalized to an equal sum across all the habitats (level of invasion after arcsin transformation, invasibility after converting residuals to positive numbers). Habitats are ranked according to the decreasing level of invasion, those with the lowest levels are not shown. Based on data reported in Chytrý et al. (2008a).

disturbed by human activities, and forest clearings are created by tree felling. Disturbance in these habitats is coupled with temporary increases in resource availability, e.g. fertilization of arable land, nutrient input into ruderal vegetation or increased light availability after opening the woodland canopy. Pulses of a high nutrient availability from external sources are typical of highly invasible habitats.

(ii) In contrast, least invasible are those habitats that are little disturbed. Most of the resistant habitats are perennial grasslands, which are also disturbed by grazing or mowing. However, such disturbances do not result in a significant temporary increase in nutrient availability, because vegetation is never disturbed completely and the resident plants respond to damage by rapid uptake of free nutrients to support their fast regrowth (Chytrý et al. 2008a). Many habitats of the low invasibility group do not experience any significant pulses of resource input from external sources, for example, alpine grasslands, mires, and oligotrophic grasslands and heathlands. However, nutrient input in these habitats (e.g. fertilizer application in an oligotrophic grassland) may cause a rapid shift towards other habitats (e.g. mesotrophic to eutrophic grassland) and such transitional habitats may indeed become highly invasible.

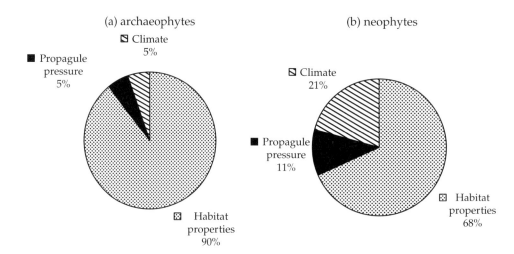

Figure 6.3 Proportional effect of the major determinants of the level of invasion by alien plants in habitats in the Czech Republic; the total proportion of variation explained by the models was 86.4% for archaeophytes (a) and 28.3% for neophytes (b); here it is recalculated to 100%. Predictor variables are in three groups, related to habitat properties, propagule pressure, and climate (see Fig. 6.1). Habitat properties also included the foliar cover of vegetation in the given habitat; this characteristic was included as an important component of community structure, with an assumed effect on community invasibility. Based on analyses presented in Chytrý et al. (2008a, their Table 2).

The results summarized in Table 6.1 support the notion that not all disturbances are necessarily conducive to invasions. In many habitats, invasions result from the alterations of the typical disturbance regime rather than from disturbances which are inherent to given habitat, for example, tree falls in forests or mowing in meadows (Hobbs and Huenneke 1992; Alpert et al. 2000). Patterns of invasibility observed across habitats are consistent with The Theory of Fluctuating Resource Availability (Davis et al. 2000). According to this theory, new species can invade a community if there are temporary pulses of unused resources. These pulses can be due to an increased supply of resources from external sources, a decreased uptake by resident vegetation, or both. Examples of the increased resource supply from external sources which increase community invasibility include fertilizer application in agricultural habitats, nutrient accumulation from atmospheric deposition, or nutrient input with flood sediments. Examples of the increased resource availability due to decreased uptake by resident vegetation are most frequently caused by disturbances, for example, herbiciding, floods, or tree felling.

6.4.2 Relative importance of factors determining the level of invasion

The analysis of the level of invasion in Central European habitats (Fig. 6.1; Chytrý et al. 2008a) quantified the effect of factors determining to what extent particular habitats are invaded. These factors act differently when archaeophytes and neophytes are examined separately. For archaeophytes, the joint effect of climate and propagule pressure is very low relative to that of habitats, and the total explained variation in their occurrence in the habitats is high. The occurrence of neophytes is less deterministic, with less variation explained, but habitat type is still the most important predictor. In spite of that, decrease in the level of invasion with increasing altitude and with decreasing proportion of urban and industrial land in the surroundings is notable. The proportional contribution of the joint effects of the three groups of predictor variables gives a different picture from that in archaeophytes. In neophytes, climate and propagule pressure are important in determining the level of invasion, accounting for about one third of the total variation explained (Fig. 6.3; Chytrý et al. 2008a).

As climate has been repeatedly shown to be one of the most important determinants of species composition or proportion of alien plants in Central European plant communities (Pyšek *et al.* 2002a, 2005; Lososová *et al.* 2004), this comparison of the relative effects of propagule pressure and climate suggests that the selected proxy variables provide a reasonable approximation to the actual propagule pressure, even though they certainly explain less variation in the proportion of aliens than would be the case if propagule immigration rate was directly measured. Interestingly, little variation is shared between habitats and climate (3.9% for archaeophytes and 3.4% for neophytes) even though different habitats occur in different climatic regions.

6.4.3 Habitat vs. propagule limitation and methodological pitfalls

Since invasions are human-mediated processes, the effect of propagule pressure on a broad geographical scale, for a variety of habitats and a large species pool of potential invaders, can be quantified through proxy variables that reflect the degree of human activity. Proxy variables representing the intensity of human activities are difficult to interpret as measures of propagule pressure in studies which focus on larger, internally heterogeneous sampling units, such as nature reserves (Macdonald *et al.* 1988; Lonsdale 1999; Pyšek *et al.* 2002a; McKinney 2004) or grid mapping cells (Deutschewitz *et al.* 2003; Kühn *et al.* 2003; Pino *et al.* 2005). Such studies usually report a positive correlation of those proxies with the number and/or proportion of alien species, but at that scale proxy variables may represent both increased propagule pressure and increased disturbance in more densely populated or urbanized areas. In our study, we controlled for disturbance effects by focusing on small, internally homogeneous plots, and on individual habitats, which themselves differ in disturbance regimes. In our models, most of the variation attributable to disturbance is therefore included in the effect of habitat. In addition, by including total vegetation cover as a predictor variable we were able to control for the variation in disturbance within individual habitats. We can therefore safely assume that the proxy variables used in this study measure the propagule pressure rather than the rate of disturbance.

The relatively low effect of propagule pressure detected here is in contrast with the results of Rouget and Richardson (2003), who reported higher importance of propagule pressure than of environmental variables for the distribution of three invasive tree species in South Africa. This difference points to the importance of the context in which invasibility is studied: while Rouget and Richardson (2003) studied recently established patterns of spread of individual populations, in which offspring usually tend to establish near to their parents, our study focused on multi-species alien assemblages which are outcomes of at least tens or hundreds years of invasion history.

6.5 Habitat-based mapping of plant invasions in Europe and prediction of future trends: the next step?

Data on the level of invasion in vegetation survey plots from Great Britain, Catalonia (NE Spain), and the Czech Republic, amassed during the European Union ALARM project (Settele *et al.* 2005), made it possible to sample a range of basic European climates, from Mediterranean to sub-continental and oceanic (Chytrý *et al.* 2008b). All three datasets consisted of several thousand vegetation plots from a range of different habitats, providing a robust assessment of the level of invasions in European habitats. The comparison of the three regions showed that (i) although there were large differences in the species composition of alien floras among the regions, patterns of habitat invasion were remarkably consistent. (ii) Extreme habitats with low nutrients were little invaded; frequently disturbed habitats with fluctuating resource availability were highly invaded. (iii) The most invaded habitats were arable land, coastal sediments, trampled areas, ruderal vegetation, sedge and reed beds, and cliffs and walls.

Inter-regional consistency of the habitat invasion patterns makes habitat a good predictor for invasion risk assessment. This assumption is supported by the result reported above (Chytrý *et al.* 2008a); for entire alien floras and the whole range of habitats across large regions, habitat properties are a much more important predictor of the level of invasion than climate or proxy measures of propagule pressure. Therefore the data from a comparative study of British, Catalonian and Czech habitats (Chytrý *et al.* 2008b) were used to produce a European map of invasions by alien plants, based on habitats (Chytrý *et al.*, 2009). This was done by translating habitat types to CORINE land-cover classes (Moss and Wyatt 1994), which had been previously mapped across Europe from the interpretation of satellite images. The data from the three regions were extrapolated to other parts of Europe, using the framework of European biogeographical regions. The overall pattern indicates high levels of invasion in industrialized western Europe and in lowland agricultural regions in the east of the continent, and with montane zones, oceanic areas in the north-west, and the boreal zone relatively little affected (Chytrý *et al.*, 2009).

Using habitats as mapping units is suitable because this approach takes into account landscape structure; habitat data allow the extrapolation of quantitative estimates of the level of invasion to other regions, based on climatic similarities—something that cannot be done with country-wise data. Finally, sampling by phytosociological method is intensive, which makes the regions studied well sampled; the results based on such data are robust. This allows reasonable precision to be achieved and provides a solid background for the assessment of risk from plant invasions, for monitoring and for modeling future changes under various scenarios of climate and land-use change.

6.6 Conclusions

1. Habitats differ considerably in their invasibility. The differences in the level of invasion between Central European habitats are mainly caused by inherent habitat properties, and to a lesser extent by propagule pressure and climatic differences between regions.
2. Patterns of habitat invasion are consistent across different regions of Europe. The same habitats usually have either high or low level of invasion despite their geographical location.
3. The most invasible habitats are those with fluctuating availability of resources, especially nutrients; most of these habitats are frequently and/or strongly disturbed.
4. The occurrence of archaeophytes is to a large extent deterministic; it mainly depends on habitat types, while propagule pressure is less important. The occurrence of neophytes is more stochastic; propagule pressure is more important than in archaeophytes, yet habitat type is still the most important predictor.
5. An approach using ecoinformatics and linking large sets of spatially explicit data from vegetation survey plots can produce robust information on macroecological patterns of plant invasions. Spatially explicit information on habitat invasions can be used to identify the areas of highest risk of invasion so as to support effective monitoring and management of alien plants; combined with scenarios of future land-use change, it may also be used for prediction of invasion risks in the future.

Acknowledgements

We thank Mark Williamson for helpful comments on the manuscript. This work was funded by the Integrated Project ALARM (GOCE-CT-2003-506675) of the FP6 of the European Union (Settele *et al.* 2005). P.P. and V.J. were also supported by the projects AV0Z60050516 (Academy of Sciences of the Czech Republic), MSM0021620828 and LC06073, M.C. by MSM0021622416 (all from the Ministry of Education, Youth and Sports of the Czech Republic).

References

Alpert, P., Bone, E. and Holzapfel, C. (2000). Invasiveness, invasibility and the role of environmental stress in the spread of non-native plants. *Perspectives in Plant Ecology, Evolution and Systematics*, **3**, 52–66.

Brown, R.L. and Peet, R.K. (2003). Diversity and invasibility of southern Appalachian plant communities. *Ecology*, **84**, 32–39.

Campos, J.A., Herrera, M., Biurrun, I., and Loidi, J. (2004). The role of alien plants in the natural coastal vegetation in central northern Spain. *Biodiversity and Conservation*, **13**, 2275–93.

Chytrý, M. (2001). Phytosociological data give biased estimates of species richness. *Journal of Vegetation Science*, **12**, 439–44.

Chytrý, M., Jarošík, V., Pyšek, P. *et al.* (2008a). Separating habitat invasibility by alien plants from the actual level of invasion. *Ecology*, **89**, 1541–53.

Chytrý, M., Maskell, L. C., Pino, J. *et al.* (2008b). Habitat invasions by alien plants: a quantitative comparison among Mediterranean, subcontinental and oceanic regions of Europe. *Journal of Applied Ecology*, **45**, 448–58.

Chytrý, M., Pyšek, P., Tichý, L., Knollová, I., and Danihelka, J. (2005). Invasions by alien plants in the Czech Republic: a quantitative assessment across habitats. *Preslia*, **77**, 339–54.

Chytrý, M., Pyšek, P., Wild, J., Pino, J., Maskell, L.C., and Vilà, M. (2009). European map of alien plant invasions, based on the quantitative assessment across habitats. *Diversity and Distributions*, **15**, 98–107.

Chytrý, M. and Rafajová, M. (2003). Czech National Phytosociological Database: basic statistics of the available vegetation-plot data. *Preslia*, **75**, 1–15.

Crawley, M.J. (1987). What makes a community invasible? In A.J. Gray, M.J. Crawley and P.J. Edwards, eds. *Colonization, succession and stability*, Blackwell Scientific Publications, Oxford, pp. 429–453.

Dark, S. (2004). The biogeography of invasive alien plants in California: an application of GIS and spatial regression analysis. *Diversity and Distributions*, **10**, 1–9.

Davies, C.E. and Moss, D. (2003). *EUNIS Habitat Classification, August 2003*. European Topic Centre on Nature Protection and Biodiversity, Paris.

Davis, M.A., Grime, J.P., and Thompson, K. (2000). Fluctuating resources in plant communities: a general theory of invasibility. *Journal of Ecology*, **88**, 528–34.

DeFerrari, C.M. and Naiman, R.J. (1994). A multi-scale assessment of the occurrence of exotic plants on the Olympic Peninsula, Washington. *Journal of Vegetation Science*, **5**, 247–58.

Deutschewitz, K., Lausch, A., Kühn, I., and Klotz, S. (2003). Native and alien plant species richness in relation to spatial heterogeneity on a regional scale in Germany. *Global Ecology and Biogeography*, **12**, 299–311.

Elton, C.S. (1958). *The Ecology of Invasions by Animals and Plants*. Methuen, London.

Essl, F. and Rabitsch, W. (eds) (2002). *Neobiota in Österreich*. Umweltbundesamt GmbH, Wien.

Firbank, L.G., Smart, S.M., Barr, C.J., *et al.* (2003). Assessing stock and change in land cover and biodiversity in GB: an introduction to Countryside Survey 2000. *Journal of Environmental Management*, **67**, 207–18.

Font, X. and Ninot, J.-M. (1995). A regional project for drawing up inventories of flora and vegetation in Catalonia (Spain). *Annali di Botanica*, **53**, 99–105.

Fridley, J.D., Brown, R.L., and Bruno, J.E. (2004). Null models of exotic invasion and scale-dependent patterns of native and exotic species richness. *Ecology*, **85**, 3215–22.

Gilbert, B. and Lechowicz, M.J. (2005). Invasibility and abiotic gradients: the positive correlation between native and exotic plant diversity. *Ecology*, **86**, 1848–55.

Haveman, R. and Janssen, J.A.M. (2008). The analysis of long-term changes in plant communities using large databases: The effect of stratified resampling. *Journal of Vegetation Science*, **19**, 355–62.

Hennekens, S.M. and Schaminée, J.H.J. (2001). TURBOVEG, a comprehensive data base management system for vegetation data. *Journal of Vegetation Science*, **12**, 589–91.

Herben, T., Mandák, B., Bímová, K., and Münzbergová, Z. (2004). Invasibility and species richness of a community: a neutral model and a survey of published data. *Ecology*, **85**, 3223–33.

Hierro, J.L., Maron, J.L., and Callaway, R.M. (2005). A biogeographical approach to plant invasions: the importance of studying exotics in their introduced and native range. *Journal of Ecology*, **93**, 5–15.

Hobbs, R.J. and Huenneke, L.F. (1992). Disturbance, diversity, and invasion: implications for conservation. *Conservation Biology*, **6**, 324–37.

Keeley, J.E., Lubin, D., and Fotheringham, C.J. (2003). Fire and grazing impacts on plant diversity and alien plant invasions in the southern Sierra Nevada. *Ecological Applications*, **13**, 1355–74.

Kennedy, T.A., Naeem, S., Howe, K.M., Knops, J.M.H., Tilman, D., and Reich, P. (2002). Biodiversity as a barrier to ecological invasion. *Nature*, **417**, 636–38.

Knight, K.S. and Reich, P.B. (2005). Opposite relationships between invasibility and native species richness at patch versus landscape scales. *Oikos*, **109**, 81–88.

Knollová, I., Chytrý, M., Tichý, L., and Hájek, O. (2005). Stratified resampling of phytosociological databases: some strategies for obtaining more representative data sets for classification studies. *Journal of Vegetation Science*, **16**, 479–86.

Kowarik, I. (1995). On the role of alien species in urban flora and vegetation. In P. Pyšek, K. Prach, M. Rejmánek, and P.M. Wade (eds) *Plant Invasions: General Aspects and Special Problems*, SPB Academic Publishing, Amsterdam, pp. 85–103.

Kühn, I., Brandl, R., May, R., and Klotz, S. (2003). Plant distribution patterns in Germany: will aliens match natives? *Feddes Repertorium*, **114**, 559–73.

La Sorte, F.A., McKinney, M.L., and Pyšek, P. (2007). Compositional similarity among urban floras within and across continents: biogeographical consequences of human mediated biotic interchange. *Global Change Biology*, **13**, 913–21.

Levine, J.M. (2000). Species diversity and biological invasions: relating local process to community pattern. *Science*, **288**, 852–54.

Levine, J.M., Adler, P.B., and Yelenik, S.G. (2004). A meta-analysis of biotic resistance to exotic plant invasions. *Ecology Letters*, **7**, 975–89.

Levine, J.M., Vilà, M., D'Antonio, C.M., Dukes, J.S., Grigulis, K., and Lavorel, S. (2003). Mechanisms underlying the impacts of exotic plant invasions. *Proceedings of the Royal Society London B*, **270**, 775–81.

Lonsdale, W.M. (1999). Global patterns of plant invasions and the concept of invasibility. *Ecology*, **80**, 1522–36.

Lososová, Z., Chytrý, M., Cimalová, Š. *et al.* (2004). Weed vegetation of arable land in Central Europe: gradients of diversity and species composition. *Journal of Vegetation Science*, **15**, 415–22.

Macdonald, I.A.W., Graber, D.M., DeBenedetti, S., Groves, R.H., and Fuentes, E.R. (1988). Introduced species in nature reserves in Mediterranean type climatic regions of the world. *Biological Conservation*, **44**, 37–66.

Maskell, L.C., Firbank, L.G., Thompson, K., Bullock, J.M., and Smart, S.M. (2006). Interactions between non-native plant species and the floristic composition of common habitats. *Journal of Ecology*, **94**, 1052–60.

McKinney, M.L. (2004). Influence of settlement time, human population, park shape and age, visitation and roads on the number of alien plant species in protected areas in the USA. *Diversity and Distributions*, **8**, 311–18.

Moss, D. and Wyatt, B.K. (1994). The CORINE Biotopes Project: a database for conservation of nature and wildlife in the European Community. *Journal of Applied Geography*, **14**, 327–49.

Pino, J., Font, X., Carbó, J., Jové, M., and Pallarès, L. (2005), Large-scale correlates of alien plant invasion in Catalonia (NE of Spain). *Biological Conservation*, **122**, 339–50.

Planty-Tabacchi, A., Tabacchi, E., Naiman, R., DeFerrari, C. and Décamps, H. (1996). Invasibility of species-rich communities in riparian zones. *Conservation Biology*, **10**, 598–607.

Pyšek, P. (1998). Alien and native species in Central European urban floras: a quantitative comparison. *Journal of Biogeography*, **25**, 155–63.

Pyšek, P., Jarošík, V., and Kučera, T. (2002a). Patterns of invasion in temperate nature reserves. *Biological Conservation*, **104**, 13–24.

Pyšek, P., Jarošík, V., Chytrý, M., Kropáč, Z., Tichý, L., and Wild, J. (2005). Alien plants in temperate weed communities: Prehistoric and recent invaders occupy different habitats. *Ecology*, **86**, 772–85.

Pyšek, P., Richardson, D.M., Rejmánek, M., Webster, G., Williamson, M., and Kirschner, J. (2004). Alien plants in checklists and floras: towards better communication between taxonomists and ecologists. *Taxon*, **53**, 131–143.

Pyšek, P. and Richardson, D.M. (2006). The biogeography of naturalization in alien plants. *Journal of Biogeography*, **33**, 2040–50.

Pyšek, P., Sádlo, J., and Mandák, B. (2002b). Catalogue of alien plants of the Czech Republic. *Preslia*, **74**, 97–186.

Rejmánek, M., Richardson, D.M., and Pyšek, P. (2005). Plant invasions and invasibility of plant communities. In E. van der Maarel (ed.), *Vegetation Ecology*, Blackwell Science, Oxford, pp. 332–55.

Richardson, D.M. and Pyšek, P. (2006). Plant invasions: merging the concepts of species invasiveness and community invasibility. *Progress in Physical Geography*, **30**, 409–31.

Roleček, J., Chytrý, M., Hájek, M., Lvončík, S., and Tichý, L. (2007). Sampling design in large-scale vegetation studies: do not sacrify ecological thinking to statistical purism! *Folia Geobotanica*, **42**, 199–208.

Rouget, M. and Richardson, D.M. (2003). Inferring process from pattern in plant invasions: a semimechanistic model incorporating propagule pressure and environmental factors. *American Naturalist*, **162**, 713–24.

Sádlo, J., Chytrý, M., and Pyšek, P. (2007). Regional species pools of vascular plants in habitats of the Czech Republic. *Preslia*, **79**, 303–21.

Settele J., Hammen, V., Hulme, P. *et al.* (2005). ALARM: Assessing LArge-scale environmental Risks for biodiversity with tested Methods. *GAIA – Ecological Perspectives for Science and Society*, **14**, 69–72.

Shea, K. and Chesson, P. (2002). Community ecology theory as a framework for biological invasions. *Trends in Ecology and Evolution*, **17**, 170–76.

Simonová, D. and Lososová, Z. (2008). Which factors determine plant invasions in man-made habitats in the Czech Republic? *Perspectives in Plant Ecology, Evolution and Systematics*, **10**, 89–100.

Sobrino, E., Sanz-Elorza, M., Dana, E.D., and González-Moreno, A. (2002). Invasibility of a coastal strip in NE Spain by alien plants. *Journal of Vegetation Science*, **13**, 585–94.

Spyreas, G., Ellis, J., Carroll, C., and Molano-Flores, B. (2004). Non-native plant commonness and dominance in the forests, wetlands, and grasslands of Illinois, USA. *Natural Areas Journal*, **24**, 290–99.

Stohlgren, T. (2007). *Measuring Plant Diversity: Lessons from the Field*. Oxford University Press, Oxford.

Stohlgren, T.J., Barnett, D., Flather, C., Kartesz, J., and Peterjohn, B. (2005). Plant species invasions along the latitudinal gradient in the United States. *Ecology*, **86**, 2298–309.

Stohlgren, T.J., Barnett, D.T., and Kartesz, J.T. (2003). The rich get richer: patterns of plant invasions in the United States *Frontiers in Ecology and Environment*, **1**, 11–14.

Stohlgren, T.J., Binkley, D., Chong, G.W. *et al.* (1999). Exotic plant species invade hot spots of native plant diversity. *Ecological Monographs*, **69**, 25–46.

Stohlgren, T.J., Bull, K.A., and Otsuki, Y. (1998). Comparison of rangeland vegetation sampling techniques in the Central Grasslands. *Journal of Range Management*, **51**, 164–72.

Stohlgren, T.J., Chong, G.W., Schell, L.D. *et al.* (2002). Assessing vulnerability to invasion by nonnative plant species at multiple spatial scales. *Environmental Management*, **29**, 566–77.

Stohlgren, T.J., Jarnevich, C., Chong, G.W., and Evangelista, P.H. (2006). Scale and plant invasions: a theory of biotic acceptance. *Preslia*, **78**, 405–26.

Stohlgren, T.J., Otsuki, Y., Villa, C.A., Lee, M., and Belnap, J. (2001). Patterns of plant invasions: a case example in native species hotspots and rare habitats. *Biological Invasions*, **3**, 37–50.

Tilman, D. (1997). Community invasibility, recruitment limitation, and grassland biodiversity. *Ecology*, **78**, 81–92.

Vilà, M., Pino, J., and Font, X. (2007). Regional assessment of plant invasions across different habitat types. *Journal of Vegetation Science*, **18**, 35–42.

Walter, J., Essl, F., Englisch, T., and Kiehn, M. (2005). Neophytes in Austria: habitat preferences and ecological effects. *Neobiota*, **6**, 13–25.

Williamson, M. (1996). *Biological Invasions*. Chapman and Hall, London.

Zobel, M. (1992). Plant species coexistence: the role of historical, evolutionary and ecological factors. *Oikos*, **65**, 314–320.

PART II
ECONOMICS

CHAPTER 7

If Invasive Species are "Pollutants", Should Polluters Pay?

R. David Simpson

7.1 Introduction

Readers of this chapter are unlikely to require an extended introduction to the damage invasive species may cause. The fact that invasive species have caused and are likely to continue to cause extensive damage does not necessarily establish the case for aggressive interventions to prevent their introduction, however. There are at least two reasons for saying this. The first is that aggressive interventions can be expensive, both in terms of the direct costs of the measures adopted and of the opportunity costs of benefits foregone in order to prevent invasions. The second reason is that it makes no sense to throw good money after bad in fighting lost causes. Economists emphasize the importance of *marginal* considerations in making policy decisions. Thus, the important consideration is generally not whether invasive species have historically caused, and do currently cause, large economic losses—they have and they do—but rather, whether the *incremental* costs of measures that might reasonably be taken to reduce the likelihood of invasions justify the *marginal* reduction in the likelihood of an invasion's occurrence they afford.

Several commentators have suggested that invasive species represent a form of "pollution" (see, e.g. Olson 2005 and Lovell *et al.* 2005, and the work they cite). This was also the position of the US District Court for the Northern District of California, which ruled that the US Environmental Protection Agency should regulate ballast water (a frequent culprit in biological invasions) under the Clean Water Act. If invasives are pollutants, then, should we apply the standard policy toolkit for dealing with pollution against them? That is, we could tax polluting activities such as intercontinental trade, require that certain technologies be employed to reduce the threat of invasion, or impose performance standards, backed by a regime of monitoring and enforcement.

When such measures are applied to other forms of pollution, such as air pollution from vehicular use or industrial water pollution, the goal of policy is generally to increase the stringency of control until marginal benefits balance marginal costs. Taxes on effluents, for example, should be imposed until the social damage caused by the last liter of waste water or ton of particulates is just balanced by the marginal cost of its abatement. Other policies seek to achieve the same outcome with different instruments.

Such approaches are typically based on the assumptions that the benefits arising from polluting activities are concave (increasing at a *decreasing* rate), while the damages are convex (increasing at an *increasing* rate). These assumptions typically assure that there is an "interior solution" characterized by the balance I have just described: the marginal benefits of control just balance the marginal costs.

Moreover, the policy variables chosen would be continuous and monotonic functions of the underlying parameters describing the threat. If there is a small probability that a little pollution will do a little damage, the optimal policy response would be to impose a little control. As any of those parameters increase, the optimal policy response would be to strengthen the controls.

Convexity is not assured in the case of biological invaders, however, and, in fact, there are reasons to suppose that expected damages would not be a convex function of the volume of shipments. I will explain this statement in more detail below by reference to several arguments and with examples from the literature, but for now let me simply note that damages are constrained by a sort of "worst case" scenario. The worst thing that can happen with regard to potential biological invasion is that the invader arrives and does its damage. Arrival is a random event, so the worst-case scenario is that the invader arrives with near certainty. With a large enough volume of shipping it may become a near certainty that an invasion will occur. The marginal expected damage caused by an additional shipment would then be negligible, since the anticipated damage from its arrival is the probability that it contains an invader *times the probability that the invader has not already arrived and become established*.

The concerns I have just recited explain why expected damages from biological invasions would be a convex function *of the number of potentially infested shipments received*. This is only half—or perhaps less than half—of the story, however. The probability with which an exotic species invades a new habitat to which it is introduced depends on two factors: how many potential opportunities for arrival it is afforded, and the likelihood that it becomes established in any given opportunity. Both of these parameters can be affected by policy. The number of arrivals is affected directly by tariffs and quotas. The probability of invasion is affected by measures to restrict the transmission of biological stowaways, such as ballast water exchange or quarantines. Moreover, to the extent that the costs of transmission-probability-reducing measures are borne by shippers, they will also act to reduce the volume of trade indirectly.

An economic analysis of the optimal choice of policies to prevent biological invasions will consider two margins. The first I have already described: balancing the incremental benefits of accepting an additional shipment—the economic gains from trade—against the incremental expected damages from biological invasion, *taking the probability of invasion arising from the "marginal shipment" as fixed*. The second consideration is to balance the incremental cost of more stringent treatment measures to prevent invasion against the induced reduction in the likelihood that an invasion will occur, *taking the number of shipments accepted as fixed*.

It is reasonable to suppose that the likelihood that an invasion will occur in any given shipment is a *concave* function of the expense incurred to prevent that event. Thus, analysis of the second consideration gives rise to conventional, well-behaved, and generally continuous policy prescriptions. Put in informal terms, then, the question of whether a policy toward invasive species should look like those adopted in response to other forms of "pollution" depends on whether damages are low enough that preventing invasions should be seen as a "lost cause" and abandoned, or if damages are high enough that prevention is a wise investment and significant resources should be put into preventing invasions.

In the remainder of this chapter I develop a simple model laying out these considerations. I find that, fixing parameters describing the demand for imports, their cost, and the cost of treating imports to reduce the likelihood of invasion, we can define optimal strategy by reference to the anticipated damages resulting from invasion. There are three possibilities. For what I will describe as low-to-medium levels of damage, the optimal policy is laissez faire. The arrival of invasive species should be regarded as inevitable, and so no costs should be incurred to fight a lost cause. There is then an abrupt change in policy for what I will describe as medium-to-high levels of damage. When the damages anticipated in the event of a successful invasion surpass a threshold value, the optimal policy involves combating invasions vigorously. Finally, if damages increase beyond a "very high" threshold, the optimal policy will involve prohibiting trade entirely.

Several features of these results are worth emphasizing. First, it would generally be true of any form of pollution policy that administrative and enforcement costs argue against action to control de minimis threats. This does not explain the result with respect to low-to-medium threats in my model, however. In some instances anticipated damages may be quite high—a significant fraction of the gains from trade in total—yet the optimal response is, essentially, to ignore them. In these instances it is not worth doing much to reduce the likelihood of invasion, and consequently, an invasion becomes a near certainty. In these cases the nonconvexity of the expected damage function is the dominant consideration, and marginal analysis shows that there is little point in incurring expenses to reduce the likelihood of an event that is likely to occur anyway.

The next point to emphasize is that there is a significant discontinuity in optimal policy response when anticipated damages in the event of expected invasion increase about the "medium" threshold. The optimal policy goes discontinuously from exerting *no* effort to reduce the likelihood of invasion to *substantial* efforts to do so, and from *very little* reduction in the volume of trade to *substantial* reductions.

Finally, when the "medium" threshold on anticipated damages is passed, the cumulative probability that invasion occurs plummets as more stringent policies are adopted. Heuristically, it is no longer prudent to treat potential invasions as a lost cause when the damage arising in the event of their occurrence becomes too great. Rather, beyond the point of "medium" damages policies should be adopted to greatly reduce, if not eliminate, the possibility of invasion. Of course, when possible damages and the cost of reducing their likelihood are high enough, trade itself become prohibitively dangerous, and would be banned.

The next section of the chapter reviews some related literature and other considerations to justify my critical assumption: that expected damages are convex in the number of shipments. I then summarize the essential findings of the chapter graphically. The fourth section introduces a formal—albeit highly schematic—model. The fifth section discusses some possible extensions of the relatively simple model I have developed. Some policy implications that the simple model *may* justify are presented in the sixth section. I have emphasized that the model *may* justify certain policy implications. It would be premature to draw strong conclusions along these lines without exploring some of the extensions discussed in the fifth section, or resolving some of the empirical issues I discuss in the seventh and concluding section.

It is worth foreshadowing one of the provisional policy implications now. One problem in proposing theoretical models of optimal policy choices is that their prescriptions often depend on empirical quantities that are extremely difficult to estimate. The model I present here suggests that the *conceptual* analysis of invasive species problems may be more complex than it is sometimes supposed. Perhaps, however, the finding that policy choices can be separated into the three regions I have identified makes the specification of optimal policy more straightforward than it first appears.

7.2 Assumptions and literature

I will suppose that the expected damage arising from the potential imports of invasive species is a convex function of the volume of imports. I base this assumption on first principles, as well as contributions to the literature on the topic. With respect to the latter, I would place this paper at the intersection of two important and influential contributions to the economic understanding of biological invasions. Richard Horan and his coauthors (2002) develop a model in which different exporters can each introduce an exotic species into a new environment. Each import shipment is regarded as a Bernoulli trial that either "succeeds" or "fails" to introduce the exotic species with a certain probability; these probabilities may vary among sources depending on the biological circumstances as well as preventative measures adopted (another paper by Christopher Costello and coauthors, 2007, presents empirical evidence that the likelihood of exotic introductions do vary by source). Hence the arrival of invasive species is a "weakest link" phenomenon; if the invasive arrives from one source, measures to prevent its introduction from another will prove futile.

I am indebted to earlier authors for many of my insights. In a sense, my chapter differs from Horan *et al.* more in emphasis than in content. The earlier paper underscores the notion that marginal expected damages vanish as the volume of imports grows and notes in passing that discontinuities in optimal policy are possible (Horan *et al.* 2002, p. 1308). However, Horan *et al.* focus much of their attention on decision making under imperfect information, an extremely important issue that I largely sweep under the rug in order to focus on the specific forms of policy interventions.

The other paper that my work resembles, and to which I am greatly indebted, is an influential and widely cited piece by Carol McAusland and Christopher Costello (2004). These authors develop a model whose structure I have mimicked. Policy makers seek to maximize gains from trade, less costs of biological controls, less expected damages from invasion. Unlike Horan *et al.*, however, McAusland and Costello assume that expected damages are linear in the volume of imports. Hence they do not identify the different types of policy approaches that may be appropriate for different characterizations of damages.

I should hasten to point out that this does not make McAusland and Costello "wrong"; on the contrary, they are almost exactly "right" in the important category of cases I have characterized above as arising when damages are medium-to-high. McAusland and Costello describe expected damages as the product of three factors: the volume of shipments, the probability that a shipment contains an invasive species (which itself depends on the efforts undertaken to detect invasives), and the damage anticipated in the event an invader becomes established. This formulation might be justified in two ways. The first would be if there were an infinite number of potential invasive species that might arrive. The second is as an approximation when there are a finite number of potential invaders but the probability with which they arrive is small enough as to reasonably ignore the possibility of redundant introductions. As I show below, the latter will be the case when damages are medium-to-high; the optimal policy calls for measures sufficient to reduce the *cumulative* probability of introduction to a relatively low level. My results *augment* those of McAusland and Costello by extending the analysis to instances of lower levels of damage, but *confirm* their findings for higher levels of potential damage.

My results are, in fact, parallel to those of McAusland and Costello in many ways. As in their work, I find that corner solutions may obtain in which it is optimal to devote no effort to reducing the likelihood of invasion.[1] The critical difference between my work and theirs is the difference between my assumption that redundant invasions cause no incremental damage and theirs that repeated invasions (or repeated inva*sives*) exhibit no diminishing damages. Thus, when McAusland and Costello find that no effort should be put into detection of invasives, they propose that a tariff take up the burden of protection. In contrast, I find that when no effort should be put into prevention, the optimal strategy is to treat the invasion as a lost cause, an event so likely to occur that it makes little sense to incur any opportunity cost to stop it.

There may be some question as to whether the assumption of nonconvex damages is appropriate. I might note in passing that other authors have pointed out the importance of nonconvexities in the analysis of invasive species (see, e. g. Olson 2005 and the work he cites in his survey paper), although the phenomenon more often cited is the sinusoidal shape of population growth functions for invaders following their introduction. However, the most compelling argument for nonconvex damages is simply the mathematical fact that probabilities cannot exceed one. If what we are concerned about is the random arrival of invasive species, the worst that can happen is that the species we are worried about arrives.

If an invasive species arrives with probability A, then the likelihood that the invader is established following S shipments is $1 - (1 - A)^S$, a function that is easily shown to be concave in S, and, of course, asymptotes to 1. Of course, we are concerned with the arrival of more than one invasive species, but the sum of concave functions is concave.

---

[1] One slight difference in approaches is that McAusland and Costello model authorities as devoting effort to *detecting* invaders and *discarding* contaminated shipments, whereas I speak of *treating* shipments. While one can appreciate formal differences between the approaches, they would appear to be identical in their formal modeling implications.

There are, however, both biological and economic reasons to suppose that expected damages could be convex over at least some range of shipments. With respect to biology, consider the simplest example: sexually reproducing species. If only *one* arrives, there cannot be an invasion. More generally, if a successful invasion requires that a critical mass of individuals become established, it may be that expected damages would be convex over some volume of shipments. Eventually, of course, the damage function would be concave, however, as with enough import events, it would become virtually certain that a viable population would be established.

It is also possible that expected damages would be convex over a range of shipments because of interaction effects between invasive species. There are examples in the biological literature in which one introduced species only became problematic when another arrives to displace the first's predators or competitors. Again, however, there must be a limit to such processes. Eventually, if an entire suite of exotic species displaces another of native species, the new ecological assemblage will be whatever it will be, but, presumably, there would be some worst-case scenario bounding damages. In fact, though, it seems one might reasonably argue that the consequences of *a few* invasions could be worse than those of *many* invasions. The problem with invasive species is often represented as being that one species explodes, unchecked by the mechanisms of predation, disease, or competition. In the fullness of time, those mechanisms would likely be reestablished by other exotic organisms.[2]

Perhaps the most problematic counterargument to the proposition that nonconvex expected damages could be an important consideration in making invasive species policy is that invasive species may make the world more homogeneous. Is this a bad thing? *De gustibus non est disputandum. If* the main reason for concern with invasive species is that they are implicated in the extinction of native species (they are[3]), *and if* the losses perceived from such extinctions are great (they may be[4]), *and if* citizens' preferences are convex in the numbers of native species preserved (surprisingly, this may be more problematic[5]), expected damages might be convex over some appreciable range. I will simply conclude that this concern for extinctions seems a weak reed to cling to in this context. I think it wise to cite relatively modest motivations for this chapter; let me just say, then, that the implications of nonconvexities for invasive species policy are worthy of consideration. It does not seem that concern for the extinction of endangered species should necessarily obviate this inquiry.[6]

---

[2] It is also worth noting that "biological control" is practised with respect to some invaders: other exotic species are introduced to prey on them. The authorities introducing the biological controls must regard their effects as positive, which would necessarily make the relationship between numbers of exotics and damages concave.

[3] See, e. g. David Wilcove, *et al.* 1998. However, other commentators (albeit with less impressive credentials as biologists) note that most such extinctions have been documented in isolated environments with relatively small suites of species to begin with (Burdick 2005). In large continental habitats, invaders seem to have been less of a concern.

[4] Various stated preference studies suggest substantial willingness to pay for species survival; see Brown and Shogren 1999.

[5] This could be the weak link. Critiques of the "embedding" or "part/whole bias" problem with stated preference studies suggest that existing studies are inconsistent with economic theory (see, e. g. Diamond 1996). This presumes convex preferences over the numbers of individuals or species preserved. One rejoinder to such criticisms is that respondents may *not* have convex preferences (Smith and Osborne 1996).

[6] One additional reference to the literature seems to be in order, although I find it somewhat difficult to categorize. Brock, Kinzig, and Perrings (2007) develop a very sophisticated model that joins a consideration of biological invasions with characterization of the biological diversity of landscapes into which invaders are introduced. As an importer's landscape becomes less diversified and more homogeneous, it becomes more vulnerable to invaders. If trade and homogenization proceed apace, it is possible that later invaders would cause more damage than did earlier ones. While the analysis in Brock, Kinzig, and Perrings is elegant, I do not believe the argument necessarily obviates the concerns I emphasize here. Only empirical work will ultimately identify the more appropriate assumptions.

I might also note in this section on assumptions and literature, and apropos of my remark concerning modest motivations, that I am addressing only one aspect of invasive species policy, and perhaps not the most important aspect at that. There is a large literature on the economics of invasive species (for recent reviews, see Olson 2005 and Lovell *et al*. 2005). Considerable, albeit controversial, work has been done attempting to value the damage that invasive species do (see e. g. OTA 1993 and Pimental *et al*. 2001). Other researchers have considered optimal measures for the control or eradication of exotic species once they have arrived (see e. g. Olson and Roy 2002). Still others consider the optimal allocation of effort to detecting potential invaders (see e. g. Mehta *et al*. 2007). While conventional wisdom has it that an ounce of prevention is worth a pound of cure, it is not clear whether the strategies I consider below, involving preventing the introduction of exotic species rather than their control once they have arrived, are always superior. Of course, the choice between the alternatives depends on the costs, or even possibility, of control; the costs of control determine the benefits of prevention (Olson 2005). In any event, suffice it to say that the policies of prevention I consider below would, in a broader analysis, be considered in the context of a wider range of policy options.

7.3 Some revealing figures

The analytical core of this chapter is in the section that follows this one. Before introducing the formal model, however, it may be useful to present some illustrations. In the formal model I suppose that the benefits from international trade are a quadratic function of the volume of trade. Formally, this arises from supposing that demand is linear and the unit costs of production and transport are constant. Expected damages are concave in the volume of trade and exponential in the probability with which any given shipment contains a dangerous invader. Each of these assumptions is justified as a reasonable schematic approximation in the following section.

Under these assumptions we can describe net social benefits—the gains from trade less the costs of treating shipments less expected damages—as a function of the two policy variables: how much trade to allow and the probability with which any shipment will result in the arrival of the invader. I designate this *probability* as the policy variable, as choosing an *expense* to incur in the treatment of shipments is equivalent to choosing the probability that a shipment gives rise to an invasion.

I want to consider what happens as the level of anticipated damages in the event of an invasion increases from an initially low level to what I characterized above as "medium", "high", and "very high" levels.

Figure 7.1a represents optimal policy when damages are at a "low" level. In the three-dimensional figure depicted there net welfare is represented on the vertical axis. The horizontal axis represents the number of shipments. The other axis—the one going "into the page"—represents the likelihood of an invasion resulting from any single shipment. This likelihood decreases with distance "away from the reader" in this depiction. Parameter values have been chosen so that the optimal level of trade, absent any invasive species concerns, would be at 50 shipments, at which point the gains from trade would be 2500.[7]

When anticipated damages in the event of an invasion are expected to be 1000, the optimal policy is, essentially, to treat invasives as a lost cause. The optimal solution is still to accept 50 shipments,[8] even though this means that there would be, in this example, a 99.5 per cent chance that the invader would arrive. To a close approximation, the welfare resulting from this strategy is the laissez faire welfare level, 2500, less the anticipated damage from invasion, 1000.

For this choice of parameters, this laissez faire/lost cause policy remains optimal so long as damages are less than about 1471. Above this threshold an interesting phenomenon arises. We

---
[7] In the notation introduced in the following section $\bar{p} = 110$, $b = 2$, $c = 10$, $k = 10$, and $\bar{\pi} = 0.10$. The probability axis is measured on a logarithmic scale, the others, linear.
[8] Actually, the optimal policy would be to accept a small fraction less than 50 shipments, but assuming that shipments are quantized, the optimum is 50, to the nearest whole number.

IF INVASIVE SPECIES ARE "POLLUTANTS", SHOULD POLLUTERS PAY?   89

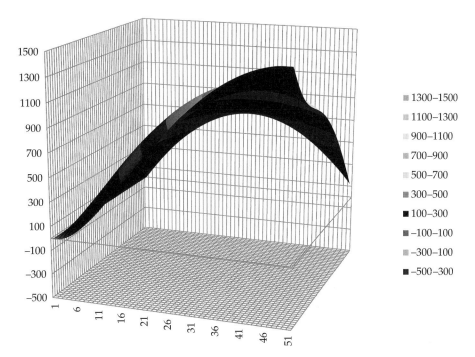

Figure 7.1a Welfare when damages are low.

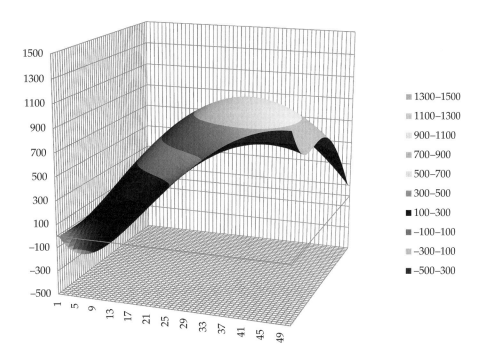

Figure 7.1b Welfare when damages are medium.

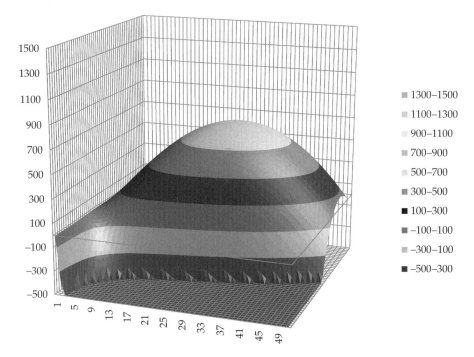

Figure 7.1c Welfare when damages are high.

would be indifferent between the laissez faire policy and one that pursues quite aggressive measures to prevent invasion. This situation is depicted in Fig. 7.1b. There are now two candidate solutions. One is the laissez faire corner solution with no restrictions on trade and no attempt to reduce the likelihood of invasion. The other solution corresponds to the top of the rounded surface depicted in Fig. 7.1b. This solution involves an order-of magnitude reduction in the likelihood of invasion from a single shipment, and a reduction in the likelihood that invasion will result from *at least* one shipment from 99.5 per cent in the laissez faire case to about 26.4 per cent.

Figure 7.1c represents a case with "high" damages. When anticipated damages are 5000, the optimal policy calls for reducing the likelihood of invasion from any one shipment by a factor of 45. This, in turn, results in roughly halving the number of shipments accepted relative to the laissez faire case. Welfare declines to about 650. It is interesting to note how this figure compares to the 2500 that would obtain if there were no invasives to worry about. Close to 1000 in expenses are incurred to reduce the likelihood of invasion but, because these expenditures are relatively effective, the expected costs of invasion decline relative to the "low" and "medium" cases (they are about 280) even though the anticipated damage in the event an invasion does occur has increased greatly.

Note in Fig. 7.1c, in contrast to 7.1b, that there is only one candidate optimal policy. When anticipated damages in the event of an invasion are high enough, the laissez faire corner solution can be rejected.

When damages exceed about 50,000 we enter the "very high" range. Then it becomes prohibitively expensive to take the measures required to reduce the probability of invasion to tolerable levels, and the optimal policy is to prohibit trade entirely.

Let us now consider in more detail the model from which these figures were drawn.

7.4 A simple model of invasive species with nonconvex damages

In this section I develop a simple model to illustrate some basic points. While this model is not at all general, I conjecture that it illustrates more general propositions. At the least, it demonstrates that important discontinuities *can* arise. In the next section I discuss some possible extensions and their likely implications.

I postulate a static model for simplicity. Such models have been useful in similar contexts (see e. g. Horan *et al.* 2002 and Costello and McAusland 2004). Transposing the model to a dynamic setting would introduce additional notation and complexity, but seems unlikely to affect the basic insights generated here.

Suppose that c. i. f. costs[9] of imports are constant at c per shipment. I will suppose that shipments may be "treated" to reduce the likelihood that they will result in the introduction of an invader. In ocean shipping, for example, common treatments in practice, or under discussion, include ballast water exchange, heating in transit, ultraviolet, filtration, ozonation, and deoxygenation (Lovell *et al.* 2005). I will suppose that treatment costs are

$$k \ln (\bar{\pi}/\pi) \qquad (7.1)$$

per unit of shipments, where $\bar{\pi}$ is the probability that an invasion will result from the arrival of an *untreated* shipment. I treat π as a control variable to be determined by policy makers.

Expression (7.1) may be motivated as follows. Suppose that a series of treatments can be required. Let each treatment cost χ, and suppose that each is equally effective, in that each eliminates any potential invaders present in the shipment with probability d. In other words, a set of T treatments can be regarded as a series of T independent Bernoulli trials, each with probability of success d. Then $\pi(T)$, the probability that an invasion occurs after T treatments have been performed, would be

$$\pi (T) = \bar{\pi} (1 - d)^T . \qquad (7.2)$$

Solving for T,

$$T = \frac{\ln \pi - \ln \bar{\pi}}{\ln (1 - d)}. \qquad (7.3)$$

---

[9] "Cost, insurance, freight": the cost of imported goods at their port of entry.

So the cost of implementing the T treatments required to reduce the probability of invasion to π is

$$K (\pi) = \chi T = \frac{\chi}{\ln (1 - d)} [\ln \pi - \ln \bar{\pi}]. \qquad (7.4)$$

If I now define $k = -\chi / \ln(1 - d)$, I have equation (7.1), which I will repeat here:

$$K (\pi) = k \ln (\bar{\pi}/\pi) .$$

Note the properties of this function. If $\pi = \bar{\pi}$ – i. e. if no treatment is conducted—then, of course, there is no treatment cost. Also,

$$\frac{\partial K}{\partial \pi} = -k/\pi; \qquad (7.5)$$

when the probability of preventing an invasion *decreases* the cost of prevention *increases*. Moreover, the rate at which treatment costs increase is inversely proportional to the probability of invasion: it would be astronomically expensive to assure that no transmission can occur.

I will use S to denote the number of shipments of imports received. Suppose that inverse demand is linear,

$$p = \bar{p} - bS. \qquad (7.6)$$

Suppose that the damages that result from an invasion are D. As discussed above, I will treat an invasion as an event that either occurs or does not occur and, if it occurs, the damage it causes cannot be exacerbated by subsequent reintroductions. Thus, the expected damage resulting from the receipt of S shipments is

$$E (L) = \lfloor 1 - (1 - \pi)^S \rfloor D. \qquad (7.7)$$

That is, the expected loss is the damage in the event of invasion, D, times the probability that the invader is introduced at least once.[10] This probability is one less the probability that it is *not* introduced in any of S opportunities.

---

[10] I may be guilty of some terminological imprecision here, using "introduce" in a way not all biologists will approve. I intend the term to mean "*introduce* an exotic species that will become *established* and do *harm*". It seems better to use the single term "introduce" for this notion than to repeat the entire phrase each time it is needed.

The probability of invasion, even absent treatment, is a relatively small number; if it were not, virtually all invasions that could have occurred would have occurred already, and our policy interest in preventing them would be obviated.[11] The following approximation is very accurate for small values of π, and simplifies several subsequent calculations:

$$1 - (1-\pi)^S = 1 - \left[\left(\frac{1/\pi - 1}{1/\pi}\right)^{\frac{1}{\pi}}\right]^{\pi S} \approx 1 - e^{-\pi S}. \quad (7.8)$$

While this approximation does make the derivation of necessary and sufficient conditions for optimal policies simpler, it is not crucial to the results. The numerical examples presented in the next section are derived using the exact form, (7.7), and they confirm that results derived using the approximation are valid more generally.

It is important to note two constraints. First,

$$\pi \leq \bar{\pi}.^{12} \quad (7.9)$$

It is not possible to "sell" additional quantities of invasion risk by increasing the probability of invasion over the value it would have absent any treatment.

Second,

$$S \geq 0. \quad (7.10)$$

shipments cannot be negative. While economists are accustomed to think of "negative imports" as "exports", it is not possible to reduce the number of invaders that have become established by exporting products to other countries.

As in Costello and McAusland (2004), I adopt a conventional welfare objective: consumer benefits less costs of production and transport, less treatment costs, less expected damages. Combining expressions (7.1), (7.6), and (7.8), and recalling that c is the unit cost of production and transport, the social objective can be written as

$$\max \int_0^S (\bar{p} - bs)\,ds - cS - k\ln(\bar{\pi}/\pi)\,S$$

$$- (1 - e^{-\pi S})\,D. \quad \text{s. t. } \pi \leq \bar{\pi}; S \geq 0 \quad (7.11)$$

The optimal policy is now found by differentiating with respect to π and S.

Differentiating (7.11) with respect to π, the first-order condition for an optimum is

$$\frac{kS}{\pi} - Se^{-\pi S} D \geq 0, \quad (7.12)$$

with equality obtaining when $\pi < \bar{\pi}$.

Differentiating (7.11) with respect to S, we find the other first-order condition:

$$\bar{p} - bS - c - k\ln(\bar{\pi}/\pi) - \pi e^{-\pi S} D \leq 0, \quad (7.13)$$

with equality obtaining when S is positive.

The case in which $S = 0$—that is, when it is optimal to prohibit imports, is straightforward. It arises when it is too risky to allow imports and too expensive to take steps to reduce the risk. In my notation, this would mean that D is large, $\bar{\pi}$ is high, and k is too large to make reducing π attractive. Of course, all comparisons of "largeness" are relative to the social gains to be realized from importing, as determined by the parameters $\bar{p}$, b, and c.

For our purposes more interesting comparisons arise in consideration of whether or not to treat imports, given that some importing is to be allowed. Consider first circumstances in which it would *not* be optimal treat imports. If $\pi = \bar{\pi}$—that is, if reductions in the probability of invasion are too costly to invest in, then $\ln(\bar{\pi}/\pi) = \ln(1) = 0$, and we may write (7.13) as

$$\bar{p} - bS - c - \bar{\pi} e^{-\bar{\pi} S} D = 0, \quad (7.14)$$

or, using the notation S^c to denote the outcome when the constraint on initial probability binds,

$$S^c = \frac{\bar{p} - c - \bar{\pi} e^{-\bar{\pi} S^c} D}{b}. \quad (7.15)$$

Using (7.12) to substitute for the last term in the numerator of (7.15), we may also write

$$S^c > \frac{\bar{p} - c - k}{b} \quad (7.16)$$

when π is constrained by $\bar{\pi}$.

---
[11] One might reasonably argue that opening trade between previously unlinked regions, and the increased speed with which trade now takes place (reducing potential invaders' mortality in transit) would call this argument into question, but as a general ecological proposition it seems true that invasions are relatively rare.

[12] The probability-reduction cost equation, (7.1), implies that $\pi > 0$ as well. This is important in the interpretation of the second-order conditions below.

Alternatively, suppose that the optimal choice of π is less than $\bar{\pi}$. It is now optimal to incur some costs to treat shipments. Then the first-order condition (7.12) is satisfied as an equality. Substituting from (7.12) into (7.13) and rearranging,

$$S^* = \frac{\bar{p} - c - k[\ln(\bar{\pi}/\pi) + 1]}{b}, \quad (7.17)$$

where I have introduced the notation S^* for the number of shipments accepted when neither constraint binds. From (7.17) we can infer that

$$S^* < \frac{\bar{p} - c - k}{b}, \quad (7.18)$$

Since $\pi < \bar{\pi}$. Combining, then,

$$S^* < \frac{\bar{p} - c - k}{b} < S^c. \quad (7.19)$$

The implication of expression (7.19) is that there can be a discontinuity in optimal policies. Note that the damages resulting from invasion, D, do not appear in expression (7.19). The damages do, however, affect the optimal policy. There are instances, then, in which damages may be small enough that it makes little sense to avoid their occurrence; that is, for low levels of potential damage it is optimal to leave π at $\bar{\pi}$.

For higher levels of damage, though, it is reasonable to incur costs to treat shipments. When damages rise above a certain threshold, then, there is a discontinuous shift in policy from laissez faire to more aggressive intervention.[13]

It is also revealing to look at second-order conditions for optimal policy. Begin again by considering the optimal policy when the probability constraint binds and $\pi = \bar{\pi}$. Differentiating (7.14) again with respect to π, the second-order condition for constrained optimization is

$$-b + \bar{\pi}^2 e^{-\bar{\pi}S} D \leq 0. \quad (7.20)$$

When neither constraint binds the second order, conditions are found by differentiating the first-order conditions, (7.12) and (7.13), again to form the

[13] More generally, the choices of policy variables can be discontinuous in the underlying parameters. Damages, D, are one of these underlying parameters, but we could make similar statements concerning discontinuities for different values of k and $\bar{\pi}$, as well as D.

Hessian matrix

$$\begin{pmatrix} -\frac{kS}{\pi^2} + S^2 e^{-\pi S} D & \frac{k}{\pi} - (1 - \pi S) e^{-\pi S} D \\ \frac{k}{\pi} - (1 - \pi S) e^{-\pi S} D & -b + \pi^2 e^{-\pi S} D \end{pmatrix} \quad (7.21)$$

Substituting from (7.12),

$$\begin{pmatrix} \frac{kS}{\pi^2}(\pi S - 1) & kS \\ kS & -b + \pi k \end{pmatrix}. \quad (7.22)$$

The sufficient conditions for unconstrained maximization are that when (7.12) and (7.13) are satisfied as equalities the diagonal terms and the determinant of (7.22) are all negative. That is,

$$1 - \pi S \geq 0, \quad (7.23)$$

$$b \geq \pi k, \quad (7.24)$$

and

$$1 - \pi S \leq \pi k / b. \quad (7.25)$$

We want to draw special attention to condition (7.23). Recall from expression (7.8) and the underlying expression, (7.7), on which that approximation was based, that $e^{-\pi S}$ is the probability with which an invasion *does not* occur. If there is an interior optimum, then, there must be *at least* a probability of e^{-1}, or about 37 per cent, that the measures taken to prevent invasion are effective.

7.5 Elaborations and extensions

I may not have emphasized firmly enough thus far that the simple model I have presented above, and on which the figures in the third section are based, is schematic and illustrative. It demonstrates what *could* happen; it does not necessarily describe what *should* happen. Having said that, though, I do believe the model provides some useful insights. These are of two types. The first is simply to illustrate that some unusual, and perhaps unanticipated, policies *may* be optimal. The second type of insight such a model *may* offer concerns more practical policy advice. While it is very difficult to say exactly when optimal policy should shift from laissez faire to aggressive interdiction, the model *may* provide some hints as to when one policy or another would be optimal. I will defer speculations regarding the second type of insights to the concluding section that follows, and focus in this section on the question of whether the specific results derived above may generalize. Before proceeding, let me explain the italicized *mays* in this paragraph. My conjectures as of this writing must be simply that: conjectures based on a simple model. Much more research would be required to confirm them.

The simple model rests on a number of simplifying assumptions, including:

- Linear demand and constant marginal costs of production and transport.
- Logarithmic costs of treatment.
- A single time period.
- A single invader that may arrive at a single point in time.
- Equal likelihood of invasion from each shipment.

Let me consider each of these assumptions in turn. The first accounts for the relatively simple form of many expressions, but does not appear to be crucial for the derivation of what should be seen as the principle result of the chapter, that optimal policies may be discontinuous (inequality [7.19]).

Presuming logarithmic costs of treatment implies that the derivative of treatment costs with respect to the probability of invasion from any given shipment is inversely proportional to that probability. Less formally, it insures that costs increase rapidly as the probability of arrival declines. This seems intuitively appealing.

Moreover, it is easily shown that the logarithmic cost specification is a special case of a more general cost function that would order treatments by their costs, call them k_t, and probabilities of deterring invasions, call them d_t. Then the probability that an invader survives T treatments would be

$$\pi(T) = \bar{\pi} \prod_{t=1}^{T} (1 - d_t). \quad (7.26)$$

If k_{T+1} is the cost of instituting the $T+1^{\text{st}}$ treatment, and d_{T+1} its effectiveness, we can approximate the derivative of treatment cost with respect to the probability of an invasive's arrival in any given shipment as

$$\frac{K[\pi(T+1)] - K[\pi(T)]}{\pi(T+1) - \pi(T)} = \frac{-k_{T+1}}{\pi(T) d_{T+1}}. \quad (7.27)$$

It seems reasonable to suppose that the ratio of k to d will increase as stringency is increased; the most cost-effective measures would be adopted first. In assuming logarithmic costs I adopted a specification whose distinguishing feature is that costs increase dramatically as the probability of invasion declines to lower levels. Expressions (7.26) and (7.27) suggest that a more realistic specification might be still more drastic in this regard.

The restriction to a static analysis *per se* seems unlikely to make a great deal of difference to the analysis. The dynamic analog to the welfare objective, (7.11), would be

$$W^0 = \int_0^S (\bar{p} - bs)\, ds - cS - k \ln(\bar{\pi}/\pi) S$$
$$- \left(1 - e^{-\pi S}\right) D + e^{-(\pi S + \delta)} W^0, \quad (7.28)$$

Where δ is the discount rate, and D should be interpreted as the net present value of all damages accruing in future periods *net* of the benefits of unrestricted trade—since, following the arrival of the invader, there would be no need to restrict trade any longer. It is easily seen that the first-order conditions for optimization differ from (7.11) and (7.12) only in that expressions in $\pi e^{-\pi S} D$ and $S e^{-\pi S} D$ would now have appended terms in $-\pi e^{-(\pi S + \delta)} W^0$ and $-Se^{-(\pi S + \delta)} W^0$, respectively. As these expressions drop out of inequality (7.19), the discontinuity result would be preserved.

It is more problematic when a dynamic model is mated with a series of potential invaders, each of which might first arrive in any time period. We could address such a problem with a stochastic dynamic programming approach, in which the number of established invaders may vary from one period to the next, but this would become extremely complicated. Suffice it to say for now that the model of a single potential invader at least gives some sense of the concerns involved, even if it does trade realism for tractability.

Finally, different potential invaders have different likelihoods of arrival, and treating all as if they were governed by the same probability is clearly unrealistic. Be that as it may, the best rejoinder is likely the same as above: building tractable models affords *some* insights, while it is not clear what more complex ones would offer to offset the opacity of their mechanics.

7.6 Implications for policy

The objective of a useful modeling exercise is not simply to illustrate possibilities, but to suggest appropriate policies. In the case of invasive species, this is greatly complicated by the difficulty one would encounter in estimating empirical magnitudes. What are the damages that would result from the introduction of an invasive species? With what probability would they occur? How effective would certain expenditures prove in reducing those probabilities? I have assumed in the model above that these quantities are known; in fact, they are not. Moreover, the caveats and qualifications I have emphasized in the previous section underscore the point that the model I have presented is as yet more illustrative than practicable.

Having made these admissions, however, it does seem that the simple model I have constructed has some potentially intriguing implications for policy formation. Rather than repeating the many caveats and qualifications required before using this structure for policy analysis, let me instead contemplate a thought experiment: *if the simple model I have developed did describe the real world, what would its implications be*? If nothing else, such an exercise will suggest whether the further work required to elaborate a more realistic depiction of real world circumstances would be likely to yield useful insights.

I believe that it would, as the simple model generates intriguing answers to several interesting questions. Let me list them.

7.6.1 When is aggressive policy appropriate?

The critical feature of the model above is the concavity of expected damages in shipments. This drives a discontinuity in policy: for "low" levels of damages a laissez faire policy response is appropriate, but above a certain threshold, a much more aggressive approach is called for. It is obvious that if anticipated damages exceed the gains from trade under a laissez faire policy, a more aggressive policy is called for. What may be somewhat surprising, however, is that it is appropriate to initiate the more aggressive policy even when damages are considerably less than would be the gains from trade under a laissez faire policy. In the numerical example I developed and illustrated in Fig. 7.1b, for example, damages above approximately 60 per cent of the maximum gains from trade would motivate an aggressive policy response.

7.6.2 Are there any easy diagnostics?

As with many economic models, a skeptic might say that the advice emerging from my analysis is "it all depends". Sometimes a laissez faire policy is appropriate, sometimes aggressive intervention. Are there straightforward ways to distinguish between prescriptions? While the answer *does* depend crucially on biological and economic parameters, if we can get some sense of what these are, we do have some relatively easy tests to guide us. For example, expression (7.12) is a weak inequality describing the optimal probability of invasion to accept per shipment.

$$\frac{kS}{\pi} - Se^{-\pi S}D \geq 0. \quad (7.12)$$

If (7.12) holds as a *strict* inequality, then no steps should be taken to reduce the likelihood of invasion; then

$$\frac{S}{D} > \pi S e^{-\pi S}. \quad (7.29)$$

Note that

$$\frac{\partial}{\partial S}\left(Se^{-\pi S}\right) = (1 - \pi S)e^{-\pi S}. \quad (7.30)$$

Hence the right-hand side of (7.29) reaches an upper bound when $S = 1/\pi$, in which case a *sufficient* condition for it to be optimal not to take steps to reduce the likelihood of invasion is that

$$S/D \geq 1/e. \quad (7.31)$$

Expression (7.30) may be useful to rule out relatively obvious instances of trivial concerns, but it is unlikely to constitute a tight bound in many instances. However, it may not be unreasonable to ask the relatively straightforward questions "what will it cost to institute the simplest measures to reduce the likelihood of invasion?" (i. e. "what is k?"); "what would the consequences of an invasion be?" ("what is D?"); and "how many times do you think the invader will arrive if we don't do anything?" ("what is πS?"). The answers to these three questions would then establish whether there is a *prima facie* case for aggressive intervention.

At the other extreme, how could we tell if aggressive measures were warranted? One reasonable question to begin with is whether *any* imports should be accepted. That is, if $\bar{p}$ is the maximum willingness to pay for the imported good, under what circumstances would it be optimal to import *at least one shipment*? This will be the case if

$$\bar{p} - c - k \ln(\bar{\pi}/\pi) > \pi D \qquad (7.32)$$

For some π. Rearranging slightly,

$$\bar{p} - c - k \ln \bar{\pi} > \pi D - k \ln \pi. \qquad (7.33)$$

The right-hand side of (7.33) reaches a *minimum* when $\pi = k/D$. Hence, a *necessary* condition for *some* trade to be optimal is that

$$\bar{p} - c > k[1 + \ln(\bar{\pi}D/k)]. \qquad (7.34)$$

Again, uncertainty regarding parameters would restrict the application of this formula, but it at least gives some direction in the formation of policy.

Perhaps the most difficult policy choice to make would arise in the case of what I have characterized as "medium-to-high" damages. How aggressive should an aggressive policy be?

Returning to the second-order conditions, recall the final condition, that the determinant of (7.22) must be negative. Rearranging expression (7.25),

$$S \geq \frac{b - \pi k}{b\pi}. \qquad (7.35)$$

Expression (7.35) provides a check on putatively optimal solutions. While inequality (7.23), the second-order condition with respect to the choice of π, requires that S be smaller than $1/\pi$, inequality (7.35) requires that S also be larger than some positive quantity (recall that $b > \pi k$ by [7.24]).[14]

7.6.3 How can abuse be prevented?

Some of the more difficult questions in thinking about invasive species policy involve political economy, rather than economics *per se*. That is, any time discussions focus on addressing externalities associated with trade, there is the possibility that measures purported to prevent "pollution" are in fact intended to protect special interests.

A number of authors have considered this possibility (see, e. g. the review in Olson 2006). Among the more provocative assertions in this literature are that certain restrictions on trade could not be justified even if invasions would have been certain absent intervention (Orden, Narrod, and Glauber 2000, as cited in Olson 2006) and that it may be impossible to distinguish optimal invasion-prevention policies from (economically) protectionist measures absent detailed information on biological parameters (Margolis, Shogren, and Fischer 2005).

It also bears mentioning in this context that trade may be the source of some *positive* externalities. Some recent economic scholarship has suggested that on balance, counter to the assertions of "anti-globalization" forces, trade may improve, rather than degrade the environment (see e. g. Bhagwati 2002, Frankel 2005). By facilitating technology transfer, trade may induce developing exporting countries to adopt cleaner production technologies. Moreover, to the extent that trade makes parties wealthier, "environmental Kuznets curve" effects may also lead to better environmental performance. Finally, in the context of invasive species specifically, there is some possibility that what is an "invasive pathway" in some respects would prove to be a "migratory corridor" in others. If stowaway species might escape climate change and habitat destruction, the effects of invasions on global biodiversity might be partially mitigated.

---

[14] It is worth noting in passing that the derivation of first- and second-order conditions for optimization implicitly assume that the probability of invasion cannot be driven literally to zero.

I cannot hope to resolve the vexing question of how we are to distinguish between legitimate concerns over biological invasions and well-disguised attempts at rent-seeking. Nor do I propose to suggest exactly how much the negative externalities of trade should be offset by potentially positive externalities. In fact, I do not intend to contribute more than a comment in passing. Perhaps it is an interesting observation, however. Gary Becker's seminal work on the "theory of competition among pressure groups [more recent usage might favor the terms "interest groups" or "special interests"] for political influence" (1983) makes the observation that the distortions introduced by such interest groups are, in a sense, self-limiting. A *small* departure from the social optimum is often easily obtained by the actions of well-organized groups with concentrated interests. Progressively larger distortions yield diminishing returns, however. The special interest advocating the change gains less for each step it induces regulators to take away from the social optimum, while the consuming public loses more with each step.

Becker's analysis assumes "well-behaved" producer and consumer benefit functions, however. As I have shown above, the introduction of a nonconvexity in damages leads to a discontinuity in policy. Relatively small potential damages should induce essentially *no* response, while damages above some threshold should induce a *big* response. The existence of such discontinuities may be a sort of blessing in disguise in the formulation of policy. By creating a burden of proof—aggressive policies should only be enacted in response to discrete externalities—they may preclude the sort of petty rent-seeking that would accomplish little for the environment while burdening consumers and administrators unnecessarily.

7.7 Conclusion

This chapter may be something of an unusual contribution to a volume such as this. Many of my colleagues have written chapters reporting work that they have been doing for several years. I, on the other hand, am reporting on work in progress, begun fairly recently, and to be entirely honest, which has been hamstrung by my own misconceptions and ignorance.[15]

Because this is work in progress, it is prudent to present many results as special cases accompanied by conjectures as to their generality. It may also be that I have not fully recognized my debt to others' work, in which the basic results of my chapter seem implicit to me, and in which my results may already be, to more perceptive readers, more apparent.

Having said this, however, I do believe that the results I have emphasized here may be useful both in understanding principles for policy toward potential invaders and in crafting policies appropriate to particular circumstances.

The latter objective is the more important, and it is appropriate to conclude with a discussion of research needs that must be met before more specific and useful policy advice can be confidently offered. While the model I have developed is very simple, even it underscores the many things that remain unknown. In addition to information concerning the gains from trade in the absence of invaders, one would need to know three basic types of information

- What is the likelihood of invasion?
- What it costs to reduce the likelihood of invasion?
- What are the consequences of invasion likely to be?

The first topic seems to lie wholly in the province of biologists and ecologists, and those of us who attempt to perform economic analyses will have to keep abreast of their progress.

Learning more about the second topic may be more of an integrated undertaking, and may involve the efforts of biologists, engineers, economists, and others. One interesting question, which I have not considered at all here, concerns the value of such information. Biological invasions are often considered irreversible, or very costly to reverse, phenomena. For this reason, it is sometimes argued that we should not run the risk of

---

[15] In this regard, I must again express my gratitude to Charles Perrings and Mark Williamson for their generous suggestions, as well as my apologies to them and other colleagues for whom my earlier efforts were likely both a distraction and a provocation.

invasion until we have a better understanding of the third category of information. There may also be something of a parallel argument, however, that improvements in invasion-prevention technology might allow trade volumes to increase over time.

It is the third topic that seems most interesting. In some instances the answers are fairly straightforward—if frightening. The consequences of introducing an epidemic disease, for example, might be characterized by (known probability distributions over) mortality and morbidity. In many cases, however, the ecological consequences of an invader's arrival probably cannot be predicted with any precision at all. Then, perhaps, we may need to focus on alternative decision rules for making choices (as in Horan *et al*. 2002).

Moreover, in many cases, one has the sense that society simply has not sorted out its preferences as regards invasive species. We may fear biological invaders because they can cause great damage to infrastructure (as in the case of zebra mussels), production (as with crop pests), or health (as with avian influenza and other diseases). Yet much of the literature on invasives emphasizes their implication in the extinction or endangerment of indigenous species. There may be substantial disagreement between individuals as to whether this is a "big" or a "small" issue, or even whether such concerns are intellectually coherent (see e. g. Sagoff 2006). It does not seem reasonable or prudent to argue this issue at the conclusion of this chapter. However, a prerequisite to developing any coherent policy is to have a clear understanding of its objectives, and puts understanding values at the head of research priorities.

Acknowledgements

I am grateful to Charles Perrings, Mark Williamson, and an anonymous reviewer for very helpful comments. I also thank participants at the workshop on "The Economics and Ecology of Invasive Species" for stimulating discussions. I am grateful to my student Ethan Arnheim for stimulating discussions of his related work. All errors are my own.

References

Becker, G. (1983). A theory of competition among pressure groups for political influence. *Quarterly Journal of Economics*, **98**(3), 371–400.

Bhagwati, J. (2000). On thinking clearly about the linkages between trade and the environment. *Environment and Development Economics*, **5**, 485–496.

Brock, W., Kinzig, A. and Perrings, C. (2007). Biological invasions, biological diversity, and trade. *Working Paper*. Arizona State University.

Burdick, A. (2005). *Out of Eden: An Odyssey of Ecological Invasion*. Farar, Srauss, and Giraux, New York.

Costello, C., Springborn, M., Solow, A. and McAusland, C. (2007). Unintended biological invasions: Does risk vary by trading partner? *Journal of Environmental Economics & Management*, **54**, 262–76.

Frankel, J.A. (2003). The Environment and Globalization. *NBER Working Paper 10090*; available at: http://www.nber.org/papers/10090.

Horan, R.D., Perrings, C., Lupi, F. and Bulte, E. (2002). Biological pollution prevention strategies under ignorance: The case of invasive species. *American Journal of Agricultural Economics*, **84**(5), 1303–10.

Lovell, S. J., Stone, S. F., and Fernandez, L. (2006). The economic impacts of aquatic invasive species: a review of the literature. *Agricultural and Resource Economics Review*, **35**, 195–208.

Margolis, M., Shogren, J. and Fischer, C. (2005). How trade politics affect invasive species control. *Ecological Economics*, **52**, 305–13.

McAusland, C. and Costello, C. (2004). Avoiding invasives: trade-related policies for controlling unintentional exotic species introductions. *Journal of Environmental Economics and Management*, 48, 954–77.

Mehta, S.V., Haight, R.G., Homans, F.R., Polasky, S. and Venette, R.C. (2007). Optimal detection and control strategies for invasive species management. *Ecological Economics*, 237–45.

Office of Technology Assessment (OTA). (1993). *Harmful Non-Indigenous Species in the United States*; Report No. OTA-F-565, Office of Technology Assessment: Washington, D.C.

Olson, L.J. (2006). The economics of terrestrial invasive species: a review of the literature. *Agricultural and Resource Economics Review*, **35**, 178–194.

Olson, L.J. and Roy, S. (2002). The economics of controlling a stochastic biological invasion. *American Journal of Agricultural Economics*, **84**, 1311–16.

Orden, D., Narrod, C. and Glauber, J.W. (2000). Least trade restrictive SPS policies: the analytical framework is there, the specific answers are not. Presented at the workshop on "The Economics of Quarantine" sponsored by Agriculture, Fisheries and Forestry-Australia, Melbourne, Australia. Available online at: http://www.usda.gov/oce/oracba/papers/orden.htm (accessed February 2, 2006).

Orden, D. and Romano, E. (1996). The avocado dispute and other technical barriers to agricultural trade under nafta. Invited paper presented at the conference "NAFTA and Agriculture: Is the Experiment Working?" San Antonio, Texas.

Pimentel, D., Zuniga, R. and Morrison, D. (2001). Update on the environmental and ecological costs associated with alien-invasive species in the United States. *Ecological Economics*, **52**(3), 273–88.

Sagoff, M. (2005). Do non native species threaten the natural environment? *Journal of Agricultural and Environmental Ethics*, **18**, 115–236.

CHAPTER 8

A Model of Prevention, Detection, and Control for Invasive Species

Stephen Polasky

> *"...you walk in, you get injected, inspected, detected, infected, neglected and selected."*
> Arlo Guthrie, Alice's Restaurant

8.1 Introduction

Though the true extent of damages from invasive species is unknown, several prior publications have summarized evidence of the potentially large costs of invasive species to society and to ecosystems (e.g. Lovell *et al.* 2006; OTA 1993; Pimentel *et al.* 2000, 2005). The large proven damages from some invasive species (Emerald Ash Borer *Agrilus planipennis*, Tamarisk *Tamarix ramosissima, Tamarix chinensis, Tamarix parviflora*, Zebra Mussel *Dreissena polymorpha*, and others) and the potential for even larger damages from future invaders calls for effective policy and management to limit introductions and control or eradicate existing invasive species. Many governments have instituted policy and management actions to do so. In recent years the US government has spent over US$1 billion dollars annually on invasive species (US National Invasive Species Council 2006).

There are many potential strategies that can be employed to limit the damage from invasive species. These strategies can be categorized into those that gather information about the presence or location of invasive species and those that control invasive species. In many instances gathering information is a prerequisite to effective control actions. These strategies can be further categorized as those aimed at preventing the introduction of the species into an ecosystem and those that to control or eradicate a species once it is in an ecosystem (see Table 8.1).

The first set of strategies to reduce damage from invasive species (stage 1) is aimed at preventing the introduction of potentially invasive species into new environments. A good example of prevention strategy is the careful inspection of airplanes landing in Hawaii to avoid the introduction of the Brown Tree Snake. If an invasive species is found during inspection, then control actions can be taken to prevent the species from being introduced into the environment, either by directly killing the individuals (as with the Brown Tree Snake), fumigating, quarantining, or rejecting the shipment of goods contaminated with the potentially invasive species. There is often a large volume of incoming material and a low rate of infestation making inspection costs per positive signal of the presence of an invasive species large. Once found, however, control actions are highly targeted and contained so the cost of control is typically small compared to the cost of inspections. In this case, inspection and control are closely tied and there is no point in doing inspection without immediately following it with control when invasive species are found. Other examples of strategies at the prevention stage do not involve inspection but go directly to control. Examples of such strategies include requirements on ballast water exchange and black lists restricting the importation of species that have the potential for being invasive. Throughout the chapter, I will refer to "prevention" strategies for both inspection and control at the first stage aimed at

Table 8.1 Categorization of invasive species strategies

Stage	Strategies	
	Information gathering	Control
Stage 1: Preventing introduction of species	Inspection	Control
Stage 2: Controlling or eliminating existing invasive species	Detection	Control

preventing invasive species from being introduced into an ecosystem.

A second set of strategies to reduce damage from invasive species involves attempting to control or eradicate invasive species after they have been introduced into an ecosystem (stage 2). Detection efforts involve monitoring ecosystems to see if an invasive species is present. Early search might be able to detect an invasive species before it becomes established and spreads, making control and possible eradication of the invasive species more effective and less costly. Search effort can also be directed at finding the extent of the invasion to help direct efforts at preventing further spread. Once an invasive species has been detected, control or eradication of the invasive species can be accomplished by biological (introduction of parasite or predator species), chemical (pesticide or herbicide applications), or physical means (e.g. mechanical or hand weeding). Throughout the chapter, I will refer to information collection strategies to find invasive species present in the environment as "detection" and strategies to reduce or eliminate invasive species from the environment as "control".

In this chapter, I construct a model to analyze optimal policy and management to reduce damage from invasive species that includes strategies for information collection and control, to prevent introduction and to control already introduced species. The three types of strategies considered (prevention, detection, and control) correspond closely to several categories used by the US National Invasive Species Council of "prevention" (prevention), "early detection and rapid response" (detection and control), and "control and management" (control). I characterize optimal strategies for this model and show how the three sets of policies interact. Using numerical simulation I explore how optimal solutions depend on the cost and effectiveness of prevention, detection, and control activities. Very strict inspection strategies and inexpensive methods to prevent species introductions make it optimal to reduce the rate of introductions. Low rates of introduction reduce the expected value of detection strategies. Stricter detection strategies can mean that invasive species are spotted early so that eradication may be possible and control and damage costs are reduced. If control of species in the environment is easy and inexpensive it will be less important to undertake aggressive prevention or detection efforts. Because of these interactions, optimal policy requires that the entire set of prevention, detection, and control strategies be jointly determined.

Much of the difficulty of designing effective strategies to combat invasive species stems from the fact that managers are often working with incomplete information. Managers may not know the rate of introduction of a species into an ecosystem or its likely survival upon being introduced. Managers may not know whether a species is present in the ecosystem or, if it is, what the extent of its range might be. Managers may also not know whether a species will become invasive should it arrive in an ecosystem or the likely damages from a species should it become invasive. In some cases, managers can form educated guesses on how likely a species is to become invasive based on its characteristics and perhaps an idea of what might be at risk if it should become invasive. Even with estimates of likelihood of damage should it become established, a manager will need additional information from monitoring, to inspect movement of items to reduce introductions, and from detection in ecosystems to ascertain whether species are present and the extent of the range of infestation. Having better information to prevent introductions, and for early warning to catch populations of invasive species while they are still small, can reduce control costs and damages.

While not dealing with all of the sources of uncertainty facing a manager, this chapter addresses several important sources of uncertainty, namely monitoring for potential introductions (inspection during the prevention stage) and monitoring the ecosystem for the presence of the species (detection). The benefits of inspection and detection come mostly from the value of information. In inspection, one can find potential invaders and use prevention strategies to keep species from being introduced into the environment. Similarly, monitoring an ecosystem for the presence of an invasive species can result in finding an invasive early, which allows subsequent control activities to be done more effectively at low cost.

There is a growing literature on the economics of invasive species (see Perrings *et al.* 2000; and see Lovell *et al.* 2006 and Olson 2006 for reviews). Several prior papers analyze optimal control of an existing population of an invasive species (e.g. Eiswerth and Johnson 2002; Eiswerth and van Kooten 2002). Olson and Roy (2002) analyze a model with nonconvexities and show when it is optimal to eradicate an existing population of an invasive species. Some of the most interesting work on invasive species control has focused on spatial aspects that arise when trying to control the spread of an invasive species (e.g. Brown *et al.* 2002; Heikkila and Peltola 2004; Sharov 2004; Sharov and Leibholtz 1998). Another set of papers have analyzed policy to prevent introductions (e.g. Horan *et al.* 2002; Sumner *et al.* 2005) and there is a related literature on trade flows and prevention (e.g. Costello and McAusland 2003; McAusland and Costello 2004). The most closely related prior papers to the present chapter are those that analyze both prevention and control in a unified model (Finoff and Tschirhart 2005; Finoff *et al.* 2007; Leung *et al.* 2002, 2005; Olson and Roy 2005). The general conclusion from these papers is that prevention is preferred to control (an ounce of prevention is worth a pound of cure).

An important aspect of the problem of invasive species that is overlooked in most of the prior literature is the problem of detection. Costello and Solow (2003) analyze a model in which species are introduced with a small population size and then grow through time. In their model, the probability of observing the species is related to the size of the population. Because of delay between introduction and detection, it is possible that the observed increase in invasive species could be real or merely an artifact of prior introductions that are only now being observed. One prior paper, Mehta *et al.* (2007), analyzes a model of detection and control. Unlike the present chapter, they do not also consider prevention.

The model of prevention, detection, and control is presented in the next section. The model is analyzed and results of the analysis are presented in Section 8.3. Section 8.4 contains a brief set of concluding remarks and a discussion of challenges needing further research.

8.2 The model

The model developed here is applied to the case with a single invasive species. At the start of each time period, $t = 0, 1, 2, \ldots$, the species is either not present in the system (N), or established and spread at a large population level (S). During the period the species may be introduced and present at a low population level (I). The species can arrive in any period. If the species is introduced in period t, then it will exist at a low population level during period t. If there is not successful control action applied during that time period, then the species will be established and spread at a high population level at the start of period $t+1$. Control actions may eliminate existing populations of the invasive species so that the population existing in period t is no longer present at the start of period $t+1$.

The ecosystem manager does not necessarily know whether the species is present in the ecosystem. From the manager's point of view, there are four potential states, each of the actual states, N, I, and S, plus an additional state in which the manager faces some uncertainty. The manager may not have detected the species in the ecosystem but not know for sure whether the species is present or not (X). If the manager has not detected the species in the ecosystem, the species may either not be present in the system (N) or introduced in the system at a low level of population (I). It is assumed that once the species has established and spread at a large population level that its presence in the environment becomes readily observable so that the manager will detect the species with probability one.

Each period the manager chooses the level of three management actions: prevention, detection, and control. Prevention effort in time period t, q_t, reduces the probability that the invasive species will be introduced in period t. Let $\lambda(q)$ be the probability that the invasive species will not arrive in the ecosystem in period t. Greater prevention effort increases the probability that the species will not arrive: $\lambda'(q) > 0$, $\lambda''(q) \leq 0$. Let the cost of prevention effort be $c(q_t)$, with $c'(q) > 0$, $c''(q) \geq 0$. I assume that $c''(q) = 0$ and $\lambda''(q) = 0$ cannot both be true and that there is sufficient curvature to guarantee an interior optimum.

After the prevention phase, the manager chooses the level of detection (search) effort in period t, s_t. Let $\gamma(s)$ be the probability of finding the invasive species if it is present in the ecosystem at a low level of population (I). Note that if the population is established and spread then it is observable even without active search. The probability of detecting the invasive species when it is established and spread is 1 for all $s \geq 0$. If the species is present at low population levels, greater search effort increases the probability of detection: $\gamma'(s) > 0$, $\gamma''(s) \leq 0$. Assume there are no false positives so that $\gamma(s) = 0$ for all s if the true state is N. Let the cost of detection effort be $d(s_t)$, with $d'(s) > 0$, $d''(s) \geq 0$. I assume that $d''(s) = 0$ and $\gamma''(q) = 0$ cannot both be true and that there is sufficient curvature to guarantee an interior optimum.

After the choices of prevention and detection the manager chooses control efforts, r_t. For simplicity, I assume that control efforts either result in eradication of the existing population or are ineffective. Further, I assume that if a species is undetected that control efforts are not undertaken. Alternatively, a manager could choose to blanket the ecosystem with control effort even though a small population has not been detected. I will not discuss this option further in this chapter. The probability of success for a given level of control differs depending on whether the population is at a low or high population level. Let the probability of eradication with a low level of population be $\psi(r)$, and with a high level of population $\varphi(r)$, with $\psi(r) > \varphi(r)$ for any level of r. Let the cost of a unit of detection effort be e, so that the total cost of detection in period t is er_t.

The manager wishes to maintain the ecosystem such that the invasive species does not become established and spread. Assume that there is a per-period benefit of W when the invasive species is either not present or introduced and present at low levels of population. Assume that there is no benefit during periods when the invasive species is established and spread at a high population level. Let δ be the discount factor between periods.

The objective of the manager is to maximize the present value of benefits minus the costs of prevention, detection, and control. The objective function for the manager can be written as follows:

$$Max E \left\{ \sum_{t=0}^{\infty} [W\eta_t - c(q_t) - d(s_t) - er_t]\delta^t \right\} \quad (8.1)$$

where $\eta_t = 0$ if the ecosystem is in state S at the beginning of period t and $\eta_t = 1$ otherwise.

The model is summarized in Fig. 8.1 and Table 8.2. Figure 8.1 illustrates the timing of the model. Table 8.2 shows the evolution of the state of the ecosystem and of the manager's information. Note that because growth is the final stage of the period, that at the beginning of each period the species must be either not be present, or established and spread so that it is readily observable. Therefore, the manager has complete information about the status of the species at the beginning of

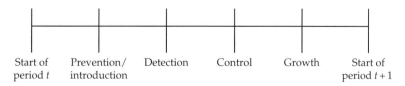

Figure 8.1 Order of events from the start of period t to the start of period $t+1$.

Table 8.2 Evolution of manager information and state of the ecosystem

	Period t						Period t + 1
Initial status	Post prevention/ introduction		Manager information		Post growth		Initial status
	Actual state	Manager information	Post detection	Post control			
N	N	X	X	X		N	N
	I	X	X	X		S	S
		X	I	N		N	N
				I		S	S
S	S	S	S	N		N	N
				S		S	S

the period, but not during the middle of the period when decisions on detection and control need to be made.

Using information on the order of events and the evolution of actual states and manager information, we can rewrite the manager's problem in terms of a stochastic dynamic programming problem. At the beginning of each period, there are two potential states that can occur {N, S}. The value function starting period t in state N is:

$$V(N) = Max_{\{q,\hat{r},s\}} W - c(q_t) - d(s_t) + \lambda(q_t)\delta V(N)$$
$$+ (1 - \lambda(q_t))(1 - \gamma(s_t))V(S) + (1 - \lambda(q_t))\gamma(s_t)$$
$$[-e\hat{r}_t + \delta(\psi(\hat{r}_t)V(N) + (1 - \psi(\hat{r}_t))V(S)] \quad (8.2)$$

where $\hat{r}_t$ is the optimal solution to the control problem given the information that the manager has at the time that control must be chosen. If the manager has detected presence of the species, and the function $\phi(e)$ is continuous and differentiable with $\phi'(r) > 0$, $\phi''(r) < 0$, then $\hat{r}_t$ solves:

$$\delta\phi'(\hat{r}_t)[V(N) - V(S)] = e \quad (8.3)$$

If the manager has not detected the species, then it is assumed that the manager should set $\hat{r}_t = 0$.

The first set of terms in equation (2), $W - c(q_t) - d(s_t)$, represents the immediate benefit of not having the invasive species present in the environment minus the costs of prevention and detection. There are two ways to prevent the invasive species from being present in the ecosystem at the start of the next period (so that $V(N)$ is the correct value function to use one period hence): a) by preventing the invasive species from being introduced, which occurs with probability $\lambda(q_t)$, and b) successfully detecting and eradicating a new colony of the species if it is introduced, which occurs with probability $(1 - \lambda(q_t))\gamma(s_t)\psi(r_t)$. There are also two ways in which the species can become established and spread during time period t (so that $V(S)$ is the correct value function to use one period hence): a) the species has been introduced but it is not detected, which occurs with probability $(1 - \lambda(q_t))(1 - \gamma(s_t))$, and b) the species is introduced and detected but not successfully eradicated, which occurs with probability $(1 - \lambda(q_t))\gamma(s_t)(1 - \psi(r_t))$. The value function starting period t in state S is:

$$V(S) = -er_t + \delta(\phi(r_t)V(N) + (1 - \phi(r_t))V(S) \quad (8.4)$$

This value function contains only control costs as there is no need for prevention and detection since the species is already established and spread in the ecosystem.

8.3 Analysis

While in principle the dynamic programming problems outlined above are solvable, finding explicit solutions to these equations can be difficult. In this section, I make one simplification regarding control technology, making it discrete rather than continuous, which makes the problem easier to handle without losing much of the power of the model. Assuming a discrete control technology focuses attention on the optimal levels of prevention and detection while still allowing for full interactions between prevention, detection, and control. In the second portion of this section I use specific functional forms and parameter values to derive exact numerical solutions. I use these solutions to perform a series of exercises that shed light on optimal policy for invasive species.

8.3.1 General solution

The problem laid out in Section 8.2 was to optimally choose prevention, detection, and control variables in each time period. Here I simplify the problem by assuming a discrete control technology. Assume that by spending an amount equal to e_I when the population is small, or e_S when the population is large, the regulator can eradicate the species with probability one. If the regulator spends less than this amount there is a zero probability of eradication. Assuming the regulator would find it advantageous to proceed with eradication under this control technology assumption, the value function equations (8.2) and (8.4) can be rewritten as follows:

$$V(N) = Max_{\{q,s\}} W - c(q_t) - d(s_t) + \lambda(q_t)\delta V(N)$$
$$+ (1 - \lambda(q_t))(1 - \gamma(s_t))V(S)$$
$$+ (1 - \lambda(q_t))\gamma(s_t)[-e_I + \delta V(N)] \quad (8.5)$$

$$V(S) = -e_S + \delta V(N) \quad (8.6)$$

Given functional forms and parameter values one can solve for the optimal choices for prevention (q^*) and detection (s^*), and the value functions $V(N)$ and $V(S)$, in each period using backward induction. By extending the time horizon to be suitably long, the optimal steady state values of choices and the value function can be derived.

8.3.2 Numerical example

In this subsection, I solve the model for a simple case with linear probability responses and quadratic costs:

$$c(q) = cq^2$$
$$d(s) = ds^2$$
$$\lambda(q) = \lambda q$$
$$\gamma(s) = \gamma s$$

These functional forms satisfy the conditions specified above and generate interior optimal solutions for suitably chosen parameter values. I solve the model using backward induction for a time horizon of 100 periods, which is sufficiently long for the model to settle to steady-state values that mimic an infinite time horizon problem.

The base case parameter values are shown in Table 8.3. The base case parameters are symmetric with respect to prevention and detection parameters (cost parameters $c = d$, and probability parameters $\lambda = \gamma$). As shown in row 1 of Table 8.4, the optimal strategy involves prevention effort that is more than twice the detection effort. In line with previous research, the results here show that it is far more efficient to try to prevent a species from being introduced than it is to search for it in the ecosystem and if it is found to spend the resources necessary to eradicate it. As the probability of successful prevention increases ("prevention efficiency") so that there is a lower chance of the invasive species being introduced into the ecosystem, the optimal prevention effort increases from 0.43 to 0.71 while the optimal

Table 8.3 Base case parameter values

Parameter name	Parameter symbol	Parameter value
Per period benefit from not having invasive species established and spread	W	10
Prevention effort cost parameter	c	5
Detection effort cost parameter	d	5
Cost of eradication with low population	e_I	4
Cost of eradication with high (established) population	e_S	8
Prevention probability parameter	λ	0.5
Detection probability parameter	γ	0.5
Discount factor	δ	0.9

Table 8.4 Optimal prevention and detection effort

Case		Optimal prevention effort (q^*)	Optimal detection effort (s^*)	$V(N)$	$V(S)$
1	Base case	0.430	0.201	21.24	11.12
2	Prevention efficiency increases by 50% ($\lambda = 0.75$)	0.710	0.137	29.61	18.65
3	Detection efficiency doubles ($\gamma = 0.75$)	0.401	0.315	22.90	12.61
4	Cost of prevention declines by 40% ($c = 3$)	0.923	0.153	27.69	16.92
5	Cost of detection declines by 40% ($d = 3$)	0.407	0.416	22.54	12.29
6	Cost of eradication of small populations declines by 50% ($e_I = 2$)	0.409	0.287	22.43	12.18
7	Cost of eradication of small populations fall to zero ($e_I = 0$)	0.380	0.380	24.09	13.68
8	Cost of eradication of for all population sizes declines by 50% ($e_I = 2, e_S = 4$)	0.325	0.211	38.08	30.27
9	Impossible to eradicate established populations (irreversible establishment and spread)	0.518	0.251	11.98	0
10	Impossible to eradicate established populations (irreversible establishment and spread) where the benefit of avoiding establishment and spread increase by 50% ($W = 15$)	0.803	0.441	20.83	0
11	Discount factor increases ($\delta = 0.95$)	0.442	0.192	38.02	28.11

Note: All cases assume base case parameter values unless otherwise specified.

detection effort declines (row 2 of Table 8.4). Having this type of efficiency increase is also quite valuable. The value function score increases nearly 40 per cent if the initial state is N and over 60 per cent if the initial state is S. Increasing detection efficiency is not nearly so valuable (row 3 of Table 8.4). A 50 per cent increase in detection efficiency lowers optimal prevention effort only slightly while raising detection effort approximately 50 per cent. The increase is the value function is small, only 8 per cent if the initial state is N and 13 per cent if the initial state is S.

Decreasing the cost of prevention (row 4 of Table 8.4) or the cost of detection (row 5 of Table 8.4) works in a similar fashion to increases in prevention or detection efficiency. A 50 per cent reduction in prevention costs results in an increase in the optimal prevention effort to over 0.923 while reducing optimal detection effort slightly. Value functions increase in a similar fashion to the increase in prevention efficiency but to a slightly smaller extent. With a 50 per cent reduction in detection costs, there is a doubling of optimal detection effort and a very slight decrease in optimal prevention effort. Similar to an increase in detection efficiency, the decline in detection costs does not have much impact on the value functions.

The results from changing parameter values on the efficiency and the cost of prevention and detection demonstrate that prevention and detection are substitutes. Prevention and detection are two alternative ways to keep the species from becoming a successful invader. High levels of prevention lead to a lower chance of the species being introduced so the marginal value of detection and control is lower. If detection is effective or cheap, this (slightly) reduces the marginal value of prevention because subsequent steps are likely to fix any problems caused by the species slipping through at the introduction stage. Overall, prevention is the more valuable tool because when it is successful, no introductions occur and there is no necessary expense or damage incurred in the second stage. Further, costs of control at the prevention stage tend to be minor compared to control costs once an invasive species has got into the environment.

When control costs for small populations decline by 50 per cent (row 6 of Table 8.4), the results are almost identical to those for a 50 per cent decline in the cost of detection. Detection and control are complements. The more effective control is for small populations the more beneficial it is to engage in detection. Also, the more effective control is for small populations, the lower is the marginal benefit of prevention. In fact, when the cost of control to eradicate a small population falls to zero (row 7 of Table 8.4) optimal detection and prevention efforts become equal. In this case, prevention and detection have identical effects in the model. When control costs for eradicating a large established population as well as for a small population fall by 50 per cent (row 8 of Table 8.4), this has the effect of lowering prevention effort as it is not so costly to have the species introduced. A fall in the cost of eradication of established populations also lowers detection effort (comparing results in row 6 with row 8). There is a large increase in the value function, because direct control costs decline should eradication be necessary and the indirect effect of reducing prevention and detection efforts also reduces costs.

Making it impossible to eradicate an invasive species once it has become established and spread (i.e. establishment and spread is an irreversible process) increases the benefits of both prevention and detection (row 9 of Table 8.4). Results are rather grim with irreversible establishment and spread under the base case scenario. The probability of a population becoming established and spread each period is 0.647. After only five years there is less than a 1 per cent chance that the species will not have become established and spread. The only way to prevent this large cumulative probability in the case of irreversible establishment and spread is if the benefits of remaining free of a large population of the invasive species are very large or the costs of prevention and detection are very small. If the benefits from not having the species established and spread increase by 50 per cent (row 10 of Table 8.4) both optimal prevention and detection efforts increase (relative to the case in row 9). The value functions show a dramatic increase as well. In the case of irreversible invasion, an increase in the benefits of 50 per cent gives a large increase in optimal prevention and detection. This increase in prevention and detection effort reduces the probability of establishment and spread of the invasive species to 0.466 per period. However, this still gives a five year cumulative probability of invasion of over 95 per cent. Only if benefits of being invasion free rise substantially or the costs of prevention and detection fall significantly relative to the base case will there be a reasonably high probability of remaining invasion free. For example, an increase in benefits from 10 to 100 results in a probability of invasion of 0.2 per cent per period.

An increase in the discount factor (decline in the discount rate) from 0.9 to 0.95 (row 11 of Table 8.4) also increases the value functions but has only a small effect on optimal policy, slightly increasing optimal prevention and slightly decreasing optimal detection. Reducing the discount rate both increases the value of future benefits, which tends to raise both prevention and detection efforts, and to increase the value of future costs, which tends to shift efforts towards prevention and away from detection.

8.4 Discussion

The problem of optimal management of invasive species is a hard problem in large part because managers must operate with limited information. Managers typically do not know when an invading species might be hitchhiking with people or goods as they move from one ecosystem to another. Managers also typically do not know when a species is introduced into an ecosystem so that they often do not know if a species is present in the ecosystem or not. Only when a species has become established and spread, leading to large damages and large control costs, is its presence readily observable. Gathering information in the form of inspections, which reduce the probability of introduction, or detection, which gives early warning of an invasion when control activities can be undertaken effectively at low cost, is valuable but also involves costly monitoring activities.

The results of the previous section show the results for the optimal effort to devote to prevention and detection, showing how these policies interact and how they are influenced also by the ease or difficulty of control policies. In general, prevention (which may include both inspection and control to prevent introductions) and detection in ecosystems are substitutes. Reducing the cost or increasing the effectiveness of one of these policies will increase the optimal effort devoted to it and reduce the optimal effort devoted to the other. Once an invasive species is present in an ecosystem, however, control and detection are complementary. Increasing the difference in control effectiveness at small versus large population levels increases the value of detection activities. When control is impossible at large population levels, making invasions that reach a certain size irreversible, then the value of both prevention and detection increase. There is greater value in stopping the invasion from reaching the critical population level, either by preventing introduction or by early detection and control.

The model considered here is quite simple in a number of regards. Useful extensions would include considerations of multiple and potentially interacting invasive species, spatial considerations on location of introduction and spread, and consideration of more states besides just having a small or large population. With multiple population states and imperfect monitoring, the problem becomes much harder. When the manager spends effort to detect but does not find the species, the manager should update probabilities on likely population sizes. Defining optimal strategies in cases where the manager learns about probabilities (or other elements of the system) introduces important additional dimensions to the model. Incorporating each of these elements would increase the realism but also decrease the tractability of the problem. Finding useful approaches that generate good solutions in realistic problems is something that deserves greater attention.

References

Brown, C., Lynch, L., and Zilberman, D. (2002). The economics of controlling insect-transmitted plant diseases. *American Journal of Agricultural Economics*, **84**(2), 279–91.

Costello, C. and McAusland, C. (2003). Protectionism, trade and measures of damage from exotic species introductions. *American Journal of Agricultural Economics*, **85**(4), 964–75.

Costello, C. and Solow, A. (2003). On the pattern of discovery of introduced species. *Proceedings of the National Academy of Sciences*, **100**(6), 3321–23.

Eiswerth, M.E. and Johnson, W.S. (2002). Managing nonindigenous invasive species: insights from dynamic analysis. *Environmental and Resource Economics*, **23**(3), 319–42.

Eiswerth, M.E. and van Kooten, G.C. (2002). Uncertainty, economics, and the spread of an invasive plant species. *American Journal of Agricultural Economics*, **84**(5), 1317–22.

Finnoff, D. and Tschirhart, J. (2005). Identifying, preventing and controlling invasive plant species. *Ecological Economics*, **52**(3), 397–416.

Finnoff, D., Shogren, J.F., Leung, B., and Lodge, D. (2007). Take a risk: preferring prevention over control of biological invaders. *Ecological Economics*, **62**(2), 216–22.

Heikkila, J. and Peltola, J. (2004). Analysis of the Colorado potato beetle protection system in Finland. *Agricultural Economics*, **31**(2–3), 343–52.

Horan, R.D., Perrings, C., Lupi, F., and Bulte., E. (2002). Biological pollution prevention strategies nder ignorance: the case of invasive species. *American Journal of Agricultural Economics*, **84**(5), 1303–10.

Leung, B., Lodge, D.M., Finnoff, D., Lewis, M.A., and Lamberti, G. (2002). An ounce of prevention or a pound of cure: bioeconomic risk analysis of invasive species. *Proceedings of the Royal Society of London, Biological Sciences*, **269**(1508), 2407–13.

Leung, B., Finnoff, D., Shogren, J.F., and Lodge, D.M. (2005). Managing invasive species: rules of thumb for rapid assessment. *Ecological Economics*, **55**(1), 24–36.

Lovell, S.J., Stone, S.F. and Fernandez, L. (2006). The economic impacts of aquatic invasive species: a review of the literature. *Review of Agricultural and Resource Economics*, **35**(1), 195–208.

McAusland, C. and Costello, C. (2004). Avoiding invasives: trade-related policies for controlling unintentional exotic species introductions. *Journal of Environmental Economics and Management*, **48**(2), 954–77.

Mehta, S., Haight, R.G., Homans, F.R., Polasky, S., and Venette, R.C. (2007). Optimal detection and control strategies for invasive species management. *Ecological Economics*, **61**, 237–45.

Office of Technology Assessment (OTA). (1993). *Harmful Non-Indigenous Species in the United States*. Publication No. OTA-F-565. Office of Technology Assessment, U.S. Congress, Washington, D.C.

Olson, L.J. and Roy, S. (2002). The economics of controlling a stochastic biological invasion. *American Journal of Agricultural Economics*, **84**(5), 1311–16.

Olson, L.J. and Roy, S. (2005). On prevention and control of an uncertain biological invasion. *Review of Agricultural Economics*, **27**(3), 491–97.

Olson, L.J. (2006). The economics of terrestrial invasive species: a review of the literature. *Review of Agricultural and Resource Economics*, **35**(1), 178–94.

Perrings, C., Williamson, M. and Dalmazzone, S., eds. (2000). *The Economics of Biological Invasions*. Edward Elgar, Cheltenham.

Pimentel, D., Lach, L., Zuniga, R. and Morrison, D. (2000). Environmental and economic costs of nonindigenous species in the United States. *Bioscience*, **50**(1), 53–56.

Pimentel, D., Zuniga, R., and Morrison, D. (2005). Update on the environmental and economic costs associated with alien-invasive species in the United States. *Ecological Economics*, **52**, 273–88.

Sharov, A.A. (2004). Bioeconomics of managing the spread of exotic pest species with barrier zones. *Risk Analysis*, **24**(4), 879–92.

Sharov, A.A. and Liebhold, A.M. (1998). Model of slowing the spread of gypsy moth (*Lepidoptera: Lymantriidae*) with a barrier zone. *Ecological Applications*, **8**(4), 1170–79.

Sumner, D.A., Bervejillo, J.E., and Jarvis, L.S. (2005). Public policy, invasive species and animal disease management. *International Food and Agribusiness Management Review*, **8**(1), 78–97.

US National Invasive Species Council. (2006). Fiscal year 2006 interagency invasive species performance-based crosscut budget. http://www.invasivespeciesinfo.gov/docs/council/FY06budget.pdf (accessed January 26, 2009).

CHAPTER 9

Second Best Policies in Invasive Species Management: When are they "Good Enough"?

David Finnoff, Alexei Potapov, and Mark A. Lewis

9.1 Introduction

Invasions of nonindigenous species and their consequences have prompted calls for the application of rapid responses and simple decision rules in policy development in response to invasions (Leung *et al.* 2005). As invasions affect both economic and ecological systems (Pimentel *et al.* 1999; Zavaleta 2000; Shogren 2000; Leung *et al.* 2002) in the formulation of simple decision rules significant amounts of biological details are typically condensed to reduce the complexity of the problem to make the framework more usable and able to provide clear rapid guidance. In some situations these simplifications have been shown to have minimal consequences. For example, in an application considering weed control in Iowa corn production (Wu 2001) found a less than two per cent improvement in the net present value of profits from the inclusion of weed dynamics in the weed control decision problem. These arguments and findings, however, are in contrast with the more general findings of (Carpenter *et al.* 1999; Perrings 1999; Ludwig *et al.* 2003). In the management of jointly determined economic and ecological systems (as are applicable with invasive species) there are tight interdependencies between economic and ecological parameters, threshold responses, and multiple time scales. These tend to make simple management policies fall short of their goals, unless the future does not matter to decision makers or the dynamical processes are slow. They point out that while it may require a significant amount of effort to construct an accurate representation of reality and to derive optimal decision rules; this effort is likely to be worthwhile.

However, there are costs to this effort. In the case of invasive species, it may be very hard to construct completely accurate representations of affected economic and ecological systems given the tremendous uncertainties that surround all components of the process. Generally, little is known ecologically or economically about an invader in the process of an invasion, usually much less than necessary to provide a perfectly accurate representation of the system. There may also be significant costs associated with gathering additional information, leaving policy makers with only a knowledge of some key characteristics of the situation. Furthermore, even when some of the characteristics are known, systems with any complexity at all tend to be difficult to optimize, and the time required to conduct the analysis may delay the implementation of policies to such dates when significant consequences of the invasion are inevitable.

Policy makers are thus faced with a quandary—they know only a little yet are charged with making the best policies they can to (attempt) controlling the system, regardless of the state of their knowledge. Given limited information, usually on only a few broad parameters (i.e. when policy makers only know a few basic parameters that characterize the situation) one aspect of the quandary boils down to how they best make their control policies? Do they attempt to dynamically optimize the system as they know it (i.e. develop control strategies that optimally transition the system as they know it to an optimal steady state) or just employ simple control strategies based on optimal steady states (and neglect the computational complexities associated with keeping the system on an optimal path)? Even in this incomplete world, there is a key trade-off to be assessed: is it worth attempting a first best policy of dynamically optimizing the system or simply settling for an easily derived second best control policy which may result in some additional costs? A first stab at answering this question can be provided by comparing the outcomes of a dynamically optimized system to those of the second best policy and assessing the additional costs of the second best policy. Situations when these additional costs are not significant provide circumstances when second best policies may provide reasonable approximations. When these additional costs are significant, second best policies will not tend to be adequate. In addition, how these additional costs depend on a few, yet critical, parameters with respect to the management of invasive species provides an understanding of the sensitivity of the trade-offs facing policy makers. The dependency of the deviations between the outcomes of these types of policies on economic and ecological characteristics can be used to characterize the circumstances when second best polices in simple systems can be reasonably expected to achieve their goals and when they will not.

Within the confines of a motivational example, we find that over an infinite time horizon second best policies always waste some resources in comparison to dynamically optimal policies, although the total amount wasted tends to be small. In situations when there are quickly spreading invasions,[1] low rates of growth in marginal damages, and high rates of growth in marginal costs, second best policies will waste less in comparison to dynamically optimal (first best) policies. However, in the opposite of circumstances, when there are slowly spreading invasions, high rates of growth in marginal damages, and low rates of growth in marginal costs, second best policies will waste more in comparison to first best policies. Although the total difference between the two policies remains small, in these situations there are large differences in costs of transition and steady state costs that together net out to a small total difference (as they work in opposite directions). In these situations the policy maker's relative concern for transitionary versus steady state periods of time becomes of upmost importance in evaluating the outcomes of second best policies.

The findings point out that there are situations when rapid responses with second best policies might work reasonably well, but their relative performance depends on critical economic and ecological parameters and some of the implications differ from extant findings. What is clear is that if invasions cause movements to new equilibria that are relatively slow, or equilibria are never attained, or regimes are constantly changing (i.e. any case or situation with long transition times or where transition times dominate the policy horizon) then second best policies will perform poorly and may result in unintended consequences. With only a basic understanding of the underlying characteristics of the invasion policy makers can have an indication of whether the effort required to derive optimal policies is worthwhile or not.

To illustrate the issues a simple model of invasive species management is developed following standard economic renewable resource management. Second best policies are derived, compared to optimal policies, and conclusions generated. Some general implications follow.

9.2 Invasion dynamics

The invasion process is modeled for the case of a network of interconnected areas/regions. An alien species has been introduced into one or several regions, has become established, and has started to spread to other regions in the network. The main vector of spread is due to internal factors within the system (movement via "hitchhiking" on human transportation and/or the natural mobility and migration of the species itself). The total number of regions N is assumed to be sufficiently large that it is possible to characterize the invasion process by a single variable, the proportion of

---

[1] In contrast to Ludwig *et al.* 2003.

regions invaded p (invaded regions N_I divided by all regions, $p = N_I/N$). It is also necessary to assume there are enough regions such that the change of p with time may be reasonably approximated by a continuous and differentiable function $p(t)$. In its simplest form, we assume that all regions are affected by roughly the same external propagule pressure and established invaders add negligibly to the total pressure. In this situation the number of newly invaded regions is proportional to the number of uninvaded regions, $dN \sim (N - N_I)$ or $dp \sim (1 - p)$ and provides a simple representation of the proportion of invaded regions over time

$$\frac{dp}{dt} = A(1 - p(t)), \quad p(0) = p_o, \quad (9.1)$$

where A represents the intensity of transport of the invader between regions (i.e. the speed of the invasion) and initial level of the invader is given by $p(0) = p_o$. Without management action the invader will spread and eventually invade all regions. The proportional representation of this simple process reflects the expectation of the invasion process. It allows for the restatement of a stochastic problem in deterministic terms. The method permits the incorporation of intersecting biology and economics at a large scale, and allows for analysis of their joint influence on decision making and optimal invasion management.

9.3 Decision models

Consider management policies which can not only slow the invasion, they are able to stop or reverse the process unlike Potapov et al. 2007, who considered the optimal management of a spreading invader with strategies that worked only to slow the spread (only delaying the timing of a fully invaded network). For simplicity these strategies are lumped into a single management variable, coined control. This refinement transforms a finite horizon problem into an infinite horizon problem and allows the possibility of long-run equilibria (steady states) other than a fully invaded system.

Let control effort with intensity $h(t)$ slow, halt, or reverse the spread and modify equation (9.1) to

$$\frac{dp}{dt} = A(1 - p(t)) - h(t). \quad (9.2)$$

The presence of the invader causes economic losses or damages. While control can reduce the flow of the invader over regions and avoid consequential damages, it is costly. Together the damage and control costs are represented by the cost function C with assumed functional form that satisfies all requirements of the maximum principle theorems (Kamien and Schwartz 1991; Seierstad and Sydsaeter 1987)

$$C(p(t), h(t)) = \frac{1}{2}[gp(t)^2 + ch(t)^2], \quad (9.3)$$

where g describes how marginal damages grow with p and c how marginal costs grow with h.

The task facing the resource manager is to determine levels of control that best balance averted damages of invasion with costs of control[2]. Given the ability of control to halt spread into perpetuity, an infinite time horizon is appropriate and relates directly to the generation of simple decision rules and optimal decision rules. Infinite horizon problems have of course been widely investigated throughout the economic literature, perhaps most obviously in the economics of renewable resources, e.g. (Brock and Xepapadeas 2003; Olson and Roy 2002).

Theoretical studies of infinite-horizon problems have been shown to possess a so-called turnpike property (Haurie 1976): their trajectories spend most of the time near a steady state. This has led to applying the Pontryagin maximum principle for their analysis, but instead of transversality conditions at the terminal state one applies the turnpike condition, that is, a requirement that the optimal trajectory ends at the steady state.

But, in many cases a complete analytical study using the maximum principle is not possible. In the face of this complexity a commonly used approach is to analyze the terminal steady states, towards which the trajectory converges as $t \to \infty$ without a consideration of the optimal transition dynamics. The technique, referred to herein as the "second best rule" to avoid confusion (Sethi and Thompson 2000 coin it the turnpike method) is just an analytical optimization across steady states. The methodology concentrates on the properties of the terminal state

---

[2] Where costs of control implicitly reflect the cost of tracking the system closely.

and not on initial transients and provides clear and simple management rules. However, the technique may implicitly alter the problem under consideration due to its neglect of these initial transients. The simplification provides a readily tractable technique that we contrast with the prescriptions of a classical dynamic optimization of the system, referred to as the "first best rule".

The two approaches considered can be summarized as:

Second best rule: Find the steady state, such that the total discounted cost of keeping the system in it from $t = 0$ to $t \to \infty$ is minimal.

First best rule: Find the terminal steady state, such that for any initial state at $t = 0$ the total discounted cost of bringing the system to the steady state at $t \to \infty$ is minimal.

In both cases the system may well spend most of the time at the steady state but the policy rules generated can be very different and their subsequent performance different. The second best rule ignores the transient regime and concentrates on the costs related to the steady state. The solution to this problem coincides with the problem of minimizing average costs, or that of an undiscounted optimal control problem.

The first best rule concentrates on the optimal transition. The terminal state, which is reached with some delay, is discounted, and therefore contributes less to the present value of the costs compared to the initial transient. This is simply a discounted infinite-horizon optimal control problem.

Both problems provide a regime of controls for the system that have important distinctions. The difference between the steady state solutions is proportional to the discount rate r, and for small r the steady state solutions converge.

9.3.1 Second best rule

The solution of this problem is straightforward as it is a static optimization across possible steady states to find the one with minimum costs into perpetuity. The problem then is to find p_s, h_s such that discounted costs J_s are minimal for $p(t) = p_s, h(t) = h_s$ where $C(p(t), h(t)) = C(p_s, h_s)$ does not change in time

$$J_s = \int_0^\infty e^{-rt} C(p_s, h_s) dt = \frac{1}{r} C(p_s, h_s). \quad (9.4)$$

where r is the discount rate. To find the minimum of (9.4) we simply have to minimize $C(p_s, h_s)$. The procedure follows from assuming that p and h do not depend on t, and the system is at the steady state, so

$$C(p(h), h) = \frac{1}{2}\left[g\left(\frac{A-h}{A}\right)^2 + c(h)^2\right]. \quad (9.5)$$

This function has its only minimum at

$$h^*\left(\underbrace{g}_{+}, \underbrace{c}_{-}, \underbrace{A}_{\pm}\right) = \frac{gA}{g + cA^2},$$

$$p^*\left(\underbrace{g}_{-}, \underbrace{c}_{+}, \underbrace{A}_{+}\right) = \frac{cA^2}{g + cA^2}, \quad (9.6)$$

$$C^*(h^*, p^*) = C^*\left(\underbrace{g}_{+}, \underbrace{c}_{+}, \underbrace{A}_{+}\right)$$

$$= \frac{gA^2}{2\left(\frac{g}{c} + A^2\right)}. \quad (9.7)$$

The comparative static dependencies of each steady state variable on parameters are given by the signs beneath parameters. A higher rate of growth in marginal damages, g, results in higher (steady state) levels of control (as control is relatively cheaper) a lower proportion of regions invaded, and higher steady state instantaneous costs. Increases in the rate of growth of marginal costs c have the opposite effects on levels of control and proportions invaded, but instantaneous costs rise regardless in the steady state. The consequences of a faster spreading invasion (increase in A) are a higher proportion of regions invaded and higher instantaneous costs, but the effect on steady state control depends on the relative magnitude of A (for given g and c). For low levels of A an increase in A results in an increase in steady state control, but the influence is reversed at high levels of A. The reason follows from optimizing (9.5). The marginal costs of control do not depend on A but marginal benefits do depend on A. When A is low, any increase in its value increases the marginal benefits of control, prompting the decision maker to increase control. But, when A is high, any increase in it lowers the marginal benefits of control, prompting the decision maker to reduce control. This is due to (9.2) which results in the productivity of h in reducing p increasing over the interval $0 \leq A \leq 1$ and diminishing for $A > 1$. Faster spreading invasions make control effort less and less effective.

Equation (9.7) provides steady state (instantaneous) costs, but to determine the total costs of transition to the steady state we need the time path of $p_s(t)$ and the convergence time of transition τ_s from the initiation of the policy h^* until $p_s(t)$ converges (approximately) to p^*. The solution for $p_s(t)$ that converges to this steady state for $t \to \infty$ for any given initial condition p_0 is

$$p_s(t) = p^* + (p_0 - p^*)e^{-At}. \tag{9.8}$$

To characterize the convergence time we set a convergence accuracy ε, such that we define that $p_s(t)$ has converged to p^* as soon as $|p_s(t) - p^*| < \varepsilon$. Assuming that $|p_s(t) - p^*| < \varepsilon$ and $p_0 = p_0^L < p^*$ (for an invasion in its early stages) or $p_0 = p_0^H > p^*$ (for an invasion in its later stages) the convergence time can be found as

$$\tau_s \begin{cases} \underbrace{g}_{-}, \underbrace{c}_{\mp}, \underbrace{A}_{\pm}, \underbrace{\varepsilon}_{-} \Big|_{p_0^L} \\ \underbrace{g}_{+}, \underbrace{c}_{-}, \underbrace{A}_{-}, \underbrace{\varepsilon}_{-} \Big|_{p_0^H} \end{cases} = \frac{1}{A} \ln\left[\frac{|p_0 - p^*|}{\varepsilon}\right] \tag{9.9}$$

where the comparative static effects for either initial condition are given by the signs lying below parameters. Regardless of initial condition, the transition time depends simply on the convergence accuracy (as ε approaches zero τ_s rises) the magnitude of difference between initial conditions and the steady state (the greater the difference ($p_0 - p^*$) the longer are transition times). In contrast the effects of the speed of the invasion process A and the growth rates of marginal damages g and marginal costs c depend on the initial conditions and their implications on (9.6) and (9.7). If an invasion is caught early and the system starts with only a slightly invaded state such that $p_0 = p_0^L < p^*$, changes in the growth rate in marginal damages

affect transition times indirectly through p^* such that an increase in the growth rate reduces p^* which reduces the time of transition. The opposite occurs for an increase in the growth rate of marginal costs which increases p^* and leads to a longer transition time.

The influence on transition times from the speed of the invasion is two-fold. There is a direct, negative effect from an increase in speed but there is also an indirect, positive effect from the affect of the speed of the spread on p^*. The overall affect depends on relative magnitudes of the counteracting effects.

The comparative statics for an invasion caught in its later stages and the system starting with a highly invaded state ($p_0 = p_0^H > p^*$) are somewhat different. In this case, as an increase in the growth rate of marginal damages reduces p^* which increases the divergence between p_0 and p^* and leads to an increase in the time of transition. The opposite occurs from an increase in the growth rate of marginal costs which increases p^* and leads to a reduction in transition time. Increases in the speed of invasion for an initially highly invaded system reduce times of transition as they lead to an increase in p^* and a reduction in the divergence between p_0 and p^*.

For any given initial condition p_0 and second best policy rule h^* the present value of total costs are a sum of present values of transition costs and steady state costs

$$J_s^* = \underbrace{\int_0^{\tau_s} e^{-rt} C(p_s(t), h^*) dt}_{T_s^* \,=\, \text{transition costs}} + \underbrace{\int_{\tau_s}^{\infty} e^{-rt} C(p^*, h^*) dt}_{SS_s^* \,=\, \text{steady state costs,}} \tag{9.10}$$

Which, using the derivations, can be written as

$$J_s^* = \underbrace{\int_0^{\tau_s} \frac{e^{-rt}}{2} [g(p^* + (p_0 - p^*)e^{-At})^2 + c(h^*)^2] dt}_{T_s^* \,=\, \text{transition costs}}$$

$$+ \underbrace{\frac{g A^2 e^{-r\tau_s}}{2r \left(\frac{g}{c} + A^2\right)}}_{SS_s^* \,=\, \text{steady state costs.}} \tag{9.11}$$

Both transition costs and steady state costs depend on all the fundamental parameters and τ_s which in turn depends on all parameters apart from r.

The dependencies are complicated and analytically ambiguous but can be evaluated numerically. Using a hypothetical benchmark dataset of $A = 1, r = 0.03, g = c = 1, \varepsilon = 0.001$, and $p_0^L = 0.001$ or $p_0^H = 0.999$ each parameter, all else constant, was varied from low to high values to observe the dependencies of τ_s, T_s^*, SS_s^* and J_s^* across each parameter. Results are summarized in Table 9.1 and those for J_s^* and T_s^* displayed in Figs 9.1 and 9.2.

If an invasion is detected early (i.e. $p_0 = p_0^L$) invasions that spread quicker are likely to have higher total costs and steady state costs, while the affect on the transition time and costs of transition depends on the magnitude of the speed of the spread. If the speed of the invasion is slow, any increase in that speed results in lower transition times but higher transition costs (as shown in Fig. 9.2). Transition costs rise in this case because the lower transition

Table 9.1 Sensitivity of second best rule variables to changes in critical parameters

Δ Parameter	Δ Variable			
	τ_s	T_s^*	SS_s^*	J_s^*
$A\vert_{p_0^L}$	±	±	+	+
$A\vert_{p_0^H}$	−	±	+	±
$r\vert_{p_0^L}$	0	−	−	−
$r\vert_{p_0^H}$	0	−	−	−
$g\vert_{p_0^L}$	−	±	+	+
$g\vert_{p_0^H}$	+	+	+	+
$c\vert_{p_0^L}$	+	+	+	+
$c\vert_{p_0^H}$	−	±	+	+
p_0^L	−	−	+	+
p_0^H	+	+	−	+

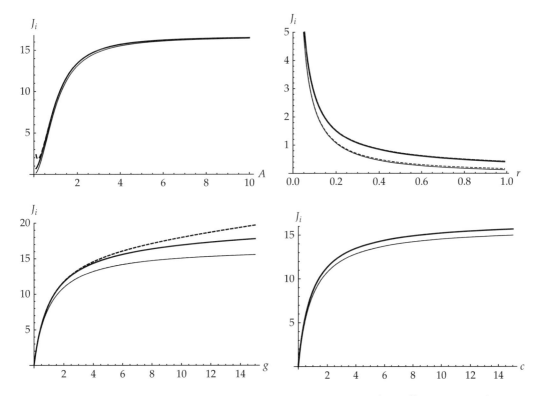

Figure 9.1 Dependencies of total costs J_i, $i = \{s, f\}$ on A, r, g, and c for initial conditions of p_0^L and p_0^H. In each panel $J_s^*\vert_{p_0^L}$ are given by light dashed lines, $J_s^*\vert_{p_0^H}$ by heavy dashed lines, $J_f^{**}\vert_{p_0^L}$ by light solid lines and $J_f^{**}\vert_{p_0^H}$ by heavy solid lines.

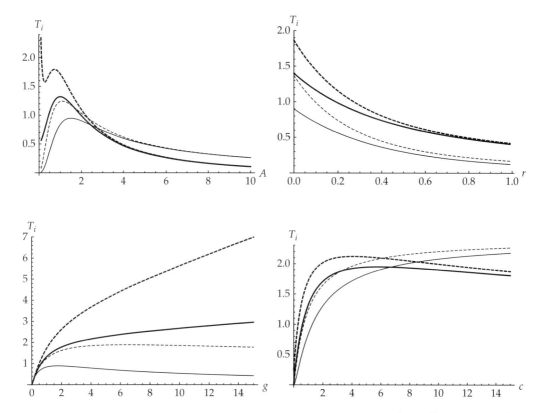

Figure 9.2 Dependencies of transition costs T_t, $i = \{s, f\}$ on A, r, g, and c for initial conditions of p_0^L and p_0^H. In each panel $T_s^*|_{p_0^L}$ are given by light dashed lines, $T_s^*|_{p_0^H}$ by heavy dashed lines, $T_f^*|_{p_0^L}$ by light solid lines and $T_f^{**}|_{p_0^H}$ by heavy solid lines.

times are accompanied by increased use of control (h^*) and eventually higher steady state proportions invaded (p^*) both with higher costs. But, if the speed of invasion is above a certain amount (corresponding to $A = 1$ in Fig. 9.2) both transition times and costs fall with any increase in speed. For any increases in speed above this level the use of control and its costs decline (over the shorter transition time) by more than the increases in instantaneous damages such that transition costs fall.

The findings are somewhat altered if an invasion is detected in its latter stages (i.e. $p_0 = p_0^H$). In this case, invasions that spread quicker are likely to have lower transition times and higher steady state costs but where the effect on costs of transition and total costs depends on the magnitude of speed. At the very lowest speeds, transition costs decline with the decline in transition times but then increase over a small range of low speeds (shown in Fig. 9.2). In this range the effect on h^* (>0) 5 dominates the reduced time of transition. However, over most of the range in speed transition, costs decline as the speed of the spread increases. This is due to p^* rising and transition times declining given a smaller divergence between the initial condition and steady state. Total costs follow the decline in transition costs only for increases in speed at the very lowest speeds of spread, but over the rest of the range total costs rise as the speed increases following the effect on steady state costs (shown in Fig. 9.1).

Higher discount rates necessarily lower the present values of all costs, regardless of initial conditions. Higher rates of growth in marginal damages and marginal costs increase each discounted cost measure apart from transition costs (shown in Fig. 9.2). If the initial conditions are an invasion in its early stages ($p_0 = p_0^L$) transition costs rise for increases in low growth rates of marginal damages but decline for increases in higher growth rates of marginal damages. Transition costs always increase for any rise in the rate of growth of marginal damage with initial conditions of an invasion in its later stages ($p_0 = p_0^H$). The opposite occurs for changes in the rate of growth in marginal costs and initial conditions. For changes in the rate of growth in marginal costs, if the invasion is detected early they affect transition costs positively. But, if the invasion is detected late in its progression and the rate of growth of marginal costs is low, any increase in this rate brings an increase in transition costs. When rates of growth in marginal costs are high, any increase in the rate causes a decrease in transition costs.

Most of these dependencies conform to intuition and the comparative statics given in (9.6) and (9.7). The reasons we observe a switching in the effects on transition costs for changes in the rate of marginal damages (with $p_0 = p_0^L$) and for changes in the growth rate of marginal costs (with $p_0 = p_0^H$) is that at high levels of either parameter the effects on steady state control and transition times dominate. For an invasion detected in its early stages, at low initial levels in the growth rate of marginal damages, any increase causes transition costs to rise as the costs from increased levels of control (9.6) outweigh the shorter transition time of their employment (9.9). However, at higher rates of growth in marginal damages, any further increase in the rate of growth will actually lower transition costs because the reductions in transition time overwhelm the increased control rates (i.e. although a higher rate of control is applied, it is applied over a shorter interval).

In a similar fashion, with an invasion detected in its later stages, transition costs only rise for an increase in rate of growth of marginal costs at lower initial levels; at higher initial levels, for any increase in this rate, transition costs decline. The decline is due in this case to transition times (9.9) falling with rising rates of growth in marginal costs (as the steady state is "pushed" towards the high initial condition) in addition to lower control expenditures.

For variations in the alternative initial conditions, if the initial proportion invaded is less than the steady state and this initial value were to be higher (i.e. the invasion is detected later) then the costs of transition might be lower (as the time of transition is lower) but steady state costs and total costs rise (as there is less time with a lower proportion of regions invaded). If the invasion is detected with an initial value greater than the steady state and this proportion were to be higher, transition costs would rise (given longer times of transition) and discounted steady state costs would fall, but total costs would rise as it would cost more to bring the system to steady state.

9.3.2 First best rule

The solution to this problem is a basic application of dynamic optimization to find the optimal path to the terminal steady state. The optimal controls on this path minimize the total discounted cost of bringing the system to the steady state J_f

$$J_f = \int_0^\infty e^{-rt} C(p(t), h(t)) dt \qquad (9.12)$$

for any p_0. Optimal controls for minimizing J_f can be obtained by the maximum principle (Kamien and Schwartz 1991). The procedure follows from writing the current value Hamiltonian (omitting time notation)

$$\mathcal{H} = -\frac{1}{2}[gp^2 + ch^2] + \mu[A(1-p) - h], \qquad (9.13)$$

where μ is the costate variable (shadow price of the spreading invader). As costs are at their lowest when $p = 0$, $\mu \leq 0$. μ should satisfy

$$\frac{d\mu}{dt} = r\mu - \frac{\partial \mathcal{H}}{\partial p} = \mu(r + A) + gp, \qquad (9.14)$$

and the evolution of the state given by (9.2). For the optimal h, $\mathcal{H}$ has to reach its maximum for each t, hence

$$\frac{\partial \mathcal{H}}{\partial h} = -ch - \mu = 0, \qquad (9.15)$$

it follows that $\mu = -ch$.[3] Time differentiation of this and substituting in (9.14) we obtain

$$\frac{dh}{dt} = (A+r)h - \frac{g}{c}p, \quad (9.16)$$

Equations (9.2) and (9.16) form a linear system with the only steady state

$$h^{**}\Big(\underbrace{g}_{+}, \underbrace{c}_{-}, \underbrace{A}_{\pm}, \underbrace{r}_{-}\Big) = \frac{gA}{g+cA(A+r)},$$

$$p^{**}\Big(\underbrace{g}_{-}, \underbrace{c}_{+}, \underbrace{A}_{+}, \underbrace{r}_{+}\Big) = \frac{cA(A+r)}{g+cA(A+r)}, \quad (9.17)$$

and instantaneous steady state costs

$$C^{**}(h^{**}, p^{**} = C^{**}\Big(\underbrace{g}_{+}, \underbrace{c}_{+}, \underbrace{A}_{+}, \underbrace{r}_{+}\Big)$$

$$= \frac{gA^2\left[\frac{g}{c} + (A+r)^2\right]}{2\left[\frac{g}{c} + A(A+r)\right]^2}. \quad (9.18)$$

Note the equivalency between the steady state results for the first and second best rules if $r \to 0$.

Comparative statics of variable on parameters are again given by the signs lying below parameters. The intuition of each follows that described for the second best rule, apart from the influence of r on all variables. Higher discount rates result in lower steady state levels of control, higher steady state proportion of regions invaded, and higher steady state instantaneous costs.

To determine the costs of transition to the steady state we need to find the optimal time paths of $p_f(t)$ and $h_f(t)$ and then the convergence time τ_f. The solution that converges to the steady state (9.17) and (9.18) for $t \to \infty$ is

$$p_f(t) = p^{**} + (p_0 - p^{**})e^{-\lambda t},$$

$$h_f(t) = h^{**} + (\lambda - A)(p_0 - p^{**})e^{-\lambda t}, \quad (9.19)$$

where

$$\lambda = \sqrt{A(A+r) + \frac{g}{c} + \left(\frac{r}{2}\right)^2} - \frac{r}{2}, \quad (9.20)$$

and $\lambda > A$.

---
[3] Sufficiency conditions require the current value Hamiltonian to be jointly concave in the control and state variables: $\mathcal{H}_{hh} < 0$, $\mathcal{H}_{pp} < 0$ and $\mathcal{H}_{hh}\mathcal{H}_{pp} - \mathcal{H}_{hp}^2 > 0$. These conditions are satisfied with the specification of the model.

To characterize the convergence time τ_f we again set a convergence accuracy ε and define that $p_f(t)$ has converged to p^{**} as soon as $|p_f(t) - p^{**}| < \varepsilon$. Assuming that $|p_f(t) - p^{**}| > \varepsilon$ and again considering the two alternative initial conditions the convergence time is

$$\tau_f \begin{cases} \Big(\underbrace{g}_{-}, \underbrace{c}_{+}, \underbrace{A}_{\pm}, \underbrace{\varepsilon}_{-}, \underbrace{r}_{+}\Big)\Big|_{p_0^L} \\ \Big(\underbrace{g}_{\pm}, \underbrace{c}_{\pm}, \underbrace{A}_{-}, \underbrace{\varepsilon}_{-}, \underbrace{r}_{+}\Big)\Big|_{p_0^H} \end{cases}$$

$$= \frac{1}{\lambda}ln\left[\frac{|p_0 - p^{**}|}{\varepsilon}\right], \quad (9.21)$$

where λ is given by (9.20). The time to convergence under the first best rule again depends on whether the initial conditions are low or high and on g, c, A, and ε as with the second best rule (although in a more complex fashion through λ and p^{**}) but also on r. Given the complexity of the comparative statics the signs given below each parameter other than ε were evaluated using the numeric example. There are a few differences with the dependencies in comparison with the second best rule. Most obviously is that higher discount rates lengthen the time to convergence. This occurs for both initial conditions because higher discount rates result in higher steady state proportions invaded and smaller steady state control rates. This rather obviously results in a longer time of transition for an initially slightly invaded state because of a greater divergence between initial conditions and the steady state. With an initially highly invaded state the divergence between the initial condition and steady state is narrowed with a higher discount rate and the reduced employment of control brings a longer transition.

Also in contrast to the second best rule, given a highly invaded initial condition, there is a switching in the dependencies on growth rates in marginal damages and marginal costs. When the growth rate in marginal damages is low and it rises, transition times rise because the divergence between the initial conditions and the steady state increases (following (9.17) and (9.18)) by more than any increase in control. However, when the growth rate in marginal damages is high and if it were to be increased then the dynamically optimal control rates would rise and bring the system to steady state quicker, even though the steady state is further from the initial condition. This is because increases in the growth of marginal damages make control costs relatively inexpensive. In a somewhat similar fashion, when the growth rate in marginal costs is low and rises, transition times rise in spite of the divergence between the initial conditions and the steady state falling. Here control is reduced by such an extent to lengthen the time of transition. But, for high growth rates in marginal costs, the divergence between the initial conditions and steady state is so small that the effect from the reduced use of control is overwhelmed and transition times fall.

For any given initial condition p_0 and first best policy rule $h_f(t)$ the present value of total costs are a sum of transition costs and steady state costs

$$J_f^{**} = \underbrace{\int_0^{\tau_f} e^{-rt} C(p_f(t), h_f(t)) dt}_{T_f^{**} \,=\, \text{transition costs}}$$

$$+ \underbrace{\int_{\tau_f}^{\infty} e^{-rt} C(p^{**}, h^{**}) dt}_{SS_f^{**} \,=\, \text{steady state costs}}, \quad (9.22)$$

which using the derivations can be written as

$$J_f^{**} = \underbrace{\int_0^{\tau_f} \frac{1}{2} e^{-rt} \left[g(p^{**} + (p_0 - p^{**})) e^{-rt} + c(h^{**} + (\lambda - A)(p_0 - p^{**})) e^{-\lambda t 2} \right] dt}_{T_f^{**} \,=\, \text{transition costs}} + \underbrace{\frac{g A^2 \left[g/c + A + r^2 \right] e^{-r\tau_f}}{2r \left[g/c + A + r^2 \right]}}_{SS_f^{**} \,=\, \text{steady state costs}},$$

(9.23)

Each cost measure again depends on all the fundamental parameters and τ_f which in turn depends on all parameters. As the dependencies again are complicated they are evaluated numerically with results summarized in Table 9.2 and illustrated for I_f^{**} and T_f^{**} in Figs 9.1 and 9.2. Most of the dependencies conform to those found under the second best rule, although consequences on transition times differ as explained in relation to equation (9.21). The magnitudes of the dependencies are somewhat different following the consequences on times of transition and result in changes in the differences of the two decision rules. Differences in the two rules provide the most meaningful contrasts and are discussed in the following comparisons.

Table 9.2 Sensitivity of first best rule variables to changes in critical parameters

Δ Parameter	Δ Variable			
	τ_f	T_f^{**}	SS_f^{**}	J_f^{**}
$A\|_{p_0^L}$	±	±	+	+
$A\|_{p_0^H}$	−	±	+	+
$r\|_{p_0^L}$	+	−	−	−
$r\|_{p_0^H}$	+	−	−	−
$g\|_{p_0^L}$	−	±	+	+
$g\|_{p_0^H}$	±	+	+	+
$c\|_{p_0^L}$	+	+	+	+
$c\|_{p_0^H}$	±	±	+	+
p_0^L	−	−	+	+
p_0^H	+	+	−	+

9.3.3 Comparison of the two rules

Comparisons of the two rules can be split into three: comparisons of the costs during transition, comparisons of costs at the steady state, and comparisons of total costs. Figure 9.3 illustrates the steady states where the interest rate in the dynamic rule is non-zero.

For any situation where $r > 0$ the instantaneous steady state costs of the first best rule are greater than those of the second best rule $C^{**} > C^*$ as the first best rule takes into account the costs of getting the system to the steady state so that $p^{**} > p^*$ and $h^{**} < h^*$. As we have noted, however, to evaluate the magnitudes of costs at the steady states

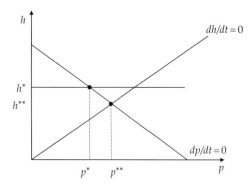

Figure 9.3 Steady States for the second best (p^*, h^*) and first best rules (p^{**}, h^{**}). If $r \to 0$ then the steady states coincide and the control isocline ($\frac{dh}{dt} = 0$) would pivot up and pass through (p^*, h^*).

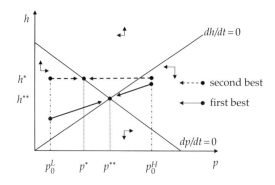

Figure 9.4 Transition paths for both rules from a low initial condition (p_0^L) or high initial condition (p_0^H). Both policies move the system to their respective steady state although the second best rule remains constant the first best rule varies with p.

the transition times must be taken into account, making direct comparisons difficult. For illustrative purposes, consider the two rules if there are no times of transition (i.e. both systems have initial conditions at their respective steady states such that $p_s(0) = p^*$ and $p_f(0) = p^{**}$. In this case the total discounted costs of the first best rule exceed those of the second best rule $I_f^{**} > J_s^*$ or $r^{-1}C^{**}(p^{**}, h^{**}) > r^{-1}C^*(p^*, h^*)$ (remembering that the comparison is over two different steady states). This result is due to the first best rule accounting for the time value of money in its optimization while the second best rule does not.

Of course this is not a realistic comparison when considering the relative performance of the two rules, which requires at least uniform initial conditions and the inclusion of the costs of transition. Figure 9.4 illustrates the phase space for transition paths for the two rules for two initial conditions, a low level of initial invasion p_0^L and a high level of initial invasion p_0^H.

The figure demonstrates that in this case, either policy will result in directing the system to their respective steady states, although the control in the second best case is invariant with the state of the invasion while control in the first best case varies. What the figure critically omits are the transition times of both policies, given analytically by (9.9) and (9.21). It is obvious for case of $r \to 0$ when the steady states coincide, the dynamic rule results in a faster transition

$$\tau_s - \tau_f = \ln\left[\frac{|p_0 - p^*|}{\varepsilon}\right]\left(\frac{\sqrt{A^2 + \frac{g}{c}} - A}{A\sqrt{A^2 + \frac{g}{c}}}\right) > 0, \quad (9.24)$$

and so lower costs of transition, and lower total costs of management (as steady state costs are the same across the two policies).

When $r > 0$ the implications are not so clear due to the divergence in steady states and the implications on dynamically optimal choices of the future being discounted. To provide some insight using the derivations generated, the numerical examples were again employed to allow a basic sensitivity analysis (or comparative static analysis) on the outcomes to variations in the fundamental parameters. Figure 9.1 depicts the dependencies of total costs under both rules across the fundamental parameters for low or high initial invasions. While the directions of changes were described with reference to Tables 9.1 and 9.2, what Fig. 9.1 makes clear are the differences in total costs between the two rules. As would be expected, total costs under the first best rule are less than those under the second best rule, but the differences are small across most parameters.

However, understanding what drives these total cost results requires looking at the results for each of the critical variables across both sets of initial conditions. To do this and maintain a decent degree of generality we focus a discussion of the results on absolute differences in times of transition ($|\tau_s - \tau_f|$) transition costs ($|T_s^* - T_f^{**}|$) steady state costs ($|SS_s^* - SS_f^{**}|$) and total costs ($|J_s^* - J_f^{**}|$) between the second and first best rules for the low (p_0^L) and high (p_0^H) initial conditions. The differences are affected by variations across the parameters and initial conditions as shown in Figs 9.1 and 9.2 and summarized in Table 9.3, where signs indicate the changes in absolute differences as each parameter is varied. Differences in total costs ($|J_s^* - J_f^{**}|$) are shown in Fig. 9.5.

The magnitude of the difference depends on the transition time. If there are large differences in times of transition there are also large differences in transition costs and steady state costs (with longer transition times less time is spent at the steady state). Overall, transition times under the second best rule are typically longer than under the first best rule. The differences in transition times between the two rules tend to be greatest with slowly spreading invasions, low discount rates, high rates of growth in marginal damages, and low rates of growth in marginal costs. For invasions that are detected

Table 9.3 Sensitivity of (absolute) differences of rule to changes in parameters

	Δ Variable			
Δ Parameter	$\|\tau_s - \tau_f\|$	$\|T_s^* - T_f^{**}\|$	$\|SS_s^* - SS_f^{**}\|$	$\|J_s^* - J_f^{**}\|$
$A\|_{p_0^L}$	±	±	±	±
$A\|_{p_0^H}$	−	−	±	−
$r\|_{p_0^L}$	−	−	−	+
$r\|_{p_0^H}$	−	−	−	±
$g\|_{p_0^L}$	±	±	+	±
$g\|_{p_0^H}$	+	+	+	+
$c\|_{p_0^L}$	±	±	±	±
$c\|_{p_0^H}$	−	±	±	−
p_0^L	−	−	−	−
p_0^H	+	+	+	+

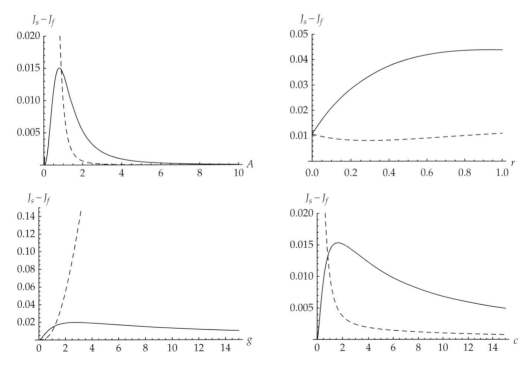

Figure 9.5 Dependencies of differences in total costs ($|J_s^* - J_f^{**}|$) on A, r, g, and c for initial conditions of p_0^L and p_0^H. In each panel $|J_s^* - J_f^{**}|\|_{p_0^L}$ are given by dashed lines, $|J_s^* - J_f^{**}|\|_{p_0^H}$ by solid lines.

early, there are maxima in the differences in transition times with very slowly spreading invasions, moderate rates of growth in marginal damages, and very low rates of growth in marginal costs. These peak differences in transition times are driven by differential effects of parameters on second best and first best steady states and first best dynamic controls. To illustrate, consider the influence of the speed of invasion on the steady states of the two policy rules

$$\frac{\partial p^*}{\partial A} = \frac{2cgA}{(g+cA^2)^2}, \quad \frac{\partial p^{**}}{\partial A} = \frac{cg(2A+r)}{(g+cA(A+r))^2}, \tag{9.25}$$

$$\frac{\partial h^*}{\partial A} = \frac{g(g-cA^2)}{(g+cA^2)^2}, \quad \frac{\partial h^{**}}{\partial A} = \frac{g(g-cA^2)}{(g+cA(A+r))^2}, \tag{9.26}$$

The magnitude of the effect is the same if the discount rate is zero, but for all else the magnitude of the influence on first best steady is less. This results in smaller changes in steady state proportions invaded, and smaller changes in steady state control effort. In addition, for very slowly spreading invasions (small A) the effects on steady state controls is positive, negative for faster spreading invasions (large A). The reason there is a switching in the effect is that when the invasion spreads very slowly, it is cheaper to control the invasion than accept the damages. But if the invasion spreads very quickly, then control costs rapidly escalate and it is cheaper to accept some more damages and control less. So, if the invasion were to spread quicker, we would always see higher steady state proportions invaded, but steady state controls would only also be higher for invasions that initially spread slowly, else they will be less (where all changes would be attenuated in the first best outcomes). In terms of transition times, considering the effects for invasions that are detected early, increases in the speed of the spread increase the distance between the initial condition and steady state proportion invaded (putting an upward force on transition time). But, at low initial speeds of spread, control rates also increase. This puts also puts an upward pressure on transition times as higher control will result in a slower transition. At high speeds of spread, control rates decrease with any increase in speed. This puts a downward pressure on transition times, at odds with the longer distance between initial conditions and the steady state. As the effects are exacerbated for the second best policy, this means that at low initial speeds of spread the increased transition times are greater than the increased transition times of the first best policy, while at high initial speeds of spread the reduction in transition time is greater than the reduction in transition time for the first best policy (bringing the two policies closer together).

For invasions that are detected in their latter stages, the difference in transition times between the two rules is at its greatest at low levels of each parameter other than the rate of growth in marginal damages, where the difference is greatest at high levels. The magnitude of the differences all decrease with an increase in each parameter, other than the rate of growth in marginal damages (which increases). Consider again the effect from changes in the speed of invasion on the (resulting) steady states (9.25 and 9.26). Under these initial conditions, increases in the speed of invasion cause the steady state proportion to move close to the initial condition. Although the steady state control may increase or decrease depending on the magnitude of the spread, the time of the transition always declines with an increase in speed of spread. With highly invaded initial conditions, faster spreading invaders approach the steady state faster (the system has to be "moved" less and any increase in control rates are overwhelmed by this and the increased speed). Again, the effects are more pronounced on the second best policy. This makes the difference between the two transition times (that is large at low speeds of spread) rapidly converge if the speed of the spread were to increase.

Differences in the costs of transition between the rules largely follow those of transition times apart from there being peaks in the differences at low rate of growth in marginal costs for both initial conditions. These peaks follow from Fig. 9.2 where the greatest difference between the two rules is found with low rates of growth in marginal costs. At these low rates of growth in marginal costs, for invasions that are caught early, second best transition times grow faster with any increase in the growth rate of marginal costs than first best transition times. For invasions that are caught later, second best transition times always fall with rising growth rates in marginal costs while first best transition times initially rise (at lower growth rates in marginal costs) then decline to converge with second best transition times. Again these are due to the differential effects of the growth rate in marginal costs on equations (9.9) and (9.21) through the parameter's influence on steady states. The growth rate in marginal costs has a smaller marginal effect on first best steady state proportions invaded than second best (with non-zero discount rates). Thus for any increase in the growth rate of marginal costs, p^* increases by more than the increase in p^{**} (which has differential effects on transition time depending on initial conditions). But, as both p^* and p^{**} are limited by a maximum of one, any increases arising from increased rates of growth in marginal costs must decline the closer p^* and p^{**} get to one. So at low growth rates in marginal costs, changes in this parameter have large effects on both transition times although by a relatively greater extent on second best transition times. At high rates, the effects on both transition times are small and the relative differences small as well.

The effects on differences in steady state costs are the (negative) mirror image of those on transition costs. Steady state costs under the second best rule are less than those under the first best rule although the difference between the two depends upon the magnitude of each parameter. Differences in steady state costs are greatest with slowly spreading invasions, low discount rates, high rates of growth in marginal damages, and low rates of growth in marginal costs. There are again single peaks in the differences in steady state costs, found with slowly spreading invasions and low rates of growth in marginal costs (under both initial conditions).

The net of differences in transition costs and steady state costs provides differences in total costs, shown in Fig. 9.5. As the two work in opposite directions, total differences are small, yet positive, implying that differences in transition costs are (slightly) greater than the differences in steady state costs. However, differences in total costs under the two rules depend quite specifically on initial conditions. If an invasion is detected in its later stages, the largest differences in total costs between the two rules are found with slowly spreading invasions, high discount rates, high rates of growth in marginal damages, and low rates of growth in marginal costs. The only dependency in differences that does not grow or decline monotonically for this initial condition is that from changes in the discount rate.

If the discount rate is low and was to be increased, the difference in total costs declines. The opposite occurs at higher discount rates where further increases result in increased differences in total costs. The switching occurs because increases in the discount rate have no effect on second best transition times or steady states (although their present value falls with an increase in the discount rate) yet increase first best transition times, steady state proportion invaded, and reduce steady state control, equation (9.17). These first best effects attenuate the reductions in the present value calculation due to the increased discount rate. In comparison, the second best transition costs fall faster than do the first best transition costs, making their differences narrow over a range of low discount rates. But, as the discount rate rises higher, total costs of the two rules again diverge. This is due to the differences in steady state costs approaching zero at higher discount rates. Thus, although the differences in transition costs have narrowed, the negative influence of differences in steady states costs tends to zero, making differences in total costs follow those of transition costs.

In contrast, if an invasion is detected early on in its progression the only dependency of differences in total costs that is not single peaked is that from the discount rate (which is increasing). All other dependencies in differences for this initial condition have a peak at low values of each parameter with the difference getting smaller at high values.

Changes in initial conditions impact the differences differentially depending on whether the variation in initial condition is for an invasion caught early (below steady states, p_0^L) or late (above steady states, p_0^H). For an invasion detected in its early stages, if the initial proportion of regions invaded were to be greater all differences decline due to the initial condition approaching the steady state. The opposite occurs for an invasion detected in its latter stages and all differences grow due to the initial condition getting further and further away from the steady state.

9.4 Discussion

From the comparisons of the two policies some insight into the debate considered can be generated. The results demonstrate that when a discount rate of zero is employed, the terminal steady states will be identical and only the transition paths will differ across the second and first best policies. In this case, the first best rule always outperforms the second best rule as steady state costs are the same and transition costs are always minimized.

However, a nonzero discount rate is probably more policy relevant. Although the second best rule will always tend to result in longer transition times, higher transition costs, and higher total costs, it can have lower steady state costs. The magnitudes of differences in total costs depend on initial conditions and the magnitude of each parameter, but tend to be small.

But one has to be careful when interpreting these results as total costs are present value calculations over infinite horizons (as typically calculated). Depending on the interest rate, the calculations may be dominated by steady state costs, as clearly seen in Tables 9.1, 9.2, and 9.3. It may be likely that transitions dominate policy horizons due to slowly moving transitions, terminal steady states never being attained, regimes constantly changing, or political myopia. In these cases, inspecting the effects on differences in costs of transition might provide more policy insight than a focus on total costs. The second best rule performs best in terms of transition costs (in relation to the first best rule) when there are fast spreading invasions, low rates of growth in marginal damages, and high rates of growth in marginal costs. The second best rule also performs relatively better if invasions are detected later(earlier) if the initial invaded proportion is below(above) the steady state. The flip side is that the second best rule performs worse (differences in transition costs increase) in the opposite of circumstances, i.e. when there are slowly spreading invasions, high rates of growth in marginal damages, and low rates of growth in marginal costs. In these cases it is likely to be important to consider dynamically optimal policies.

These are largely the same conditions as found with total costs, but the magnitude in the differences are larger for transition costs. This has the implication that what might seem a close approximation in terms of total costs is not so close in terms of transition costs. The debate boils down to the relative importance of transitionary periods and their costs. If the transitions are minor to the policy maker, the second best rule tends to work well, the point being to get a policy up and running quickly. But if periods of transition have a high degree of policy importance, there are circumstances when dynamically optimal policies will likely be worth the effort they require.

9.5 Conclusions

Given the rapid and often shocking consequences of the spread of invasive species there have been calls for the application of rules of thumb in their management. This contrasts significantly with arguments that in these types of situations, effort intensive characterizations of optimal rules of management are likely to be worthwhile. To provide some insight into this debate facing policy makers, we weighed the trade-offs and characterized critical parameters of applying simple decision rules in contrast to dynamically optimal decision rules for a simple example of the management of invasive species.

We focused on a specific component within the debate and posed the question of whether or not managers can use second best policies that do not vary over time nor as conditions change and not significantly waste resources in comparison to first best programs of dynamically optimal management that might vary tremendously over time or as conditions change? In the example considered, we find that over an infinite time horizon, second best policies always waste some resources in comparison to first best policies but the magnitude of the waste tends to be small. In situations with fast spreading invasions, low rates of growth in marginal damages, and high rates of growth in marginal costs, second best policies will waste less than in comparison to dynamically optimal policies. In the opposite of circumstances, with slowly spreading invasions, high rates of growth in marginal damages, and low rates of growth in marginal costs, second best policies will waste more in total than in comparison to dynamically optimal policies. Furthermore, in these circumstances the differences in transitions costs and steady state costs of the two policy rules is greatest, although in the opposite direction, making the difference in total costs small. In these circumstances the relative importance of periods of transition in comparison to steady states will be of utmost importance to assessing whether a first best rule is worth the effort. The performance of the policy rule thus critically depends on the specifics of the invasion, making any policies designed in a "one size fits all" format likely to result in wasted resources. Of course, even in this most simple of situations, there are five parameters that must be known: the initial conditions, speed of invasion, relevant discount rate, rates of change in marginal damages, and marginal costs. While policy makers may know one or a few of these parameters, it is likely that there will remain significant unknowns over these parameters making research in these areas critical. For example, at this point, perhaps what might be most well known are spread rates of known invaders. With regards this parameter, and assuming all else constant, for a quickly spreading species, like water hyacinth, our model would imply that the extra costs incurred from applying a second best rule will not be excessive and will manage the system well enough. But, for a slowly spreading species, such as gypsy moth, all else being constant it will likely be worth the effort to dynamically optimize the system.

An important caveat to our findings relates to the assumptions employed in the analytical example. While the linear dynamics provided an example that provided (largely) analytical solutions, it abstracts significantly from reality in that it allows for single steady states for each rule. In reality one can likely expect highly nonlinear systems with multiple steady states. In these situations second best policies are likely to perform worse than found here. For example, if one were to employ the standard Levins model of metapopulation dynamics (Levins 1969) extended to include background

propagule pressure, then there are situations when a constant level of control will have three potential steady states, one slightly invaded, one highly invaded, and an intermediate value. In this situation if the invasion is detected at any point after the slightly invaded steady state, and the corresponding constant level of control is applied as a second best policy rule, the resulting steady state will be heavily invaded with correspondingly high damages. The point is that with nonlinearities the performance of simple rules is likely to deteriorate unless invasions are caught very early in their spread, making the effort to characterize optimal policies likely worthwhile.

Acknowledgements

The authors would like to thank Charles Perrings and an anonymous reviewer for many helpful suggestions and ISIS group members for useful discussions. This research has been supported by a grant from the National Science Foundation (DEB 02-13698). ML also acknowledges support from a Canada Research Chair, and NSERC Discovery and Collaborative Research Opportunity grants.

References

Brock, W.A., and Xepapadeas, A. (2003). Valuing biodiversity from an economic perspective: a unified economic, ecological, and genetic approach. *The American Economic Review*, **93**(5), 1597–1614.

Carpenter, S.R., Brock, W., and Hanson, P. (1999). Ecological and social dynamics in simple models of ecosystem management. *Conservation Ecology*, **3**(2), 4 [online].

Haurie, A. (1976). Optimal control on an infinite time horizon. The turnpike approach. *Journal of Mathematical Economics*, **3**, 81–102.

Kamien, M.I., Schwartz, N.L. (1991). *Dynamic Optimization: The Ccalculus of Variations and Optimal Control in Economics and Management*. North-Holland, Amsterdam.

Leung, B., Lodge, D.M., Finnoff, D., F. Shogren, J., Lewis, M.A., and Lamberti, G. (2002). An ounce of prevention or a pound of cure: bioeconomic risk analysis of invasive species. *Proceedings: Biological Sciences (formerly Proceedings Royal Society London B)*, **269**, 2407–13.

Leung, B., Finnoff, D., Lodge, D.M., and Shogren, J.F. (2005). Managing invasive species: Rules of thumb for rapid assessment. *Ecological Economics*, **55**, 24–36.

Levins, R. (1969). Some demographic and genetic consequences of environmental heterogeneity for biological control. *Bulletin of the Entomological Society of America*, **15**, 237–40.

Ludwig, D., Carpenter, S., and Brock, W. (2003). Optimal phosphorus loading for a potentially euthrophic lake. *Ecological Applications*, **13**(4), 1135–52.

Olson, L.J., and Roy, S. (2002). The economics of controlling a stochastic biological invasion. *Amer. J. Agr. Econ.*, **84**(5), 1311–16.

Perrings, C. (1999). Comment on "Ecological and Social Dynamics in Simple Models of Ecosystem Management" by S.R. Carpenter, W.A Brock, and P. Hanson. *Conservation Ecology*, **3**(2), 10[online].

Pimentel, D., Lach, L., Zuniga, R., and Morrison, D. (1999). Environmental and economic costs of nonindigenous species in the United States. *Bioscience*, **50**, 53–65.

Potapov, A.B., Lewis, M.A. and Finnoff, D.C. (2007). Optimal control of biological invasions in lake networks. *Natural Resource Modeling*, **20**(3), 351–379.

Seierstad, A., and Sydsaeter, K. (1987). *Optimal Control Theory with Economic Applications*, North-Holland, Amsterdam.

Sethi, S.P., and Thompson, G.L. (2000). *Optimal Control Theory: Applications to Management Science and Economics*, 2nd edition. Kluwer Academic, Norwell, MA.

Shogren, J.F. (2000). Risk reductions strategies against the explosive invader. In C. Perrings, M. Williamson, S. Dalmazzone, ed. *The Economics of Biological Invasion*. Edward Elgar, Northhampton, MA, pp. 56–69.

Wu, J. (2001). Optimal weed control under static and dynamic decision rules. *Agricultural Economics*, **25**(1), 119–30.

Zavaleta, E. (2000). The economic value of controlling an invasive shrub. *Ambio*, **29**(8), 462–67.

CHAPTER 10

Optimal Random Exploration for Trade-related Nonindigenous Species Risk

Michael Springborn, Christopher Costello, and Peyton Ferrier

10.1 Introduction

Driven by international trade liberalization and falling transportation costs, US food imports have grown at a mean annual rate of 4.5 per cent by volume since 1989 (US Department of Agriculture 2007). Against this backdrop of expanding trade, environmental safety concerns regarding tainted food imports have attracted attention in the wake of well-publicized incidents of contaminated pet food and toothpaste and discoveries of shipments infested with invasive pests and pathogens. Environmental threats in the form of non-indigenous species (NIS) such as citrus canker, bovine spongiform encephalopathy, and plum pox can have substantial consequences for trade and agricultural productivity. For example, citrus cuttings intercepted by Customs and Border Protection in 2005 carrying citrus canker could have caused between $173 and $890 million in damages if the disease became established in California (CBP 2005). In addition to the potential for NIS to cause both economic and ecological harm, they are costly to control once introduced. Annual spending on emergency eradication programs grew in the 1990s from $10.4 million to $232 million (Lynch and Lichtenberg 2006).

Despite a large scale effort to mitigate NIS introduction risk through border inspections, only a small percentage of goods are physically examined. Given notable breakdowns in imported food safety, the multiple US government agencies responsible for import safety have emphasized the need for improvement in the targeting of inspections based on risk (Interagency Working Group on Import Safety 2007). The need to support targeting inspections by gathering pertinent data is clear. The task, though, is complicated by imperfect information about the risks posed by particular sources.

Allocating limited border inspection resources for NIS risk resembles the type of natural resource problem for which adaptive management has been advocated. The targeting of inspections involves both learning by doing and decision-making repeated over time. Each inspection provides information about the probability that goods from a given source are infested. Over time we can learn about risk, either confirming our initial beliefs or correcting them when inaccurate. Such learning improves future targeting. For example, learning that a given source of trade is more risky than initially thought allows for redirecting future inspections from lower risk sources to focus on the greater risk. This type of learning mainly occurs through the intentional use of management actions (inspections) as experiments, in this case exploration through the diversion of some portion of risk-based inspections to shipments which would otherwise not have been examined to uncover surprises. This approach is the essence of adaptive management in the tradition of Holling (1978) and Walters and Hilborn (1978).

Each inspection provides expected benefits of two types: the immediate potential to intercept an infested shipment and information of value for improving future targeting. The myopic approach is to focus on exploiting current information for the best immediate performance. Adopting intentional exploration acknowledges future payoffs but comes with the immediate opportunity cost of forgone inspections of likely riskier sources. A central challenge of adaptive management, and of this chapter, is balancing this exploitation–exploration tradeoff. One approach, discussed in greater detail in Section 10.4, is to jointly estimate the immediate benefit and value of information to identify when exploration is worth the opportunity cost. While appealing in its directness, such valuation approaches are data-intensive and computationally demanding.[1]

A simpler strategy is to implement exploration by allocating a limited portion of inspections randomly. Under the US Agricultural Quarantine Inspection Monitoring (AQIM) program, a limited random sample of agricultural cargo (and other pathways) is comprehensively inspected for quarantine materials such as NIS. The need for a tractable exploratory program which is sensitive to the opportunity costs of redirecting risk-based inspections motivates our central research question: what factors influence the optimal intensity of an active adaptive management approach to inspections? Specifically, we aim to identify variables from the port inspection setting which influence the gains to exploration via random inspections.

We begin by describing a Bayesian learning model of trade-related NIS risk in Section 10.2 which captures uncertainty over the true probability that trade from a given source is infested and provides a framework for updating these beliefs as observations accrue. The formal inspection allocation decision problem is expressed in Section 10.3, where the computational demands of the central task are made clear. The analysis in Section 10.4 begins with the simplest possible nontrivial version of the problem. We subsequently add elements of real-world complexity to build intuition for the ultimate task of exploring random inspection policy in an empirically-based setting. Our workhorse method applies Monte Carlo simulation under various policies to characterize performance in terms of interceptions and to identify optimal choices for design and intensity of exploration.

The empirical analysis uses both monthly inspections and import volume data for the period 1996 through 2006. Our inspections dataset, provided by the Animal and Plant Health Inspection Service (APHIS) of the US Department of Agriculture (USDA), includes the outcomes of examinations of fruit and vegetable imports for exotic pests at US ports. Over seven million shipments were inspected across 144 ports during the time period, with approximately 0.8 per cent (62,000 shipments) found to be infested with a pest species of concern. Our trade dataset is comprised of monthly import volumes published by the US Department of Commerce (1996–2006). The merger of these two datasets is nontrivial because of significant differences in the way agricultural commodities are identified and grouped. We take advantage of a concordance, recently developed by the USDA, which maps commodity definitions in the inspections data to those in the trade data.[2]

Empirical features of the combined dataset motivate much of our focus on how particular system variables affect the optimal random inspection policy. The proportion of shipments inspected and found to be infested differs significantly for different commodities from different countries. We will think about a particular commodity from a particular country as one import "source" or pathway. As the number of different sources subject to a random inspection policy increases, we might expect the returns to such a program to dissipate

---

[1] Work is ongoing to extend such a method to the scale of the port inspections problem (see Springborn 2008).

[2] We gratefully acknowledge the work of the Economic Research Service at USDA in facilitating the merger of these two datasets, in particular the efforts of Donna Roberts and collaborators.

since more ultimately irrelevant (low probability of infestation) sources will garner inspections. If we are most uncertain about the source(s) perceived to be the riskiest, this uncertainty will simply be resolved through the myopic or "greedy" approach of targeting sources of high expected risk, making additional exploration unattractive. However, if uncertainty is higher for sources not targeted in a myopic approach, greedy behavior may well ignore potentially important alternatives. Thus, we examine the role of the particular distribution of uncertainty across sources.

If shipments from the riskiest source are simply not available in a particular period due to the types of seasonal variability observed in the trade record, there will be some degree of "forced" exploration of alternative sources of the next greatest perceived risk. We might expect this to reduce returns to additional random exploration. However, such availability constraints mean that the pool of alternative shipments, and therefore the set of choices, is frequently shifting. This could increase the scope of useful information and therefore the level of optimal random exploration. A further port-level constraint on allocating national inspections arises since inspection resources are not easily transferrable between entry points. Finally, since the value of learning depends on the potential to use the information in the future, the size of the inspection pool, reflecting current and future demand for targeting, is likely to influence the returns to exploration.

We find that the optimal policy is indeed sensitive to the number of import sources, the distribution of uncertainty amongst them and the size of the inspection budget. Constraints on targeting imposed by both shipping variation and port-level inspections budgets are also important. In settings where the optimal level of random inspection is positive, we present two cautionary results. The first points to the potential danger of just dabbling with (under-allocating to) random inspections. The second provides insight into forming realistic expectations over program rewards when the mean and median payoffs are of different sign. We conclude the paper with a discussion of extensions of the risk and inspection process modeling for potential future study.

10.2 A Bayesian learning model of trade-related NIS risk

To implement a risk-based inspection targeting approach with learning, we first specify a conceptual model of the shipment infestation process and a framework for updating the assessment of risk as observations accrue. We make the assumption that production, sanitation, and packaging techniques from a particular trade source j lead to the infestation of a fixed fraction, p_j, of shipments. Each source j represents a particular commodity from a particular country, such as Mexican avocados, Mexican tomatoes, or Australian tomatoes. The ability to further differentiate pathways, for example by producer within each country, while potentially useful is not possible with available data and not considered here. By assuming that p_j is fixed, we are abstracting away from the possibility that the true infestation probability evolves over time due to exogenous natural factors or explicit decisions like producer changes in phytosanitary practices.[3]

Whether or not a particular shipment is infested (regardless of whether or not it is inspected) is modeled as the outcome of a Bernoulli trial with parameter p_j. The decision maker has incomplete information about the true level of this probability measure for each source. We model subjective beliefs on this parameter p_j with a beta distribution. This flexible density is commonly used for similar processes, restricted to the interval [0,1] and easily updated in a Bayesian way as observations are made.

---

[3] This is a reasonable assumption if exporting parties are atomistic. However, if individual exporters are relatively large or inspections are targeted at the level of an exporting firm, behavioral responses could critically influence optimal exploration. Other sources of temporal variation in p_j might include seasonal changes in weather and pest population dynamics. Springborn (2008) observes that reported infestation rates of US agricultural imports can change significantly from month to month. The simple beta-Bernoulli model is extended using a hierarchical Bayesian approach to allow for such variation. While this model could be implemented in the current setting, we use the simpler model for parsimony and clarity of results.

A particular outcome from this Bernoulli infestation process is only observed when a shipment is inspected. For simplicity we assume that inspections perfectly identify infestations when present. Since truly exhaustive physical inspections are typically cost-prohibitive, this is an abstraction from reality.[4] Using the terminology of Bernoulli trials, a "success" corresponds to an infestation while a "failure" denotes a clean shipment. Let s_j^t and f_j^t represent the two "hyperparameters" of the beta distribution, describing beliefs over the probability p_j at time t: $p_j \sim \text{Beta}(s_j^t, f_j^t)$. The initial prior distribution for beliefs over p_j at time zero, here referred to as the "sampling prior", is specified by the pairing (s_j^0, f_j^0) and captures all available information on infestation likelihood at the time. We describe our specific methodology for establishing such priors in the empirical application in Section 10.4.2.3.

Bayesian updating of this subjective distribution with inspection outcomes takes place as follows. If a Bernoulli trial results in a success *and* is observed, then s_j is incremented by one: $s^{t+1} = s^t + 1$. An observed failure augments f_j in the same fashion. This updating short-cut is evident upon applying Bayes rule to calculate the posterior density function given the prior distribution and the observation made (see Gelman *et al.* 2004, p. 34). We take advantage of the fact that, given a beta distribution prior and a Bernoulli trial, the posterior distribution incorporating the new information is also beta.[5] As the sum of the two hyperparameters increases with observations, the variance of beliefs over p_j typically falls and uncertainty is reduced. The sum of the beta hyperparameters $(s_j^t + f_j^t)$, is often loosely referred to as the "sample size" because each observation of a Bernoulli trial augments the sum by one. In the remainder of the chapter we will use this workhorse Bayesian framework to capture the learning component of our formal decision models.

10.3 The inspection allocation decision problem

The objective in our inspection allocation decision problem is to maximize the expected averted damage from interceptions. A choice is made each period to apportion a fixed number of inspections at each of several different ports to shipments from various sources. We define the following:

- J: The number of import sources. Individual sources are particular exporter–commodity pairings and will be indexed by j.
- Z: The number of receiving ports where inspections may occur. Individual ports are indexed by z.
- K_z: The total number of inspections available in a given period at port z. Let K represent the total number of inspections available in a given period across all ports: $K = \sum_{z=1}^{Z} K_z$.
- $x_{j,z}^t$: The scalar decision variable indicating the number of times a shipment from source j is inspected at port z in period t. The vector of controls for a given port z is given by $\mathbf{x}_z^t = (x_{1,z}^t, \ldots, x_{J,z}^t)'$. The $J \times Z$ dimensional matrix of controls for *all* ports at period t is given by $\mathbf{X}^t = (\mathbf{x}_1^t, \ldots, \mathbf{x}_Z^t)$. The level of the decision variable is naturally constrained by the number of shipments received from source j at port z at time t: $x_{j,z}^t \leq \bar{x}_{j,z}^t, \forall j = 1, \ldots, J$ and $\forall z = 1, \ldots, Z$.
- $y_{j,z}^t$: The binomial random variable capturing the number of shipments found to be infested ("successes") from $x_{j,z}^t$ inspections.

Next we specify the decision problem in a dynamic programming framework which includes the state, control variables, randomness, dynamics, and rewards:

- *State*: The state of the system in period t is fully described by the vector of hyperparameters (of the beta distribution) for all sources: $(\mathbf{s_t}, \mathbf{f_t}) = (s_1^t, \ldots, s_J^t, f_1^t, \ldots, f_J^t)$. We assume that the infestation probability is independent of port of entry: $p_{j,z} = p_j$.

---

[4] This assumption is important to keep in mind when applying our results to a practical setting, particularly in the US. For example, the AQIM program, in addition to uncovering surprising risks, initiates inspections of specified intensity to provide data to evaluate the performance of standard targeted inspections. We do not consider this performance assessment value to random inspections in this chapter.

[5] In Bayesian terminology, the beta distribution prior is said to be *conjugate* to the Bernoulli likelihood function since the resulting posterior is also beta.

- *Control variables*: The decision or control is given by the matrix $\mathbf{X^t}$ (defined above). The total number of inspections at port z in time t is constrained by the inspection budget constraint: $\sum_{j=1}^{J} x_{j,z}^{t} \leq K_z$.
- *Randomness*: Let $\sum_{z=1}^{Z} x_{j,z}^{t} = x_{j}^{t}$ represent the total number of shipments from source j inspected across all ports at time t. Inspecting x_{j}^{t} shipments results in the discovery of y_{j}^{t} infested shipments. Thus y_{j}^{t} is a beta-binomial random variable with parameters x_{j}^{t} and p_j.
- *Dynamics*: Let the total number of inspections for each source at time t be given by $\mathbf{x^t} = (x_1^t, \ldots, x_J^t)$. Also, let vector of inspection outcomes at time t be represented by $\mathbf{y^t} = (y_1^t, \ldots, y_J^t)$. Bayesian updating of the state of the system is given by: $\mathbf{s^{t+1}} = \mathbf{s^t} + \mathbf{y^t}$ and $\mathbf{f^{t+1}} = \mathbf{f^t} + \mathbf{x^t} - \mathbf{y^t}$.
- *Reward*: The expected payoff is equal to the expected averted damage which, in this case, refers to the expected number of pests intercepted times the expected damage per accepted infested shipment. Let c_j represent the constant expected damage from failing to intercept an infested shipment from source j. The expected value of a beta random variable with parameters s_j^t and f_j^t is $s_j^t/(s_j^t + f_j^t)$. The expected averted damage for period t is therefore given by:

$$E\left[\sum_{j=1}^{J} c_j y_j^t \mid x_j^t, s_j^t, f_j^t\right] = \sum_{j=1}^{J} c_j x_j^t E\left[p_j^t \mid s_j^t, f_j^t\right]$$

$$= \sum_{j=1}^{J} c_j x_j^t \left(\frac{s_j^t}{s_j^t + f_j^t}\right). \quad (10.1)$$

The relationship between propagule pressure (here accepted infested shipments) and damage from successful establishments of NIS is a poorly understood element of biological invasions. There is reason to expect that c_j could be nonlinear over the cumulative number of accepted infested shipments. However, there is not a clear consensus on the likely form of this nonlinearity. Increasing marginal damages (a convex function) might be driven by the Allee effect. Alternatively it might be the case that after a species successfully establishes, further introductions have relatively little impact (a concave function). In the absence of accepted guidance we assume that expected cumulative damages are linear. Only the relative values of c_j are needed as input to the decision model since we assume the inspection budget is fixed. We also assume that c_j is independent of the port z where goods are inspected, a simplification that is more reasonable when the areas around ports have similar vulnerability to infestation. While the expected damage from an infestation is an important uncertain decision parameter in its own right, in the absence of suitable data we will set $c_j = 1$ and concentrate on the infestation probability p_j.

In each period t, the decision maker chooses how many shipments from each source j at each port z to inspect $(\mathbf{X^t})$ to maximize the total expected payoffs over an infinite horizon, given beliefs described by $(\mathbf{s_t}, \mathbf{f_t})$ and available shipments $(\bar{x}_{j,z}^t)$. Letting U represent the payoff of the system, the decision problem is defined as

$$EU^* = \max_{\mathbf{X^t}} \left\{ \sum_{j=1}^{J} \sum_{z=1}^{Z} x_{j,z}^t \left(\frac{s_j^t}{s_j^t + f_j^t}\right) \right. \quad (10.2)$$

$$\left. + \sum_{w=t+1}^{\infty} \beta^{w-t} \sum_{j=1}^{J} \sum_{z=1}^{Z} E\left[x_{j,z}^w \left(\frac{s_j^w}{s_j^w + f_j^w}\right)\right] \right\}$$

$$\text{s.t.} \sum_{j=1}^{J} x_{j,z}^t \leq K_z, \, x_{j,z}^t \in \left\{0, 1, 2, \ldots, \min\left\{\bar{x}_{j,z}^t, K_z\right\}\right\},$$

$$\forall j = 1, \ldots, J \quad \text{and} \quad \forall z = 1, \ldots, Z.$$

The expectation in the second summand is taken over all possible future paths for the state and control variables. The constraint shows that inspections are limited by the port-specific inspection budget (K_z) and by the number of available shipments from a given source $(\bar{x}_{j,z}^t)$.

The first summand in the objective function could be combined with the second for a more concise expression which sums over time $w = t, \ldots, \infty$. The current period t expected payoff appears separately in Equation 10.2 to highlight the distinction between immediate and future returns to the current choice of inspections $(\mathbf{X^t})$. The separation also makes clear that while the time t state describing the decision maker's beliefs $(\mathbf{s_t}, \mathbf{f_t})$ and shipment levels $(\bar{x}_{j,z}^t)$ are given, future states and shipments enter in expectation.

This dynamic programming problem may, in principle, be solved over a finite horizon, T, using backwards induction. In the final period T, the optimal action is to exhaust the inspection budget by inspecting shipments with the greatest expected probability of infestation, $E[p_j^T | s_j^T, f_j^T]$, conditional on expected shipping. Then one can back up through time and solve each period conditional on the state $(\mathbf{s}^t, \mathbf{f}^t)$ and shipping projections. For our problem, and most of practical importance, such a direct solution method is computationally prohibitive given the immense number of future possible states.

10.4 Allocation method and results

Various methods have been proposed to approximate an optimal solution to the problem specified in Equation 10.2. Our structure is similar to a class of decision problems referred to as multi-armed bandits (MAB). The canonical MAB problem involves repeatedly choosing one slot machine ("one-armed bandit") to play amongst a set, with incomplete information over the probability of a success for any given machine. For our problem, each slot machine might be thought of as representing a particular source of trade with only a partially understood probability of "success" if inspected or "played". Springborn (2008) addresses a similar inspection problem using a solution approach from the MAB literature known as Lagrangian decomposition, a form of approximate dynamic programming. While this methodology has the appeal of estimating and internalizing the value of information gleaned from a given observation, it remains computationally intensive and requires extended expertise in dynamic programming to implement.

A much simpler approach involves allocating available inspections to the shipments with the highest expected payoff (expected risk) given current beliefs. This strategy, which ignores the value of information of an observation, is commonly referred to as a "myopic" or "greedy" approach since it maximizes expected immediate payoffs and ignores the future. Because information is incomplete, the greedy strategy runs a long-run risk of concentrating on sources which are not the most rewarding. If the sources believed to be of high risk are given priority in inspections, we would expect to eventually uncover and correct mistaken assessments of risk which overstate the danger of a source. However, such an approach is not effective for identifying the opposite error—sources *mistakenly* believed to be of low danger will continue to be ignored. To the extent that high risk sources have yet to be prioritized as such, a strictly myopic strategy might forgo significant potential improvements in performance.

A modification to this greedy approach, which is only modestly more involved but acknowledges the potential benefits of exploration, is to allocate a fixed fraction (ϵ) of inspections randomly while the remainder is allocated greedily. First described by Watkins (1989), this so-called "ϵ-greedy" approach is a commonly used algorithm for bandit problems Vermorel and Mohri (2005). The analytical task is therefore vastly simplified to identifying the best choice of ϵ, that is, the share of inspections allocated randomly which results in the greatest number of expected interceptions. Despite this simplification, the optimal ϵ-greedy policy for the fully dynamic and spatially disaggregated problem considered in Equation 10.2 is not analytically tractable and requires a simulation-based analysis. We therefore begin by exploring a simpler single port, two-period, two-source version of the problem for which direct numerical solutions are possible.

10.4.1 The tractable inspection allocation decision problem

A single port manager is faced with the task of maximizing interceptions of infested shipments from two sources, A and B, over two periods using a simple strategy: a fixed share of greedy inspections are allocated to the source with the highest expected risk (payoff) while the remaining "exploratory" inspections go to the less risky source. Without loss of generality, we assume that subjective initial beliefs (sampling prior), given by $\{(s_A^0, f_A^0), (s_B^0, f_B^0)\}$, are such that the expected value of the probability of infestation for B is greater: $E[p_B] > E[p_A]$. Recall that K represents the number of inspections available in a given period and the number of exploratory inspections is given by ϵK. Let $E[p_j] = \hat{p}_j$ and represent the posterior value of $\hat{p}_j$ by $\hat{p}'_j$. If g is the index of the greedy source in a given period ($E[p_g] \geq E[p_{\bar{g}}]$), the decision problem is given by:

$$EU^* = \max_{x^1, x^2} \; x^1_A \hat{p}_A + x^1_B \hat{p}_B + E\left[x^2_A \hat{p}'_A + x^2_B \hat{p}'_B\right]$$

$$\text{s.t. } x^t_g = (1-\epsilon)K, \; x^t_{\bar{g}} = \epsilon K, \; \epsilon K \in \{0, 1, 2, \ldots, K\}. \tag{10.3}$$

This problem may be re-expressed as:

$$EU^* = \max_{\epsilon} \; (1-\epsilon)K\hat{p}_B + \epsilon K \hat{p}_A \tag{10.4}$$

$$+ Pr(\hat{p}'_A > \hat{p}'_B)\left((1-\epsilon)K\hat{p}'_{A,\hat{p}'_A > \hat{p}'_B} + \epsilon K \hat{p}'_{B,\hat{p}'_A > \hat{p}'_B}\right)$$

$$+ Pr(\hat{p}'_A \leq \hat{p}'_B)\left((1-\epsilon)K\hat{p}'_{B,\hat{p}'_A \leq \hat{p}'_B} + \epsilon K \hat{p}'_{A,\hat{p}'_A \leq \hat{p}'_B}\right)$$

$$\text{s.t. } \epsilon K \in \{0, 1, 2, \ldots, K\}.$$

Several important features emerge from this simplification. First, the immediate opportunity cost of shifting an inspection in the first period from the greedy choice (B) to the exploratory choice (A) is clear: the greater expected payoff ($\hat{p}_B$) must be forgone in order to learn about the alternative. We also see that the action taken in the second period depends on observations from the first period. With likelihood $Pr(\hat{p}'_A > \hat{p}'_B)$ we observe inspection outcomes in the first period which cause us to *reorder* the sources in terms of perceived payoff and allocate the greedy portion of inspections to A in the second period. This may occur because our updated beliefs involve a lower expected infestation rate for B ($\hat{p}'_B < \hat{p}_B$) and/or a higher expected rate for A ($\hat{p}'_A > \hat{p}_A$). Finally, it is apparent that, if ϵ was not fixed we would optimally set this to zero in the second period as there are no further periods in which to take advantage of exploratory information gathered in this last period.

Given the constraint on ϵK in Equation 10.4, we are essentially faced with an integer programming problem for which the typical first order conditions are inappropriate. Additionally, the second period expected payoff depends on $Pr(\hat{p}'_A > \hat{p}'_B)$ and thus on the expected probability that one binomial random variable is larger than another, which generates a complex nonlinear equation in the choice variable. We therefore solve for the optimal level of ϵ numerically for a set of illustrative cases. Maintaining the assumption $\hat{p}_B < \hat{p}_A$, we vary: (1) the spread between $\hat{p}_B$ and $\hat{p}_A$, (2) the initial sample size, $n^0_j = s^0_j + f^0_j$, and (3) the total number of possible inspections, K.

We first consider how the initial sample size $\{n^0_A, n^0_B\}$ and expected rate spread ($\hat{p}_B - \hat{p}_A$) might influence the optimal level of exploratory inspections. Insight from the MAB literature (e.g. Brezzi and Lai (2002)), which decomposes the value of a given observation into the immediate value and informational value, provides a prediction. As n_j increases with observations, the variance of our beliefs (uncertainty) typically falls and the less informational value there is in the next observation. Thus if n_A in particular is relatively small, we would expect the informational value and expected returns to exploration to be relatively high, all other things being equal. We find that this is indeed the case. The objective function from Equation 10.4 is plotted in Fig. 10.1 for selected combinations of parameters. The horizontal axis is presented in units of exploratory inspections (ϵK). Dedicating a nonzero share to exploratory inspections is optimal when the sample size of the target of those exploratory inspections is relatively small compared to the sample size of the (greedy) alternative. The influence of the second component of the value of a given observation, the immediate payoff, is more straightforward. As the spread between $\hat{p}_B$ and $\hat{p}_A$ increases, the immediate opportunity cost of exploration (forgone immediate expected rewards) increases and we expect the optimal exploration rate to fall. This is observed in Fig. 10.1 where $\epsilon^* K$, if positive, is largest when the spread is lowest.

When we consider the sensitivity of optimal exploration with respect to the inspection budget K, both the MAB literature and simple model results provide little guidance. While the *absolute* number of optimal exploratory inspections ($\epsilon^* K$), if positive, clearly increases with K the share ϵ^* itself does not follow a discernible pattern. Table 10.1 shows $\epsilon^* K$ over various cases for the particular example of low spread ($\{\hat{p}_A, \hat{p}_B\} = \{0.7, 0.8\}$). As the sample size of the non-targeted source (A) increases and as K decreases, $\epsilon^* K$ falls, eventually reaching zero. For

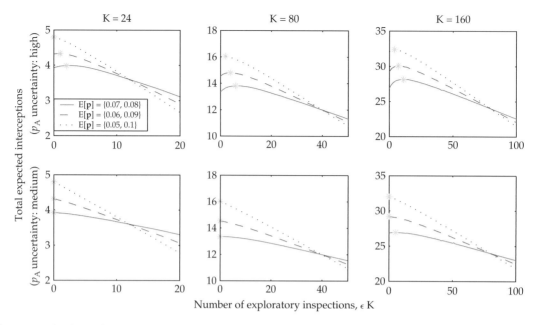

Figure 10.1 The objective function or expected performance in terms of interceptions for the two-period, two-source problem. The cases include various levels of the inspection budget (K) and infestation rate ($E[\mathbf{p}] = \hat{\mathbf{p}} = \{\hat{p}_A, \hat{p}_B\}$). In both rows, beliefs over p_B reflect low uncertainty via a large initial sample size ($n_B^0 = 50$). The top row presents the case where beliefs over p_A reflect high uncertainty ($n_A^0 = 10$) and the bottom row the case where beliefs over p_A reflect medium uncertainty ($n_A^0 = 30$). The optimal number of exploratory inspections in each case is marked by a star.

cases with even greater spread between probabilities, $\epsilon^* K$ is lower and reaches zero sooner.

Next, we look at how the inspection budget influences the gain from replacing a myopic strategy with an optimal random inspection approach. When performance is considered in terms of percentage improvement relative to $\epsilon = 0$ this improvement is increasing in K (given $\epsilon^* > 0$). The simple model then suggests that returns to random exploration, relative to the strictly greedy strategy, will be greater when the resource budget is larger.

Table 10.1 The optimal number of exploratory inspections ($\epsilon^* K$) for various levels of sample size and inspection budget given $\{\hat{p}_A, \hat{p}_B\} = \{0.7, 0.8\}$.

	K		
$\{n_A^0, n_B^0\}$	24	80	160
{10, 50}	2	6	11
{30, 50}	0	0	5
{50, 50}	0	0	1
{50, 30}	0	0	0

While these simple model results suggest a potential role for random exploration under certain cases, the real-world setting is richer and includes factors which are difficult to examine in the two period model. The number of sources we are concerned about is much larger and the time horizon is vastly extended. Empirically we also see that the *availability* of sources for inspection is constrained by both seasonal variation in trade volume and the division of imports across numerous ports of entry. If the most rewarding source is not available because no shipments are forthcoming, either in a particular period or at a particular port, then there will be in effect *forced* exploration of the next most rewarding sources. Implications for the right level of exploratory inspection are not immediately clear, however. One might expect forced exploration to reduce the need for additional exploration, random or otherwise. However, it might be the case that the need to identify not just the single most rewarding source over the entire pool of sources but over *each* sub-pool of sources that might arise could have the opposing effect of increasing the scope of useful information and hence random exploration. We turn next to the optimal level of random inspections in this richer setting.

10.4.2 A fully dynamic model and non-constant exploration policy

Given that uncertainty will typically fall with additional observations over time, it is reasonable to suggest that an approach which allows the rate of random allocation to attenuate over time might outperform the fixed-ϵ strategy.[6] Early treatments of this "ϵ-decreasing" approach Cesa-Bianchi and Fischer 1998; Auer *et al.* 2002 involve setting an initial random allocation rate, ϵ_0, which decays according to some specified function of time. While other variations of non-constant, ϵ-frequency random selection have been developed, we restrict our attention here to ϵ-decreasing since experimental evidence suggests this family performs as good as non-constant alternatives, given carefully chosen parameters Auer *et al.* 2002; Vermorel and Mohri 2005. The ϵ-decreasing function we consider is given by $\epsilon_t = \epsilon_0(1/t)$ proposed by Auer *et al.* (2002). We also explored the performance of the policy function $\epsilon_t = \epsilon_0(\ln(t)/t)$ (Cesa-Bianchi and Fischer (1998) but do not report the results here as they are qualitatively similar but show weaker improvements in performance relative to the $(1/t)$ weighting.

The analytical task lies in finding the best choice of ϵ for the constant ϵ-greedy approach and ϵ_0 for the ϵ-decreasing strategy. We identified the optimal parameters for these alternatives in the port inspection setting using a Monte Carlo simulation analysis. Performance, in terms of cumulative infested shipments uncovered, was assessed over 100 month-long periods ending in 2006. We will first describe and present results for a set of stylized cases before considering the empirically derived setting. In either case, our starting point involves establishing the beta sampling prior which captures beliefs over p_j for each of the sources at the beginning of the first decision period. In the stylized set-

---
[6] Recall that we have assumed that p_j is fixed. If this assumption does not hold, all else equal, as the rate of change of p_j increases, the case for an attenuating exploratory strategy is lessened.

ting we selected sampling priors for a set of illustrative cases. For the empirical setting, described in more detail in Section 10.4.2.3 below, we used a subset of the data record to inform the sampling priors. Each Monte Carlo iteration began with a random draw, p_j^0, for each source from its particular sampling prior. This parameter value was fixed for the duration of the iteration (100 periods), acting as the "true" Bernoulli parameter driving the stochastic infestation process, which the decision maker learns about if observations are made.

At the beginning of each month-long period the inspection budget for each port was apportioned, according to a particular allocation algorithm, across the available number of shipments at each port. In the empirical setting, availability is given by the US Department of Commerce import data record. Each inspection for a given source j uncovered either an infestation or no infestation according to a random Bernoulli trial with parameter p_j^0. Inspections were assumed to be perfect and deterministic—the presence or absence of an infestation was the stochastic element. At the end of the period subjective beliefs over the parameters were updated given the total number of inspections conducted and infestations found. This process was repeated for all 100 months to complete one iteration of the Monte Carlo simulation. In each case presented below, unless otherwise noted, we conducted 1000 Monte Carlo iterations.

10.4.2.1 RESULTS FOR THE STYLIZED SETTING
In the stylized setting we selected a series of cases to develop intuition. The panels in Fig. 10.2 present the performance of the simplest policy, ϵ-greedy, for a case of two-sources $\{\hat{p}_{lo}, \hat{p}_{hi}\} = \{0.06, 0.09\}$ under various levels of the inspection budget and uncertainty over the probability measure. (We also considered larger and smaller spreads between $\hat{p}_{lo}$ and $\hat{p}_{hi}$ as in Section 10.4.1 but the results were similar and not presented here.) To ease comparison between cases, performance is plotted in terms of percentage improvement from the number of interceptions under a baseline of no exploration, $\epsilon = 0$. First, we find that results from the simple model typically hold under the extended time horizon. Again, the level of optimal exploration (ϵ^*, marked

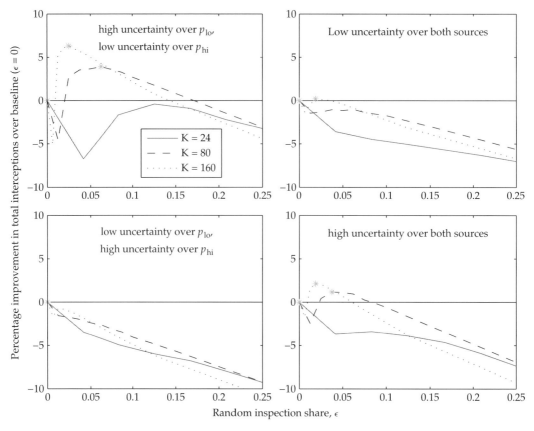

Figure 10.2 Percentage improvement in total expected interceptions relative to the baseline ($\epsilon=0$) over levels of a constant ϵ-policy for two sources, $\{\hat{p}_{lo}, \hat{p}_{hi}\} = \{0.06, 0.09\}$. Each panel presents mean Monte Carlo simulation results for three levels of inspection budget (K). Low (high) uncertainty indicates a sampling prior sample size of $n_j^0 = 50$ ($n_j^0 = 10$). The optimal number of exploratory inspections in each case is marked by a star. The range of ϵ presented below 0.25 since performance in each case declines monotonically to the upper bound at one.

by a star) tends to be positive when uncertainty is high over p_{lo}, the source believed less likely to be infested (top left and bottom right panels). Also, when optimal random inspection is positive, relative benefits of ϵ^* over the baseline are increasing in K.

The actual number of optimal exploratory inspections ($\epsilon^* K$) is not necessarily increasing in the inspection budget. Across the range of inspection budgets (24, 80, 160), the optimal number of exploratory inspections per period over the cases considered lies in a range from two to five. While $\epsilon^* K$ (when positive) does not appear to vary widely or scale consistently with the inspection budget, a larger budget does allow for greater exploitation of the information gleaned.

While affirming key results from the two-period model, the extended time horizon model also illustrates interesting additional outcomes. One such result is the danger in conservatively "dabbling" in random exploration, which may provide relatively little new information relative to its costs in terms of foregone targeted inspections. In the simple two-period setting, when ϵ^* was greater than zero, each value of ϵ between zero and ϵ^* also led to better performance than the baseline. Put another way, when optimal random inspection was positive, dabbling with some smaller level of ϵ was not a danger in the sense of leading to expected outcomes worse than no exploration. In the full dynamic model this was not always the case. The performance function between zero and ϵ^* was poten-

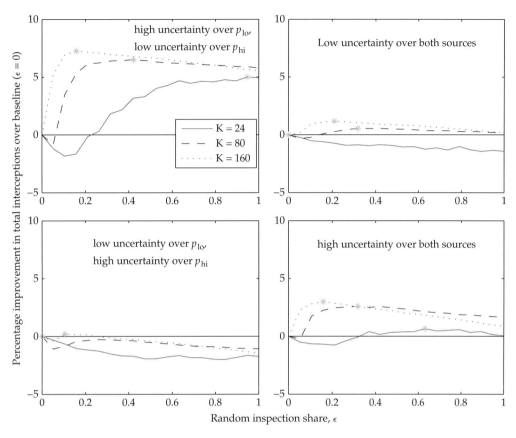

Figure 10.3 Percentage improvement in total expected interceptions relative to the baseline ($\epsilon=0$) over levels of an ϵ-decreasing policy for two sources, $\{\hat{p}_{lo}, \hat{p}_{hi}\} = \{0.06, 0.09\}$. Each panel presents mean Monte Carlo simulation results (6000 iterations) for three levels of the inspection budget (K). Low (high) uncertainty indicates a sampling prior sample size of $n_j^0 = 50$ ($n_j^0 = 10$). The optimal number of exploratory inspections in each case is marked by a star.

tially non-monotonic and reached negative levels. This is apparent in Fig. 10.2: in each case where we found that $\epsilon^* > 0$, there was a smaller level of ϵ for which we observed a *reduction* in performance roughly similar in magnitude to the improvement in performance at ϵ^*.

Recall that one observation highlighted in the two-period analysis was that maintaining a constant level of exploratory inspections can be simply a burden with no expected payoff when the opportunity to influence future decisions becomes limited. In Fig. 10.3 we present the performance of the ϵ-decreasing strategy where the share of exploratory inspections at any time t is given by $\epsilon_t = \epsilon_0/t$. We found that performance under an efficient ϵ-decreasing strategy performed consistently better than an optimized constant-ϵ approach. Not surprisingly then, the optimal level for random inspections can be nonzero under the ϵ-decreasing policy in a wider range of cases, e.g. even when uncertainty for the low infestation rate source is not relatively greater and the value of information from an exploratory observation dampened.

Before leaving behind the simple two-source setting, it is instructive to consider the nature of the improvement when ϵ^* is positive. Consider a shift in the distribution of payoffs over Monte Carlo simulations for the case $\hat{p} = \{.05, .10\}$, $n = \{10, 50\}$ and $K=160$ as the applied policy is shifted from $\epsilon_0 = 0$ to $\epsilon_0 = \epsilon_0^*$. The top panel in Fig. 10.4 shows the empirical probability mass functions (pmfs) for performance in terms of interceptions under the baseline

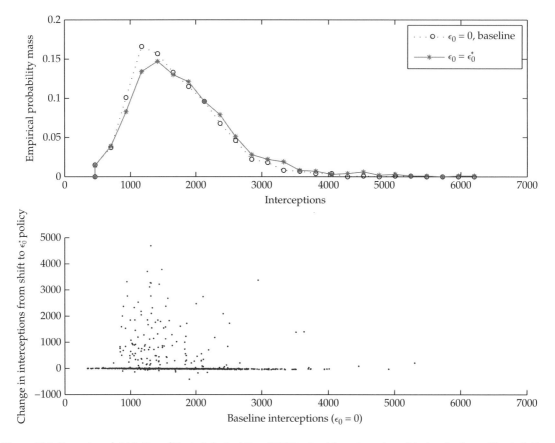

Figure 10.4 Comparison of distributions of Monte Carlo simulations (1000 iterations) for ϵ-decreasing policies baseline ($\epsilon_0 = 0$) and ϵ_0^*. The case includes two sources with high uncertainty over the lower rate ($\hat{\mathbf{p}} = \{0.05, 0.1\}$, $\mathbf{n} = \{10, 50\}$).

($\epsilon_0 = 0$) and ϵ_0^* policies. The shift to ϵ_0^* reduced the likelihood of modal outcomes, with mass moving towards the righthand tail (better performance).

The second panel in Fig. 10.4 shows the change in performance from the shift to the ϵ_0^* policy for individual Monte Carlo iterations. While most changes were clustered near zero, optimal random exploration sometimes led to vast improvements in realizations where the strictly myopic baseline approach performed especially poorly, for example, in the range of zero to 2200 on the horizontal axis. In fact, the modal difference in Monte Carlo outcomes was a *negative* number relatively close to zero. This result is instructive in forming realistic expectations about the performance of an exploratory policy. Even when performance is improved (in expectation) under optimal random inspections, the most likely outcome may be slightly *worse*

performance than the strictly greedy approach. Thus, exploration in this case acts like a moderately costly insurance policy protecting against the potential for missing large payoffs (realizing heavy damages).

To this point we have ignored the issue of shipment availability by assuming that at least K shipments of each source are available each period. An important difference between our setting and the typical MAB problem is that, in reality, the availability of shipments from different sources can be highly variable, especially over a seasonal time scale. Figure 10.5 shows monthly shipment volume time series for four different sources in our dataset that convey a sense of the seasonality.

Can constraints on the availability of shipments significantly affect the returns to random inspec-

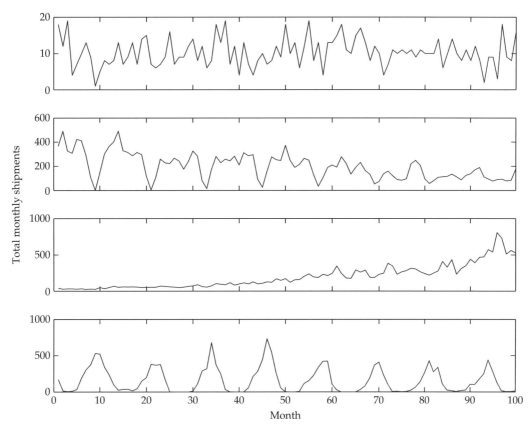

Figure 10.5 Four examples of monthly shipment volume variability for sources (commodity-exporter pairs) from 1999–2006.

tions? To explore this question we considered a simple constraint where, for each source, stylized shipment levels for one month each year were set to zero. Monte Carlo simulations showed that relative expected gains from random exploration were dissipated under this scenario for both the constant and the ϵ-decreasing strategy. A strictly greedy approach with no exploration was optimal for each of the cases considered. Essentially, the constraint forced the observation of each source at least once every 12 periods. This compelled learning provided information to reveal when targeting should be adjusted in a sufficient number of cases to erode the gains to *additional* exploration to the point where such a policy was no longer worth the opportunity cost. Because this result was not particularly surprising in a two-source setting where the constraint forces inspection of both alternatives in at least one period each year, we turn now to results from an expanded set of ten sources.

We explore several cases for optimal random inspection over a ten-source set, however the general results for the two central questions of interest are consistent.[7] Of interest is whether the relative

---

[7] In addition to the various levels of inspection budget (K), we intersect two specifications for initial expected infestation probability ($\hat{p}$) and two specifications for initial sample size (**n**). For $\hat{p}$ we include two vectors with levels for the expected probability ranging from 0.01 to 0.16, roughly corresponding to the observed empirical range. The first vector is a simple linear spacing over the range while the second specification includes a cluster of low levels (0.01 to 0.05) and a cluster of high levels (0.12 to 0.16) a stylized representation of the bi-modal grouping observed in the data. The vector of initial sample sizes (**n**) is a linear spacing between 10 and 50, either with high uncertainty associated with the high infestation probability sources or the reverse.

returns to random exploration are dissipated under a larger number of sources (some with a much lower infestation rate) each of which garner a share of random inspections. For the constant ϵ policy this was indeed the case to the degree that a strictly greedy policy was optimal for all of the cases considered. Some positive level of random inspection under the ϵ-decreasing strategy was optimal only when the inspection budget was large ($K = 160$). With a large number of sources the expected benefit of an exploratory inspection was diluted by the fact that the non-targeted pool contained many sources perceived to be of medium and extremely low threat where the likelihood of uncovering a surprising risk was very low. Essentially, too much effort was wasted confirming that sources with a low probability of infestation were not *much* worse than initially believed.

Our second question concerns whether returns to exploration are dissipated by shipping variability in a larger set of sources. In the two-source case, the effect of shipping constraints is straightforward—when one source is not available the other is the default choice. In the ten-source set, the effect is harder to predict. When a targeted source is not available a further choice must be made between the remaining sources, resulting in a broader set of decisions to be made which could incentivize additional exploration. However, we observed the opposite. Relative returns to random inspection were *further* dissipated when shipping availability was constrained to be zero once per year for each source. Optimal random inspection was essentially zero for each case and policy considered.

The stylized settings explored above suggest predictions for the empirical setting we turn to next. While there may be gains to be had from allocating some portion of resources to random exploration, especially when the inspection budget is large and the uncertainty over non-targeted sources is high, these gains potentially dissipate as the number of sources grows large and the availability of shipments is constrained. Further, a nonconstant ϵ strategy (e.g. ϵ-decreasing) may out-perform a constant ϵ approach since uncertainty (and the likelihood of a surprise) typically falls as observations accrue over time.

10.4.2.2 USE AND INTERPRETATION OF THE DATA

To empirically characterize import pest risk, we merged inspections data described in the introduction with monthly measures of total imported shipments published by the US Department of Commerce (USDOC, 1996–2006). Our "empirical set" of trade sources was constructed from the consideration of eight agricultural commodities (Asparagus, Cauliflower/Broccoli, Celery, Cherries, Pineapple, Spinach, Strawberry, Tomato) received across nine US ports of entry (Boston MA, Chicago IL, Laredo TX, Los Angeles CA, Miami FL, Nogales AZ, Philadelphia PA, San Diego CA, San Francisco CA). Restricting our attention to sources which were inspected at least once and which had at least 15 total shipments received, we were left with a set of 50 commodity–origin pairs.

Since our objective was to develop realistic theoretical insight, we used inspections data to establish model parameters. To do so we made a couple of simplifying assumptions. While the unit of inspections at US ports is typically a bill of lading or manifest which may include multiple commodities, we focused on inspections of single-commodity shipments. When inspections are conducted for exotic species risk there are several different ways in which imports may be identified as problematic. The container or product may be found to be contaminated with unwanted material, such as leaves or soil. There may be problems with import documents including discrepancies with the accompanying commodity. For this analysis, we restricted attention to the sole category of finding an actionable pest in, on, or with the product as the simplest direct measure of infestation probability.

10.4.2.3 RESULTS FOR THE EMPIRICAL SETTING

To conduct Monte Carlo simulations in this setting we first identified reasonable levels for the inspection budget and for the initial (prior) uncertainty over the probability of infestation (p_j) for each source. The inspection budget unique to each port (K_z) was set equal to the average monthly port inspection rate, for a total of 344 ($\sum_{z=1}^{9} K_z$) inspections each month across all nine entry points.

To construct the beta sampling prior describing beliefs over p_j for each of the 50 sources, we started with a common beta "fitting prior" uniquely specified by two assumptions. We set the mean of the common fitting prior equal to the overall average raw observed infestation rate across all sources (0.04). The parameters (s, f) were chosen to be as small as possible to generate a "diffuse" distribution reflecting high uncertainty sensitivity to the data.[8]

Finally, to generate a specific sampling prior for each of the 50 sources, we updated the fitting prior with inspection outcome data (i.e. observed Bernoulli trials) for each source over some number of months (M) immediately prior to the start of the Monte Carlo policy assessment period. For example, if a given source j was inspected frequently over the M training periods, its initial sample size (n_j^0) was relatively large and the uncertainty low. If many infestations appeared in the data over this period for this source relative to the number of inspections, the initial expected probability of infestation was high. We subjectively set $M = 10$ to achieve an initial set of sources which diverged from the fitting prior (shared by all sources) yet still featured some uncertainty. We considered alternative lengths of the training period M as well as sampling these M periods from different time frames, neither of which significantly changed the results discussed below.

The Monte Carlo simulations were run using the empirical shipping record of the 100 monthly periods ending in 2006. Availability of shipments for inspection was therefore determined by the historical variation in shipping, both over time and across ports. When random exploration policies are considered against this full set of 50 sources across nine points of entry, we found that no positive level of random allocation was optimal for either the constant or the ϵ-decreasing strategies. This outcome was foreshadowed by results in the stylized setting. Gains from random inspection can dissipate when the availability of shipments are constrained, as in our empirical case where we observed both seasonal variability over time and spatial variability across ports.

Such dissipation may also occur as the number of sources grows large. Inspection of the empirical set of sources revealed that the vast majority (40 of 50) either had a low expected infestation probability (a sampling prior with mean less than 0.01) or were low volume (a mean shipping rate of less than ten per period). We might expect that including these sources in the pool subject to random inspection would dilute the returns to such exploration given the low infestation likelihood or low potential for using inspection information from limited availability. However, even when these sources were excluded, there was no relative gain to any level of random inspection.

We did, however, find evidence that the lack of a positive return to a random inspection regime was, at least in part, driven by the seasonal and spatial availability of sources across time and ports. To verify a role for port constraints and seasonality, we considered the empirical set of sources collapsed to a single hypothetical "mega-port" without any seasonal variability. Since each source in this scenario was available each period for as many inspections as desired (up to $K = 344$) shipments were effectively "unconstrained". Even with the large set of sources ($J = 50$), the optimal level of random inspection in this hypothetical setting was no longer zero and generated substantial gains (both in the fixed epsilon and decreasing epsilon policy). The relative returns to random exploration over the baseline ($\epsilon_o = 0$) for this unconstrained imports and single port scenario for the ϵ-decreasing strategy are plotted in Fig. 10.6 as case 2 (the top line).

What factors then are important in dissipating positive returns to optimal random inspection? We reversed the mega-port aggregation and consider unconstrained imports across all nine ports. In this scenario each port has its own inspection budget (K_z) and every source is available for inspection each period ($\bar{x}_{j,z}^t \geq K_z$). The relative returns to this unconstrained imports, multiple ports scenario are plotted in Fig. 10.6 as case 3 (second line from the top). While returns are dampened relative the the single port setting (2), optimal random inspection

---

[8] Specifically, since the mean ($E[p] = 0.04$) was less than one-half, the f parameter must be larger than the s parameter. We set $f = 1$ to be as low as possible while avoiding the U-shaped beta pdf with increasing mass as p approaches one, which arises when f is less than one. Given $E[p]$ and f, s is determined by $E[p] = s/(s + f)$.

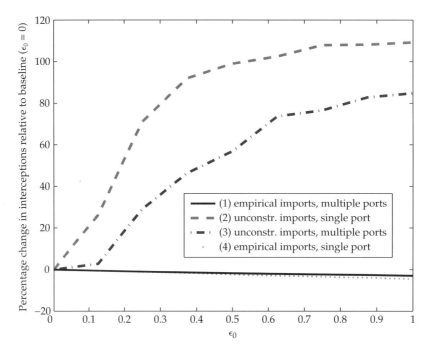

Figure 10.6 Percentage improvement in total expected interceptions relative to the baseline ($\epsilon_o=0$) under an ϵ-decreasing strategy given the empirical set of sources. Results for this set (1) reflecting empirical imports and multiple ports, are presented in comparison to (2) unconstrained imports at a single port, (3) unconstrained imports at multiple ports, and (4) empirical imports collapsed to a single port.

($\epsilon_o^* = 1$) was positive and provided substantial gains over no exploration. This finding left seasonal variation as the likely lead source of dissipation. This was indeed the case. When the single port setting was maintained but shipping levels set by the empirical record, optimal random search fell to zero (Fig. 10.6, case 4). While the distribution of shipments and inspection resources across multiple ports slightly dampened the (still positive) returns to random inspection, seasonal variation in shipping levels severely undercut the benefits of random inspection due to "forced" exploration of alternative sources when leading sources were not available to inspect.

From the discussion of stylized setting results in Section 10.4.2.1, risk averse preferences emerged as potentially important—losses from exploration were frequent but relatively small while averted losses from exploration were less common but of potentially large magnitude (see Fig. 10.4). Given this stylized result we examined outcomes in the current empirical setting for evidence of a similar "long tail" of potential averted damages to be had from exploration. We found that the forced exploration from seasonal shipping variability effectively "mined" the existing opportunities to uncover surprisingly risky sources. Because there was no evidence of this long tail of averted damage levels over repeated Monte Carlo simulations, there appears to be limited, if any, potential for consideration of risk aversion to reverse the conclusion in this particular empirical setting.

10.5 Discussion

Our analysis considers the use of random inspections to maximize pest interceptions over time. The optimal level of random inspection is more likely to be positive when uncertainty over the otherwise nontargeted sources is relatively high, the number of sources is low, and when the inspection budget is large. Assuming a fixed probability of infestation, we observed that a rate of random inspection which decreased over time performed better than a constant share of random exploration. To the extent this stationary assumption holds, the ϵ-decreasing strategy expands the range of cases in which exploration can provide positive expected returns.

When the case for some degree of random inspections is strong, two cautionary results are of interest. First, we note that "dabbling" can be dangerous; sometimes a conservative allocation to random inspections can perform much worse than no random inspections at all. Second, even when the mean payoff to exploration is positive, it could very well be that the median payoff is negative. Thus the learning approach considered here may be thought of as a form of insurance which typically leads to moderate costs but protects against infrequent extreme losses. Finally, we observe that seasonal import variability observed in the data can lead to dissipation of the returns to random inspection.

We have focused on one objective of an inspection regime with learning—improving future returns through enhanced targeting. Application of our findings to existing programs should be tempered by the degree to which multiple objectives are served. For example, US Customs and Border Protection relies on random inspection of uniform intensity in the AQIM program to evaluate the efficacy of its broader targeted inspections. Because both the quality and opportunity cost of the information generated from these random inspections may be quite different, this modified model could have interesting implications for an exploratory policy.

We have also abstracted from modeling a time-varying pest presence, for example, shifts in the likelihood of infestation. Springborn (2008) observes that infestation probabilities may not be constant across months as assumed here. A hierarchical Bayesian model could be used to capture the idea that the infestation probability for any given subset of time (e.g. month) is itself a stochastic draw from some fixed population. The scope for learning is thus expanded to include short-term learning about the current threat which is then used to inform beliefs over the long-term average and monthly variability. It may still be the case that the very distribution from which seasonal risk emerges is shifting over time. This would result in a decay in the value of existing data data over time and

undermine the case for an attenuating exploration strategy (e.g. ϵ-decreasing).

Finally, we assumed that the expected damage from accepting an infested shipment was independent of the region and season in which it was received. A better understanding of how propagule pressure maps into establishments—including spatial and temporal dependencies—would allow for the design of optimal learning strategies better tuned to the scales at which NIS damages are manifested. The level of expected damage from an introduction from a particular source is a poorly understood variable worthy of study in its own right. Source differences in expected damage would have strong implications for the optimal allocation of inspection resources.

Acknowledgements

The authors thank Robert Deacon, Bruce Kendall and two anonymous referees for helpful comments on an earlier draft. The views expressed here are those of the authors, and may not be attributed to the Economic Research Service or the U.S. Department of Agriculture.

References

Auer, P., N. Cesa-Bianchi, and P. Fischer (2002). Finite time analysis of the multiarmed bandit problem. *Machine Learning* **47**(2), 235–56.

Brezzi, M. and T. L. Lai (2002). Optimal learning and experimentation in bandit problems. *Journal of Economic Dynamics and Control* **27**, 87–108.

Cesa-Bianchi, N. and P. Fischer (1998). Finite-time regret bounds for the multiarmed bandit problem. In *Proceedings of the 15th International Conference on Machine Learning*, San Francisco, CA, pp. 100–108. Morgan Kaufmann.

Gelman, A., J. Carlin, H. Stern, and D. B. Rubin (2004). *Bayesian Data Analysis* (2 ed.). Chapman and Hall/CRC: Washington, District of Columbia.

Holling, C. (Ed.) (1978). *Adaptive Environmental Assessment and Management*. Wiley.

Interagency Working Group on Import Safety (2007). Action plan for import safety. A Report to the President, available online, URL: www.importsafety.gov.

Lynch, L. and E. Lichtenberg (2006). Foreword: Special issue on invasive species. *Agricultural and Resource Economics Review* **35**, iii–v.

Springborn, M. (2008). Bayesian adaptive management with learning. Working paper, University of California at Davis, Department of Environmental Science and Policy.

U.S. Department of Agriculture (2007). Foreign Agricultural Trade of the United States (FATUS) database.

U.S. Department of Commerce, Bureau of the Census, Data User Services Division (1996-2006). *U.S. Imports of Merchandise* (monthly). CD-ROM.

Vermorel, J. and M. Mohri (2005). Multi-armed bandit algorithms and empirical evaluation. In *Proceedings of the 16th European Conference on Machine Learning*, pp. 437–48.

Walters, C. and R. Hilborn (1978). Ecological Optimization and Adaptive Management. *Annual Review of Ecology and Systematics* **9**(1), 157–88.

Watkins, C. (1989). Learning from Delayed Rewards. Ph. D. thesis, Cambridge University.

CHAPTER 11

The Role of Space in Invasive Species Management

Julia Touza, Martin Drechsler, Karin Johst, and Katharina Dehnen-Schmutz

11.1 Introduction

With the increasing political attention given to the problems posed by invasive species, economic studies on invasive species management have developed rapidly in recent years. An important branch of this literature focuses on prevention and control policies against biological invasions (see Olson 2007 and Touza 2007 for recent reviews). Invasive species are often viewed as pests that cause damage, and the literature evaluates economic choices related to prevention and/or control. Issues so far studied include the interactions between expenditures on prevention and on control (e.g. Finnoff *et al.* 2007; Leung *et al.* 2006; Heikkila and Peltola 2004), the influence of environmental uncertainties surrounding the invasion process, and the irreversibility of the damage (e.g. Horan *et al.* 2002; Olson and Roy 2002), and the efficiency of economic instruments such as import tariffs, inspections, tradeable permits, or ambient tax (e.g. Costello and McAusland 2003; McAusland and Costello 2004; Horan and Lupi 2005; Jones and Corona 2007). However, little attention has so far been paid by economists to the spatial dimension of invasion processes. Spatial factors are identified as highly relevant by ecologists, and have attracted considerable attention in ecological research (e.g. Higgins and Richardson 1996; Shigesada and Kawasaki 1997). Indeed, most ecologically-based management decision support tools mainly concentrate on the evaluation of measures in a space-less context.

Bockstael (1996), in a discussion of spatial heterogeneity in economics, argued that there are important economic-environmental interdependencies with specific spatial dimensions that have not received the attention they deserve. Similarly, Deacon *et al.* (1998) wrote:

> The spatial dimensions of resource use may turn out to be as important as the exhaustively studied temporal dimensions in many contexts. Curiously, the profession (environmental economics) is only now beginning to move in this direction.

Since then, an increasing number of studies consider decisions made over space, such as dispersed pollution, fisheries management, and conservation planning (establishment of protected areas) (e.g. Antle *et al.* 2003; Neubert 2003; Sanchirico and Wilen 2001, 2005; Brock and Xepapadeas 2005; Smith *et al.* 2007). Results have shown that optimal management strategies can change significantly if the analysis includes spatial linkages and/or heterogeneities between interacting systems (habitats, fishing grounds, etc.). Sanchirico and Wilen (1999), for instance, model a fishery industry exploiting a patchy resource over space and show that the structure of the diffusion system among fishing grounds affects the spatial distribution of fishing effort and fish stock. They show that the inclusion of the spatial dimension has relevant implications for policy development.

This chapter focuses on the spatial characteristics of invasive species management. Specifically, it focuses on control of the spread of a species introduced to an area. Since this is influenced by landscape features (Jules *et al.* 2002), management of the spread of invasive species should be conditioned by those features. This means that control of the spread of an introduced species is a spatially dynamic process, that is, decisions on where and when to start management may often be crucial for the invader's success. For example, the probability of seedling establishment in suitable habitats has been shown to be significantly affected by the distances from the closest seed sources, and may therefore influence control strategies as well (Stephenson *et al.* 2007). The timing of pesticide application may impact differently on landscape-wide distributions of the invader population depending on spatial process such as wind speed and direction, temperature, and species' dispersal ability (Parry *et al.* 2006). On a larger scale, the success of prevention and control policies at an international level depends on individual countries' efforts (Perrings *et al.* 2002).

In this chapter we first review the emerging literature on the spatial economics of invasive species. The studies are reviewed in terms of their use of space: the inclusion of local heterogeneity and the analysis of a network of areas (i.e. two-patches versus multiple-patches). We show that most of the studies either focus on heterogeneity in simple landscapes with two patches, or on complex landscapes with multiple areas but ignoring spatial heterogeneity. Following the presentation of the results of the review, we develop a simple model to explore the role of spatial heterogeneity in invasive species management in a landscape with multiple areas. A final section recapitulates the main points and offers our conclusions.

11.2 Spatial aspects of the economics of biological invasions

Analysis of spatially explicit policy decisions depends on how space is considered. In order to facilitate this review, we extend the classification developed by Drechsler *et al.* (2007) and distinguish four types of studies that take into account space to varying degrees (Fig. 11.1): (1) Spatially implicit studies: they consider different sites which may differ by state (invaded versus non-invaded), but which are otherwise identical in terms of local properties or distances between them. (2) Spatially differentiated studies: the areas analyzed have different properties (next to possible differences in state), e.g. different sizes, ecological qualities, institutional issues, etc. (3) Spatially explicit studies: the locations of sites are considered explicitly and distances between sites differ among pairs of sites. (4) Spatially explicit and spatially differentiated studies: the locations of the sites are known explicitly and the areas' properties as in (2) vary across space. We term models that are both spatially explicit and spatially differentiated as *spatially heterogeneous*. In each of the four classes we further distinguish studies according to how many sites are considered.

11.2.1 Spatially implicit models

Spatially implicit studies do not explicitly analyze space (i.e. distance, direction, shape are irrelevant), and the environment is considered homogeneous. The surveyed studies focus either on social policy decisions based on the dynamics of invaded and non-invaded areas (Perrings 2002; Potapov *et al.* 2006); or on private individual decisions subject to spatial interactions with adjacent agents due to invaders' migration effects (McKee 2006; Grimsrud *et al.* 2008). In the first case, space is modeled as implicitly consisting of many areas but aggregated on two variables, invaded and non-invaded sites. In the second case, the analysis has a two areas setting (i.e. the studied landscape is assumed to be composed of two areas). All these studies evaluate control measures, although the definition of control varies among them. In Perrings (2002) control efforts reduce the space occupied by invaders; in Potapov *et al.* (2007) the efforts applied aim to reduce further spread to non-invaded areas; and in Dehnen-Schmutz *et al.* (2004), McKee (2006), and Grimsrud *et al.* (2008) control efforts imply a reduction in the level of infestation.

Perrings (2002) and Potapov *et al.* (2007) look at the invasion management problem from the perspective that space can be differentiated uniquely according to the state of the system: invaded and non-invaded. The dynamics of the invaded and non-invaded areas follows a metapopulation

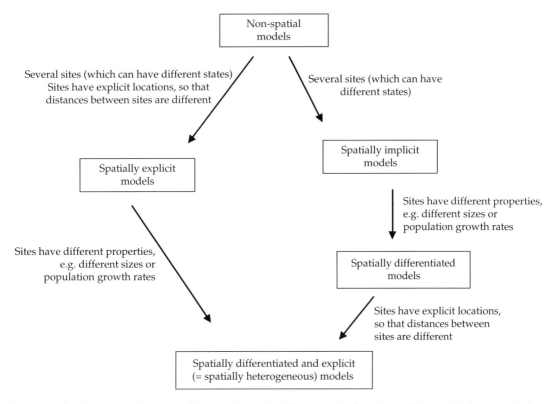

Figure 11.1 Classification of models in terms of how space is considered. The arrows describe which spatial characteristic has to be added to move from one model type to another, for instance, if we make the sites in a spatial implicit model different, we obtain a spatially differentiated model.

model characterized by invaders' colonization rate (i.e. invasion rate) and no extinction events. In Perrings (2002) the discounted net social benefits from exploiting the native species is maximized by choosing the rate of invaders' control. This study shows that there is an establishment threshold after which the invasive species spreads, and this threshold depends on the cost of control and the invasion impact: the cheaper the act of clearing the space, and the higher the impact of the invader on social welfare, the higher the threshold. Once the species has established, the optimal level of control increases with the marginal benefit of native species and the marginal costs of invasive species impact, and decreases with the marginal cost of control. Potapov *et al.* (2007) assume a network of lakes with no differences in size or distances between the lakes. The invader propagule pressure is assumed constant and defined as the average number of propagules that can be transported from any invaded lake to any given non-invaded lake. The authors study control measures that can limit the propagules transported between lakes and their survival (e.g. establishing checkpoints and disinfecting boats and fishing materials) by controlling the outflow of invaders at infected lakes and the inflow of invaders at uninfected lakes. They conclude that there are three types of optimal policies that minimize the invasion costs (damage plus control costs): control of invaded (donor) lakes, control in non-invaded (receiving) lakes, and no control. At any point in time of the invasion development only one of these prevention policies is optimal, which one depends on the efficiency of prevention measures and the unit costs of control in invaded and non-invaded lakes. The higher the invasion damages and the efficiency of prevention measures are, the higher is the optimal level of control. As expected, discounting reduces early investment and allows invasion to progress quickly.

McKee (2006) focuses on the control of Glasshouse Whitefly (*Trialeurodes vaporariorum*) infestations in strawberry crops in California. This work analyzes the effect of migration of the invader on pesticide use. Migration occurs when the crops of the adjacent field are removed and the glasshouse whiteflies search for alternative suitable hosts nearby. The analysis computes numerically the optimal timing of pesticide use on the Glasshouse Whitefly population (egg, nymph, and adult) and consequently on the growing season's profits. Migration is simulated as a positive increase of the invader population in one period during the growing season as a consequence of crop removal in an adjacent field. The results indicate that there are times in the season when the only way for a grower to optimize the timing of pesticide use is to receive information from adjacent growers about intended host plant harvesting times. This is because the application of pesticides has to be done before the adjacent owner removes his crop. Coordination among growers is thus necessary to control the Glasshouse Whitefly population and to increase producers' profits.

Grimsrud *et al.* (2008) analyze the effect of short-distance diffusion between adjacent pasture lands in the control of a weed population. Weeds reduce growth of grass via competition, affecting thus the livestock production and the benefit to the rancher. Diffusion is assumed to be a constant proportion of the weed stock in the neighboring pasture. The results for Yellow Starthistle (*Centaurea solstitialis* L.) in fields of New Mexico indicate that regardless of the initial levels of infestation and regardless of the effort by an adjacent rancher, without some form of incentive payment the rancher will not exert any effort to control the weed because it is cost-prohibitive. Furthermore, coordination efforts between ranchers are necessary to minimize impacts, that is, control efforts should be exerted by both ranchers otherwise efficient control is not possible. This coordination is more likely when weed infestation levels are low. However, if the adjacent rancher free-rides, that is, he decreases his efforts as his neighbor's effort increases, then the marginal benefits of weed removal for his neighbor decrease, and as a consequence, he will also decrease his efforts.

The effectiveness of current control levels in a spatially implicit approach is analyzed by Dehnen-Schmutz *et al.* (2004). The study focuses on the invasive non-native plant *Rhododendron ponticum* in the British Isles. Data on the origin of infestation (whether planted or spread from neighboring sites) at 67 locations are used to get an estimate of the probability that one site is invaded from neighboring sites. The model assumes that landowners controlling the species on their own sites reduce the probability of invasion in neighboring sites, thus creating a social benefit. At the socially optimal level of control, control expenditure equal to the expected benefits from damage avoided in contiguous sites would be sufficient to prevent further spread of the species. The model shows that current control effort is only sufficient to prevent damage in 2.2 contiguous sites. To achieve the socially optimal control for the average number of 5.6 neighboring sites in the study, social support for private control efforts would have to be more than doubled.

11.2.2 Spatially differentiated models

A number of economic studies are spatially differentiated, that is, they incorporate space by focusing on two areas that have different properties (besides differences in state, such as invaded or non-invaded) but do not consider the geographical location of these areas (Drechsler *et al.* 2007). The environmental literature related to the problem of managing invasions involves the analysis of agents who undertake collective harvest of migratory natural resources (e.g. Munro, 1990). This literature investigates cooperative and non-cooperative strategies between neighboring landowners/countries. Noncooperative behavior encourages free-riding behavior, because species will migrate from the unmanaged into the managed area. Even though there are no spatially differentiated studies on invasive species, there are studies considering populations of a pest species in two adjacent areas, where the species causes damage to other uses of the areas (e.g. browsing damage). Our survey includes three studies, Huffaker *et al.* (1992), Skonhoft and Olaussen (2005), and Bhat and

Huffaker (2007). The differences between the two areas considered in these studies are, for instance, the level of damage inflicted by the pest (e.g. because the climatic characteristics may vary, making them more or less productive to forestry), or the pest populations' carrying capacity (because the areas may have different sizes). In the first paper only the pest population in one area is subject to control, and it is shown that even in this case populations in both sites should be taken into account when taking control decisions because of the migration pressure from the adjacent population. This pressure depends on the within-group competition for vital resources, and therefore on the population level in the adjacent land.

In Skonhoft and Olaussen (2005), the species also has economic value associated with hunting activities. This paper studies a unified management regime under a public plan that sets hunting levels of a moose population inhabiting two areas. The population in area one causes browsing damage in area two, but not vice-versa. The results indicate that the steady-state welfare maximum is achieved by equating the value of the marginal growth in each location to marginal pest damage. It is also shown that by taking migration into account when setting the harvest (control) measures, there is a higher degree of accordance between hunting income and browsing damage between the landowners in the two areas (e.g. by giving hunters of area two hunting licenses in area one, and reducing the stock levels of area one, there is less migration and less damage in area two).

Bhat and Huffaker (2007) analyze transfer-payments between agents to sustain a self-enforcement cooperative control of two adjacent populations of a pest species. Given that the levels of pest populations, and as a consequence the benefits of cooperation, vary with time, the authors suggest a flexible contractual mechanism that allows landowners to negotiate adjustments in payments, to encourage compliance with the cooperative agreement.

11.2.3 Spatially explicit models

We found only one study on the explicit analysis of space on invasion management: Sharov and Liebhold (1998). They analyze the present net value of benefits to establishing barriers at the front of an invading population in order to slow the spread of that population. The spatial aspects included are: the direction of the invasion, the shape of the area that can be potentially invaded, and the distance from the introduction point. The area to be potentially invaded is nevertheless assumed not to be spatially differentiated—that is, the landscape considered is homogenous. They define the benefits of creating a barrier as the delay in the impacts of the invader in the uninfested area. The analysis is carried out using three scenarios where the population expands along an infinite strip, along a rectangular area, and as small colonies. Numerical simulations are carried out for the example of the spread of the Gypsy Moth (*Lymantria dispar*) in the US. The study concludes that slowing down the spread by eradicating small, isolated colonies of Gypsy Moth within barrier zones is not only feasible, but also economically justified (because it generates positive net benefits). In addition, considerable benefits from slowing population spread may exist even if only a small portion of potential range remains uninfested. This seems to contradict those who argue that eradication should not be done in a country/area with a large proportion already invaded because of the high risk of recolonization.

11.2.4 Spatially explicit and differentiated models

Studies that are both spatially explicit and differentiated analyze environment–invader interactions explicitly by considering location and distance along with heterogeneity in the landscape. Our survey includes just two study of this type: Higgins *et al.* (2000) and Potapov and Lewis (2008). The first integrates the demography of alien plant populations with information on the spatial structure and dynamics of the ecosystem being invaded using a computer-based simulation approach. The authors model the rates and patterns of invasive plant spread using GIS information and successional vegetation models. Control strategies for two established plant invaders (*Acacia cyclops* and *Pinus pinaster*) in the Fynbos ecosystems in Cape Peninsula, South Africa, are analyzed. In addition, Higgins *et al.* (2000) study the impacts on native plant diversity and the cost effectiveness of alternative clearing strategies. They conclude that the most cost effective and rapid strategy is to start by clearing low-density stands of juvenile plants, then clear higher density stands of juvenile plants, and leave high-density stands of adult plants until last.

Very recently, Potapov and Lewis (2008) formulated an optimal control problem for a system of heterogeneous lakes that are connected via traffic of recreational boats. As in Potapov *et al.* (2007), described above, they study control measures that can limit the propagules transported between lakes, such as washing boating and fishing equipment. They assume an Allee effect in the local invaders populations and are able to derive the "colonization threshold", a critical flow in terms of arriving boats per year in a lake, below which the invader population cannot establish. They numerically solve the problem of optimal spatial resource allocation for a given configuration of lakes, and a problem of optimal invasion stopping configuration. For the first problem they ignore spatial heterogeneity and show that the optimal control allocation is located at the invasion front, and the intensity of control declines with the distance from the invasion boundary. For the second problem, their illustration shows that if the lake system has clusters, it is more efficient to stop invasion between the clusters. Once a cluster has several invaded lakes, they suggest it is better to abandon the cluster and invest the resources in preventing other clusters from becoming invaded.

11.3 Modeling invasions in spatially heterogeneous systems

In this section we develop a simple model of the spread of an organism through a spatially heterogeneous network of habitat patches, in order to better understand the role of spatial heterogeneity of the landscape in invasion control. We compare the cost-effectiveness of spatially homogenous and heterogeneous control strategies. Cost-effectiveness here is measured in terms of the time needed for the organism to invade the whole network (which is to be maximized) for a given budget. We show that the time to invade the network depends on the distribution of patch properties, such as the extinction rates of local populations. Assuming that the network structure can be affected by management we further show that the cost-effective allocation of resources depends on the current level of spatial heterogeneity in the network and on the function relating costs and management effort.

11.3.1 The management problem

We consider the spread of an introduced organism through a model landscape of rectangular shape. This landscape is divided into 12 strips ($m = 0, \ldots 11$), each containing an equal number of $n = N/12$ patches, where N is the total number of patches (Fig. 11.2).

We assume that initially, the n patches in the first strip are all occupied while all other patches are empty. The objective of a regulator is to maximize the expected time T that it takes until the first patch in the last strip ($m = 11$) has become occupied. Assuming that the damage caused by invasive species increases with the time an area has been invaded, this management objective contributes to the overall objective of minimizing costs of invasive species to the society. Spatial network heterogeneity is included assuming that patches of the same strip have the same size, but patch sizes may differ among strips. For simplicity we assume that the patch sizes alternate from strip to strip so that the patches in the uneven-numbered strips ($m = 1, 3, 5, 7, 9$) have size A_{10} each, and those

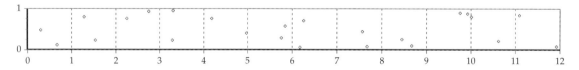

Figure 11.2 Invasion landscape with 24 patches evenly distributed over the strips. Each strip has length and width d. Invasion starts in strip 0 with two patches occupied. The time a population needs to arrive in strip 11 is monitored as invasion time.

in even-numbered strips have size A_{20} each. Spatial network heterogeneity can thus be measured by the ratio $h_0 = A_{10}/A_{20}$ where $h_0 = 1$ ($h_0 \neq 1$) represents a spatially homogenous (heterogeneous) network.

We assume that a budget B is available to reduce the sizes (= carrying capacities) of the patches. These control measures makes the treated patches less suitable for the survival of local populations (cf. eq. 11.4 below) and reduce the risk (increase the expected time) of invasion. For simplicity, we assume that the probability that each strip will become colonized is not affected by management (though see the discussion in Section 11.3.4). We are interested in the impact of spatial network heterogeneity (h_0) on the cost-effective spatial allocation of control efforts. Focusing on the role of space alone, we do not consider state-dependent management (e.g. management that depends on the current occupancy of patches). Therefore we consider the problem in a static setting: the budget is available and has to be spent once and for all, and the effects of management last forever, so that the patch sizes after management do not change in time. We assume that all patches in the uneven-numbered strips receive the same management effort and the same for the patches in the even-numbered patches. This means that each patch in an uneven-numbered strip receives a budget of b_1, and each patch in the even-numbered strips receives

$$b_2 = b - b_1 \quad (11.1)$$

where

$$b \equiv \frac{B}{5n} \quad (11.2)$$

After management each patch in an uneven-numbered (even-numbered) strip has size A_1 (A_2). The decision problem is to determine the level of b_1 that maximizes the invasion time T, where $b_1 = b/2$ ($b_1 \neq b/2$) corresponds to a spatially homogeneous (heterogeneous) allocation of resources.

11.3.2 The model

The analysis is based on a stochastic metapopulation model (Drechsler and Wissel 1998; Frank and Wissel 2002; Johst et al. 2002) in which the dynamics of an invading species is modeled as a spatially explicit Markov process. We consider an ensemble of N patches each of which is inhabited by a local invasive species population or not. The state of the species in the network is characterized by a vector **x** which is composed of the x_i ($i = 1 \ldots N$) and tells which of the N patches are occupied by a local invasive population ($x_i = 1$) and which are empty ($x_i = 0$). Starting from an initial occupancy vector $\mathbf{x}_0$ (where $x_i = 1$ for the patches in the first strip, $m = 0$, and $x_i = 0$ for all other patches) we are interested in the stochastic dynamics of **x** and particularly in the average time T it takes until the system is found in a final state $\mathbf{x}_f$ (where at least one patch in the last strip, $m = 11$, has been invaded).

The stochastic transition from a state **x** to a state **x**' is described by transition rates μ_i and λ_{ij}. The first rate μ_i represents the stochastic extinction of a local population on patch i and is modeled as

$$\mu_i = e^{(i)} \quad (11.3)$$

According to metapopulation theory (e.g. Drechsler 1998; Hanski 1999), the extinction parameter $e^{(i)}$ is a function of patch carrying capacity $A^{(i)}$:

$$e^{(i)} = \varepsilon \left(A^{(i)} \right)^{-y} \quad (11.4)$$

where ε and y are species-specific parameters. In particular, large (small) y means that the expected life time of a local population ($1/e^{(i)}$) increases strongly (weakly) with increasing carrying capacity, and corresponds to species whose population growth rate is subject to weak (strong) environmental fluctuations.

The management of the spread of the introduced species is by control of extinction probabilities through regulation of the size of patches. The patch sizes after management are given by $A^{(i)} = A_1$ if patch i is located in an uneven-numbered strip ($m = 1, 3, 5, 7, 9$) and $A^{(i)} = A_2$ if patch i is located in an even-numbered strip ($m = 2, 4, 6, 8, 10$). The corresponding local extinction rates of patches in uneven- and even-numbered strips are denoted as e_1 and e_2, respectively.

The second transition rate, λ_{ij}, describes the colonization of an empty patch j by a local population on patch i:

$$\lambda_{ij} = c(1 - x_j) \exp(-\alpha d_{ij}) \quad (11.5)$$

$(1 - x_j)$ comes from allowing only an empty patch to become colonized, $\exp(-\alpha d_{ij})$ has the probability of patch i colonizing patch j decaying exponentially with the distance d_{ij} between the two patches at a rate α. The parameter c contains, among others, the number of emigrants from a local population (e.g. Hanski 1999; Frank and Wissel 2002).

This completes the ecological component of the model. We now turn to the economic side and define two cost functions that relate budget b_s to patch size A_s ($s = 1, 2$: uneven- and even-numbered strips, respectively). We believe these cost functions should encompass a wide range of real situations. In the first case, we assume that the cost of patch size reduction is related to the *proportion* of area removed:

$$A_s(b_s) = A_{s0} q(b_s) \quad (s = 1, 2) \quad (11.6)$$

where q is a function of b_s only, with $q(0) = 1$ and $dq/db_s < 0$. In the second case, cost is a function of the *amount* of area removed:

$$A_s(b_s) = A_{s0} - k(b_s) \quad (s = 1, 2) \quad (11.7)$$

where k is a function of b_s only, with $k(0) = 0$ and $dk/db_s > 0$.

11.3.3 Analysis and results

We analyze the model in two steps. In the first step we numerically determine the invasion time T as a function of the local extinction rates e_1 and e_2 of the patches in the uneven- and even-numbered strips. This measures the impact of spatial network heterogeneity on the rate at which the invasive species spreads through the system. For this we simulate the metapopulation dynamics 1000 times each over 5000 time steps and calculate the average over the times T recorded for each simulation run. In the second step we use the result of the first step to analytically determine the cost-effective level of b_1 for the cost functions, eq. (11.6) and (11.7).

In the first step we find that T can be related to spatial network heterogeneity via the geometric mean of the local extinction rates, $(e_1 e_2)^{1/2}$. Figure 11.3 shows T as a function of the product

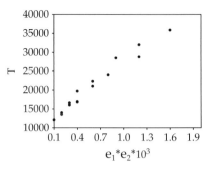

Figure 11.3 Invasion time as a function of the product $(e_1 e_2)$. Parameters are: colonization parameter $c = 0.08$ per time step, patch number $N = 24$. Mean dispersal distance is $1/\alpha = 2d$ where d is the width of a strip (Fig. 11.2).

$e_1 e_2$. All points are located more or less on a single (curved) line which indicates that in good approximation, the local extinction rates e_1 and e_2 affect T only via the product $e_1 e_2$. This means that multiplying e_1 with a certain factor and dividing e_2 by the same factor leaves T nearly unchanged.

Since network heterogeneity is measured by the product $e_1 e_2$, we can readily discuss the cost-effective allocation of control efforts. Since invasion time is a function of $e_1 e_2$ only, the optimal amount of the budget allocated to the uneven-numbered patches is given by

$$b_1{}^* = \arg\left[\max_{b_1}(e_1 e_2)\right] = \arg\left[\min_{b_1}(A_1 A_2)\right] \quad (11.8)$$

(read the equation as: "maximize $e_1 e_2$ as a function of b_1 and take the corresponding argument which is the level of b_1 that maximizes $e_1 e_2$"). In the second equality we used that e_s ($s = 1, 2$) is inversely related to A_s via a simple power law (eq. 11.4).

For the cost function eq. (11.6), eq. (11.8) becomes

$$b_1{}^* = \arg\left[\min_{b_1}(q(b_1) q(b - b_1))\right] \quad (11.9)$$

Due to the special property of the exponential function that $\exp(a + b) = \exp(a)\exp(b)$ we can conclude that the product $q(b_1)q(b - b_1)$ is independent of b_1 if $q(.)$ is the exponential function. We slightly generalize from the exponential form and choose

$$q(b) = a \exp\left\{-(\beta b)^{1+\delta}\right\} \quad (11.10)$$

with constants $a, \beta > 0$ and δ a real number.

For $\delta = 0$, q declines exponentially with increasing budget while for $\delta > 0$ ($\delta < 0$) it declines more (less) than exponentially.

Using some straightforward algebra we find that:

$$\frac{b_1^*}{b} = \begin{cases} \in \{0, 1\} & \delta > 0 \\ \in [0, 1] & \delta = 0 \\ 0.5 & \delta < 0 \end{cases} \quad (11.11)$$

This means that if q declines more than exponentially with increasing budget spent, control efforts should focus either on the uneven-numbered or on the even-numbered patches to maximize invasion time T, not on both. If q declines at an exponential rate, all solutions lead to the same T; and for q declining less than exponentially control efforts should be equal across patches.

In economic terms, one would expect the cost-effective strategy to equalize the ratios of marginal benefits and marginal costs of control in all patches. Using eq. (11.4), the marginal benefit and the marginal cost of changing A_s are given by

$$\frac{\partial T}{\partial A_s} = \frac{\partial T}{\partial (e_1 e_2)} \frac{\partial (e_1 e_2)}{\partial A_s} = -\frac{\partial T}{\partial (e_1 e_2)} e_{3-s} \varepsilon y A_s^{-y-1}$$

$$= -\frac{y e_1 e_2}{A_s} \frac{\partial T}{\partial (e_1 e_2)} \quad (11.12)$$

and

$$\frac{\partial b}{\partial A_s} = -\frac{1}{\beta(1+\delta) A_s} (\ln(A_s/A_{s0}))^{-\delta/(1+\delta)} \quad (11.13)$$

and the ratio of the two is:

$$\Phi(A_s) = \frac{\partial T}{\partial A_s} \left(\frac{\partial b}{\partial A_s}\right)^{-1} = \frac{y e_1 e_2 \beta (1+\delta)}{(\ln(A_s/A_{s0}))^{\delta/(1+\delta)}} \frac{\partial T}{\partial (e_1 e_2)}$$

$$(11.14)$$

A necessary condition for cost-effectiveness is that $\Phi(A_1) = \Phi(A_2)$. For $\delta = 0$ the ratio $\Phi(A_s)$ is independent of A_s so that all solutions of b_1 are cost-effective, which confirms eq. (11.11). For $\delta \neq 0$, equality of the two ratios $\Phi(A_1) = \Phi(A_2)$ is obtained if $A_1/A_{10} = A_2/A_{20}$, that is, if the budget is spent evenly among uneven- and even-numbered patches ($b_1 = b_2 = b/2$). For $\delta > 0$ this interior solution is dominated by the corner solutions, $b_1 = 0$ and $b_1 = b$, identified in eq. (11.11); for $\delta < 0$ it equals the solution found in eq. (11.11). To summarize, for the first type of cost function the cost-effective allocation of the budget is independent of the initial patch sizes A_{10} and A_{20}—and thus the degree of network heterogeneity—but depends on the shape of the cost function. The observation that the initial patch sizes do not affect the cost-effective allocation of control measures is due to the fact that control costs are related to the proportion of habitat removed (eq. 11.6) and that the increase in the invasion time caused by a proportional reduction of habitat size is independent of the initial patch size (eq. 11.8).

Now consider the second type of cost function, eq. (11.7). Inserting eq. (11.7) into eq. (11.8) leads to

$$b_1^* = \arg\left[\min_{b_1} ((A_{10} - k(b_1))(A_{20} - k(b - b_1)))\right]$$

$$= \arg\left[\min_{b_1} \left(\ln(1 - k(b_1)/A_{10})\right.\right.$$

$$\left.\left. + \ln(1 - k(b - b_1)/A_{20})\right)\right] \quad (11.15)$$

Equation (11.15) can be solved analytically if the changes in patch size (k) are small compared to initial patch sizes A_{s0}. Expanding the logarithm to first order of (k/A_{s0}), eq. (11.15) becomes

$$b_1^* \approx \arg\left[\min_{b_1} (-k(b_1)/A_{10} - k(b - b_1)/A_{20})\right]$$

$$= \arg\left[\max_{b_1} (k(b_1)/A_{10} + k(b - b_1)/A_{20})\right] \quad (11.16)$$

For constant marginal costs, $k(b_s) = \beta b_s$ with constant $\beta > 0$, we obtain

$$b_1^* \approx \arg\left[\max_{b_1} (b_1/A_{10} + (b - b_1)/A_{20})\right]$$

$$= \begin{cases} b & A_{10} < A_{20} \\ \in [0, b] & A_{10} = A_{20} \\ 0 & A_{10} > A_{20} \end{cases} \quad (11.17)$$

Equation (11.17) states that it is cost-effective to allocate the entire budget to the smallest patches (if all patches have initially equal size all solutions are cost-effective). The reason for this is that the magnitude of the marginal benefit, $|\partial T/\partial A_s|$ decreases with increasing patch size so the ratio of marginal benefits and marginal costs is highest for the small patches (note that we assumed small changes in the patch size so that the largest patches always remain the largest ones: $A_{10} > A_{20} \Rightarrow A_1 > A_2$). If marginal costs decrease,

$$k(b_s) = \beta b^\gamma \quad (11.18)$$

with constant $\gamma > 1$, the same is true: the entire budget should flow into the smallest patches (if all patches have equal sizes the budget should flow either entirely into the uneven- or into the even-numbered patches). If marginal costs increase, eq. (18) with $\gamma < 1$, a non-trivial solution appears:

$$b_1^* \approx \arg\left[\max_{b_1}(b_1^\gamma/A_{10}) + (b-b_1)^\gamma/A_{20})\right] \quad (11.19)$$

In the case of identical patch sizes, $A_{10} = A_{20}$, where marginal benefits are identical for all patches, an even allocation of the budget, $b_1 = b_2 = b/2$ is cost-effective. For uneven patch sizes, $A_{20} > A_{10}$, the cost-effective allocation can be expected to shift towards small patches ($s = 1$). Solving the first order condition for the cost-effective budget yields

$$b_1^* \approx \frac{b}{1 + (A_{10}/A_{20})^{1/(1-\gamma)}} \quad (11.20)$$

which exceeds $b/2$ if $A_{10} < A_{20}$. Note that the same conclusion might be drawn from eq. (11.20), even assuming decreasing marginal costs ($\gamma > 1$), but in this case we would expect a corner solution. To summarize, for the second type of cost function, the optimal allocation of the budget depends on the shape of the cost function but also, critically, on the sizes of the patches. In general, if patch sizes are heterogeneous more of the budget should be allocated to the smaller patches, because the marginal benefit of control is inversely related to patch size. The decline in marginal benefits with increasing patch size is due to the fact that a proportional change in the size of a patch affects the invasion time in the same way, regardless of whether it occurs in a small or in a large patch (eq. 11.8).

To summarize, if control costs are proportionate to the *share* of the patch area removed, the cost-effective allocation of control efforts depends only on the shape of the cost function (eqs. (11.10) and (11.11)). Control effort does not depend on the initial sizes of the patches. If control costs are proportionate to the *amount* of patch area removed and if all patches have identical sizes, all allocations are equally cost-effective. If patch sizes differ in size, small patches receive a greater share than larger ones. If marginal costs do not increase the entire budget is allocated to the small patches; if they do increase less than the entire budget, but more than half of it, is allocated to the small patches (eq. 11.20).

11.4 Discussion and conclusions

The model analysis above shows that spatial heterogeneity can affect the cost-effective allocation of resources to delay invasion. We considered a case where control measures affect the sizes of habitat patches for invader populations and control costs may be related to the *proportion* or to the *amount* of habitat area removed. In the former case spatial heterogeneity did not affect the cost-effective allocation of resources, but it did in the latter. These results are largely driven by the fact that spatial heterogeneity in local extinction rates affects the invasion time only via the geometric mean of these extinction rates (Fig. 11.3). This finding is novel, but also quite plausible—being consistent with both metapopulation theory and stochastic processes. Frank and Wissel (2002) develop a formula for the expected life time of a spatially heterogeneous metapopulation. They show that this life time depends on the number of patches (N) and an appropriate spatially aggregated ratio of colonization and extinction rates of the local populations. This ratio is formulated as the geometric mean over all patches of the local (patch-specific) ratios. Since these patch-specific ratios contain the local extinction rates, the paper of Frank and Wissel (2002) leads us to conclude that local extinction rates enter the lifetime of the metapopulation via their geometric mean only.

Frank and Wissel (2002), as well as papers they build on, use elements of the theory of stochastic processes (e.g. Goel and Richter-Dyn 1974) which view population dynamics as Markov processes of birth and death events. In Goel and Richter-Dyn (1974), the life time of a metapopulation is represented by the so-called mean-first-passage-time M_{0N} of a Markov process starting from a state with N occupied patches and ending at a state with zero occupied patches. The formula provided by these authors for the mean first passage time is a complicated combination of colonization:extinction ratios. The same book also provides a formula for the mean-first-passage-time M_{N1} for a Markov process starting from a state with one occupied patch and ending at a state with N occupied patches—which can be identified with the invasion time of the present analysis. The formula for M_{N1} is somewhat more complex than that for M_{0N} but has a very similar structure. The main difference is that M_{N1} depends on inverse ratios of colonization and extinction rates. This strongly indicates that metapopulation life time and invasion time are tightly but inversely related. Given our finding that the metapopulation life time is a function of the geometric mean of local extinction times, this strongly supports and generalizes that finding (Fig. 11.3).

Since the discussed role of the geometric mean of the local extinction times is a fairly general feature of metapopulation dynamics and independent of the particular spatial arrangement of the habitat patches, our results can be expected to hold also in more complex spatial settings than the one assumed in this chapter.

In this chapter we have focused on control measures that affect the local extinction rates (e) of the invasive population. One could, however, focus on the emigration of invaders from local populations. That would mean that c in eq. (11.5) is affected by management and may vary among patches. As described above, Frank and Wissel (2002) show that the metapopulation life time is a function of the geometric mean over local colonization:extinction ratios. We hypothesize that invasion time can also be written as a function where the c values of the individual patches enter only via their geometric mean. For control measures that affect c the general results can therefore be expected to be very similar to ours obtained for control measures affecting e.

Invasion processes usually occur over large heterogeneous areas, where spatial features have been shown to affect the susceptibility of areas to establishment of local invasive species population and the impact of control practices (e.g. Stephenson et al. 2006; Parry et al. 2006). This implies that management of a locally invasive population cannot be studied in isolation from the habitat structure and population dynamics at the landscape scale. Both the control costs and the vulnerability of the landscape to invasion depend on the level of spatial heterogeneity in the landscape. The spatial dimension is thus an important aspect of policies for the prevention and control of biological invasions. Our simple static spatial model indicates the need to identify where there are clusters of larger suitable patches/habitats for establishment in the landscape, because these need to be treated differently than those regions with smaller patches (provided that control costs are sensitive to the amount of area removed). Thus, areas with smaller patches should receive a greater share of control resources than areas with large ones because the marginal benefits of control practices decline with patch size. Our results are consistent, for example, with programs where corridors of unsuitable areas are established among vulnerable areas. This landscape management approach is not new in the literature, Hulme (2006) suggests altering the surrounding habitat "matrix" to limit invasions, and Gosper et al. (2005), in the context of invasive plants dispersed by birds, propose landscape manipulations, for example by placing invasive plant seed sinks in habitats that are unsuitable for seedling recruitment.

We conclude that more studies are needed that integrate the spatial and the temporal dimension of invasion processes, considering spatial heterogeneity in a landscape context (whatever the spatial scale), as well as allowing control measures to be dynamic and state-dependent. Potapov and Lewis (2008) is an example of such studies. Furthermore, the role of spatial heterogeneity raises new challenges at the institutional level particularly when there are many landowners/regions/countries involved in the invasion process. Therefore, the study of mechanisms to induce cooperation over time and space is an important challenge in the economics of invasive species.

References

Antle, J., Capalbo, S., Elliott, E., and Paustian, K. (2003). Spatial heterogeneity, contract design, and the efficiency of carbon sequestration policies for agriculture. *Journal of Environmental Economics and Management*, **46**, 231–50.

Bhat, M.G. and Huffaker, R.G. (2007). Management of a transboundary wildlife population: A self-enforcing cooperative agreement with renegotiation and variable transfer payments. *Journal of Environmental Economics and Management*, **53**, 54–67.

Bockstael, N.E. (1996). Modeling economics and ecology: the importance of a spatial perspective. *American Journal of Agricultural Economics*, **78**, 1168–80.

Brock, W.A. and Xepapadeas, A. (2005). Optimal control and spatial heterogeneity: pattern formation in economic-ecological models. Working papers 11, Wisconsin Madison—Social Systems.

Chabrerie, O., Roulier, F., Hoeblich, H., *et al.* (2007). Defining patch mosaic functional types to predict invasion patterns in a forest landscape. *Ecological Applications*, **17**(2), 464–81.

Costello, C. and McAuslandm C. (2003). Protectionism, trade, and measures of damage from exotic species introductions. *American Journal of Agricultural Economics*, **85**, 964–75.

Deacon, R.T., Brookshire, D.S., Fisher, A.C., *et al.* (1998). Research trends and opportunities in environmental and natural resource economics. *Environmental and Resource Economics*, **11**(3–4), 383–97.

Dehnen-Schmutz, K., Perrings, C., and Williamson, M. (2004). Rhododendron ponticum in the British Isles: an economic analysis. *Journal of Environmental Management*, **70**, 323–32.

Drechsler, M. and Wissel, C. (1998). Trade-offs between local and regional scale management in metapopulations. *Biological Conservation*, **81**, 31–41.

Drechsler, M. and Wätzold, F. (2001). The importance of economic costs in the development of guidelines for spatial conservation management. *Biological Conservation*, **97**(1), 51–59.

Drechsler, M., Grimm, V., Mysiak, J., and Wätzold F. (2007). Differences and similarities between ecological and economic models for biodiversity conservation. *Ecological Economics*, **62**, 232–241.

Finnoff, D., Shogren, J.F., Leung, B., and Lodge, D. (2007). Take a risk: preferring prevention over control of biological invaders. *Ecological Economics*, **62**, 216–222.

Foxcroft, C.L., Rouget, M., and Richardson, D.M. (2007). Risk assessment of riparian plant invasions into protected areas. *Conservation Biology*, **21**(2), 412–21.

Frank, K. and Wissel, C. (2002). A formula for the mean lifetime of metapopulations in heterogeneous landscapes. *American Naturalist*, **159**(5), 530–52.

Gosper, C.R., Stansbury, C.D., and Vivian-Smith G. (2005). Seed dispersal of fleshy-fruited invasive plants by birds: contributing factors and management options. *Diversity and Distributions*, **11**, 549–58.

Grimsrud, K.M., Chermak, J.M., Hansen, J., Thacher, J.A., and Krause, K. (2008). A two-agent dynamic model with an invasive weed diffusion externality: An application to Yellow Starthistle (*Centaurea solstitialis* L.) in New Mexico. *Journal of Environmental Management*, **89**(4): 322–335.

Hanski, I. (1999). *Metapopulation Ecology*. Oxford series in ecology and evolution. Oxford University Press, USA.

Heikkila, J. and Peltola, J. (2004). Analysis of the Colorado potato beetle protection system in Finland. *Agricultural Economics*, **31**, 343–352.

Higgins, S.I. and Richardson, D.M. (1996). A review of models of alien plant spread. *Ecological Modelling*, **87**, 249–265.

Higgins, S.I., Richardson, D.M., and Cowling, R.M. (2000). Using a dynamic landscape model for planning the management of alien plant invasions. *Ecological Applications*, **10**(6), 1833–1848.

Holland, D. and Schnier, K.E. (2006). Individual Habitat Quotas for Fisheries. *Journal of Environmental Economics and Management*, **51**, 72–92.

Horan, R.D. and Lupi, F. (2005). Tradable risk permits to prevent future introductions of invasive alien species into the Great Lakes. *Ecological Economics*, **52**, 289–304.

Horan, R.D., Perrings, C., Lupi, F., and Bulte, E.H. (2002). Biological pollution prevention strategies under ignorance: the case of invasive species. *American Journal of Agricultural Economics*, **84**, 1303–10.

Huffaker, R.G., Bhat, M.G., and Lenhart, S.M. (1992). Optimal trapping strategies for diffusing nuisance-beaver populations. *Natural Resource Modeling*, **6**, 71–97.

Hulme, P.E. (2006). Beyond control: wider implications for the management of biological invasions. *Journal of Applied Ecology*, **43**, 835–47.

Jones, K.R. and Corona J.P. (2007). An ambient tax approach to invasive species. *Ecological. Economics*, **64**: 534–543.

Johst, K., Brandl, R., and Eber, S. (2002). Metapopulation persistence in dynamic landscapes: the role of dispersal distance. *Oikos*, **98**, 263–270.

Jules, E.S., Kauffman, M.J., Ritts, W.D., and Carroll, A. (2002). Spread of an invasive pathogen over a variable landscape: A nonnative root rot on Port Orford cedar. *Ecology*, **83**(11), 3167–81.

Leung, B., Lodge, D.M., Finnoff, D., Shogren, J.F., Lewis, M.A., and Lamberti, G. (2002). An ounce of prevention or a proud of cure: bioeconomic risk analysis of invasive species. *Proceedings of the Royal Society: Biological Sciences*, **269**, 2407–13.

Leung, B., Finnoff, D., Shogren, J.F., and Lodge, D.M. (2006). Managing invasive species: rules of thumb for rapid assessment. *Ecological Economics*, **55**, 24–36.

Lunney, D., Pressey, B., Archer, M., Hand, S., Godthelp, H., and Curtin A. (1997). Integrating ecology and economics: illustrating the need to resolve the conflicts of space and time. *Ecological Economics*, **23**, 135–143.

McAusland, C. and Costello, C. (2004). Avoiding invasives: trade related policies for controlling unintentional exotic species introductions. *Journal of Environmental Economics and Management*, **48**, 954–977.

McKee, G.J. (2006). *Modelling the effect of spatial externalities on invasive species management*. Agribusiness and Applied Economics Report No. 583. Agricultural Experiment Station, North Dakota State University.

Munro, G.R. (1990). The optimal management of transboundary renewable resources. *Canadian Journal of Economics*, **12**, 355–377.

Neubert, M.G. (2003). Marine reserves and optimal harvesting. *Ecology Letters*, **6**(9), 843–49.

Olson, L.J. and Roy, S. (2002). The economics of controlling a stochastic biological invasion. *American Journal of Agricultural Economics*, **84**, 1311–1316.

Olson, L.J. (2007). Economics of Terrestrial Invasive Species: A review of the literature. *Agricultural and Resource Economics Review*, **35**(1), 178–94.

Parry, H.R., Evans, A.J., and Morganm D. (2006). Aphid population response to agriculture landscape change: A spatially explicit, individual-based model. *Ecological Modelling*, **199**, 451–463.

Perrings, C. (2002). Biological invasions in aquatic systems: the economic problem. *Bulletin of Marine Science*, **70**(2), 541–52.

Perrings, C., Williamson, M., Barbier, E.B., *et al.* (2002). Biological invasion risks and the public good: an economic perspective. *Conservation Ecology*, 6(1), 1 [online]. URL: http://www.consecol.org/vol.6/iss1/art1

Potapov, A.B., Lewis, M.A., and Finnoff, D.C. (2007). Optimal control of biological invasions in lake networks. *Natural Resource Modeling*, **20**(3), 351–79.

Potapov, A.B. and Lewis, M.A. (2008). Allee effect and control of lake system invasion. *Bulletin of Mathematical Biology*. **70**(5): 1371–1397.

Sanchirico, J. and Wilen, J. (1999). Bioeconomics of spatial exploitation in a patchy environment. *Journal of Environmental Economics and Management*, **37**, 129–150.

Sanchirico, J. and Wilen, J. (2001). A Bioeconomic Model of Marine Reserve Creation. *Journal of Environmental Economics and Management*, **42**, 257–76.

Sanchirico, J. and Wilen, J. (2005). Optimal Spatial Management of Renewable Resources: Matching Policy Scope to Ecosystem Scale, *Journal of Environmental Economics and Management*, **50**, 23–46.

Sharov, A.A. and Liebhold, A.M. (1998). Bioeconomics of managing the spread of exotic pest species with barrier zones. *Ecological Applications*, **8**(3), 833–45.

Shigesada, N., and Kawasaki, K. (1997). *Biological Invasions: Theory and Practice*. Oxford University Press, Oxford.

Skonhoft, A. and Olaussen, J.O. (2005). Managing a migratory species that is both a value and a pest. *Land Economics*, **81**(1), 34–50.

Smith, M.D., Sanchirico, James N., and Wilen, J. (2007). The economics of spatial-dynamic processes: Aplications to renewable resources. Resources for the Future Discussion Paper No. 07-27-REV.

Stephenson, C.M., Kohn, D.D., Park, K.J., Atkinson, R., Edwards, C., and Travis, J.M. (2007). Testing mechanistic models of seed dispersal for the invasive *Rhododendron ponticum* (L.). *Perspectives in Plant Ecology, Evolution and Systematics*, **9**, 15–28.

Touza, J., Dehnen-Schmutz, K., and Jones, G. (2007). Economic analysis of invasive species policies. In W. Nentwig, ed. *Biological Invasions, Ecological Studies vol. 193*. Springer, Berlin, pp. 353–66.

PART III
MANAGEMENT AND POLICY

CHAPTER 12

The Impact of Invasive Alien Species on Ecosystem Services and Human Well-being

Liba Pejchar and Harold Mooney

12.1 Introduction

12.1.1 Invasive alien species and global change

Invasive alien species, defined as those non-native species which threaten ecosystems, habitats, or species (CBD 2008), are key drivers of human-caused global environmental change (Vitousek et al. 1997; Lonsdale 1999; Williamson 1999; D'Antonio et al. 2004). Widely heralded as the second greatest agent of species endangerment and extinction after habitat destruction, particularly on islands (Wilcove et al. 1998), invasive alien species (IAS) are also inflicting serious impacts on the ecosystem processes that are the life-line of the human enterprise (Mack et al. 2000; Zavaleta 2000; Mooney 2005). These changes have global consequences for human well-being including the wholesale loss or alteration of goods (agricultural and forest products, fisheries) and services (clean and plentiful drinking water, climate stabilization, pollination, culture, and recreation) (Daily 1997).

Much effort has gone into understanding what makes a species invasive (Lonsdale 1999; Sakai et al. 2001) and on documenting ecological effects of the invasion (Parker et al. 1999). Although invasion-driven changes to the structure and functioning of ecosystems are well-documented (Mooney and Hobbs 2000), little is known about the mechanisms linking invasive species to ecosystem services (Charles and Dukes 2007). Additionally, the economic impact of invasive species on these services is often not quantified (Mack et al. 2000) nor incorporated into economic impact assessments (Colautti et al. 2006). As such, the impacts of invasive species can result in an "invisible tax" on ecosystem services that is rarely included in decision making.

12.1.2 Quantifying impacts of invasive alien species

There have been several attempts to quantify the economic impact of IAS at a national level (OTA 1993; Pimentel et al. 2000; Pimentel et al. 2005; Xu et al. 2006). In these cases, impacts are staggering (e.g. $14.45 billion in China) but largely anecdotal and wide-ranging. For example, OTA (1993) and Pimentel et al. (2000, 2005) figures for the total cost of invasive species in the US range from $131 billion cumulative to $128 billion annually but do not use systematic empirical methods of estimating costs and do not consider benefits, making it difficult to judge the validity of the estimates (Perrings et al. 2000; Born et al. 2005; Lovell et al. 2006). In addition, many effects of invasive species on non-market based ES are overlooked (Costello and McAusland 2003; Duncan et al. 2004; Born et al. 2005; Hoagland and Jin 2006).

Because current calculations of economic impacts are hampered by missing data on non-market services and thus are probably underestimates (Eiswerth and van Kooten 2002; Perrings et al.

2002; Keller *et al.* 2007), policy responses to-date have been based on very rough estimates of economic damages (Hoagland and Jin 2006) that may be unsuitable for developing a national strategy for IAS (Born *et al.* 2005). Filling the economic assessment gap may be very worthwhile if more data demonstrates that today's investments in prevention and eradication could save us millions of dollars in diminished losses to human health, agriculture, forestry, and the preservation of natural systems and the services they provide in the future (Simberloff 2003).

12.1.3 Defining ecosystem services

Ecosystems are life-support systems that provide a suite of goods and services that are vital to human health and livelihood. These assets are referred to as "ecosystem services" and are defined as benefits that people obtain from ecosystems (MA 2005). In this review we discuss three of the four categories of ecosystem services as defined in the Millennium Ecosystem Assessment:

- Provisioning services: food, fuel, fiber, medicine, and fresh water.
- Regulating services: pollination, climate, water, natural hazards, air quality, and disease regulation.
- Cultural services: recreation and tourism, aesthetic, and cultural heritage values.

We excluded supporting services from this review because the impacts of IAS on these services (nutrient cycling, photosynthesis, etc.) have been reviewed extensively elsewhere (Vitousek 1990; Dukes and Mooney 2004). With the exception of some of the provisioning services, most ecosystem services have been traditionally viewed as free "public goods" (e.g. clean water, climate mitigation, and clean air). Lacking a formal market, these natural assets are usually overlooked in public, corporate, and individual decision-making (Daily *et al.* 1997).

In the following sections we review the literature to understand the significance of making decisions about the prevention and/or control of IAS that ignore impacts on ecosystem services. For the most part, we report damage costs associated with IAS in monetary terms. The costs we present for various provisioning, regulating, and cultural services may be roughly comparable since most of the literature mostly clusters around the early 2000s. Whether damage costs of any magnitude will change the way we manage IAS will naturally depend on the benefits of the activities that lead to the introduction and spread of each species. We suggest that identifying potential damage costs and estimating their magnitude is a positive first step towards properly accounting for the full impact of IAS.

12.2 Mechanisms

IAS are most well known for their impact on biodiversity but can also act as keystone species, precipitating massive changes to ecosystem structure and function that, in turn, affect the provision of ecosystem services. Much research shows that alien species that are most likely to invade differ from native species in some trait behavior or function that can be either discrete (e.g. nitrogen fixation) or continuous (e.g. growth rate). The mechanisms by which these changes occur, however, are not always well understood (Levine *et al.* 2003). Some possible mechanisms include exploitation competition (e.g. resources), interference competition (e.g. allelopathy), and direct predation, herbivory, and parasitism (Vitousek 1990; Crooks 2002).

IAS affect ecosystem services using these mechanisms at four levels of complexity: 1) species, 2) communities, 3) ecosystems, and 4) the atmosphere (Levine *et al.* 2003; Charles and Dukes 2007). At the species level, competition and other impacts of IAS can lead to the decline or loss of economically valuable species for food, fiber, forage, fuel, and medicine. When invasions result in extinctions, there is also a loss of option value—the opportunities forgone as a result of loss of evolutionary or exploitation potential (Perrings *et al.* 2000). Extinctions can also lead to a decrease in resilience to disturbance and increased vulnerability to more invasions which could magnify impacts on ecosystem services.

At the community level alien invasions may lead to a loss of aesthetic value (e.g. kudzu covers forests and houses in the southeastern US) (McNeely 2001a). IAS may also disrupt mutualisms, compromising pollination and pest control services for agriculture (Traveset and Richardson 2006). Finally, IAS can change the physical environment by increasing soil salinity (Zavaleta 2000; Vivrette and Muller 1977) or through allelopathy (Orr et al. 2005).

At the ecosystem level IAS can alter trophic interactions and decrease water quality (e.g. golden apple snail; Naylor 1996). Changes in nutrient cycling from invasive plants that fix nitrogen can benefit IAS and alter fire frequencies (e.g. *Myrica faya* in Hawaii; Mack and D'Antonio 1998). The loss of native "fire-resistant" forests or grasslands means the loss of a "fire-prevention" ecosystem services to neighboring human and natural communities. Finally, several IAS substantially alter hydrological cycles through high evapotranspiration rates, by lowering the water table, and by changing the timing and magnitude of runoff (e.g. *Tamarisk* in southwestern US, Zavaleta 2000; acacias and pines in South Africa, Le Maitre et al. 1996).

The mechanisms by which IAS impact climate and atmospheric composition are least studied. IAS may alter the composition of the atmosphere by changing rates of carbon dioxide sequestration, emitting gases or volatile organic compounds that have adverse health effects. Across all of these levels of complexity, IAS affect populations, community interactions, ecosystem processes, and abiotic variables. This cascading effect of biological invasions can lead to impacts on multiple ecosystem services.

12.3 Ecosystem services

In the following discussion, we present examples and synthesis regarding the effects of IAS on particular provisioning, regulating, and cultural services.

12.3.1 Provisioning

12.3.1.1 FOOD, FIBER, AND FUEL
Introduced species are both a blessing and a curse for agriculture and food security worldwide. For instance, most food crops are non-native species, yet introduced "weedy" species or IAS can also reduce crop yields by billions of dollars annually (OTA 1993).

The impacts of a few alien invasive plants on agriculture are very well documented. For instance, widespread invasive annual herbs and grasses such as Leafy Spurge (*Euphorbia esula*), knapweeds (*Centaurea* spp.), and Yellow Starthistle (*Centaurea solstitialis*) reduce income from grazed lands (Duncan et al. 2004). Leafy Spurge alone costs western US landowners up to $130 million and 1433 jobs annually (Leitch et al. 1994) and the invasion of knapweed has a $42 million annual impact on Montana's economy (Hirsch and Leitch 1996). Eagle et al. (2007) calculated that Yellow Starthistle, which is unpalatable to cows and can be toxic to horses (Cordy 1978), costs California $7.65 million annually in lost livestock forage value and costs ranchers $9.45 million in out-of-pocket expenditures. These numbers amount to 7 per cent of all revenue from harvested pasture in California.

Australia faces similar economic threats from IAS. The weeds *Chondrilla juncea*, *Heliotropium europaeum*, and *Avena fatua* cost an estimated A$10m, A$40m, and A$42m respectively in lost crop revenue and *Echium plantagineum*, *Onopordum* spp., and *Vulpia* spp. cost A$30m, A$20m, and A$30m respectively in lost revenue from pastureland (Watkinson et al. 2000). These losses are particularly significant where farming is precarious and droughts are not uncommon.

Cost-effective management of IAS involves weighing costs from damage against costs of control (eradication or containment). Because IAS can have such substantial impacts on food, fiber, and fuel, the costs of control can be lower than the damage. For example, Tansy Ragwort (*Senecio jacobaea*) poisons livestock and results in lost livestock forage. Coombs et al. (1996) found that biological control of tansy ragwort generates a benefit/cost ratio of 15:1 with a 7 per cent discount rate. In the case of severe infestations, however, the private costs of removal may overwhelm the private benefits. If a social benefit cost analysis favors control, public subsidies may be needed to implement a biological control program (Jetter et al. 2003).

Despite the above exceptions, comprehensive economic impact data are lacking for most IAS in agricultural systems (Eagle et al. 2007). In addition, environmental and societal costs are often not included in analyses of even the best-documented invasive species (Duncan *et al.* 2004), even where they have been demonstrated to have serious impacts on ecosystem services such as increased fire risk (Lodge *et al.* 2006: see Subsection 12.3.2.6). Accounting for all effects frequently changes the benefit-cost analysis. For instance, Johnson *et al.* (1999) demonstrate that controlling Redberry Juniper (*Juniperus pinchotii*) in Texas rangelands is economically feasible over a 30-year time horizon because of increased livestock production resulting from control. The net benefits of controlling this species would be even higher if other services such as increased water available to recharge aquifers were included in the analysis (Texas Soil and Water Conservation Board 1991).

Vertebrate invaders are just as damaging as IAS plants to agricultural goods and services. European rabbits (*Oryctolagus cuniculus*) cost Australia A$600m annually in pasturelands (Wilson 1995), and European starlings and mynas devastate grain and fruit crops in North America and Asia (Somers and Morris 2002; Long 1981). Insects and pathogens can also do serious damage to crops and farm animals, resulting in widespread hunger and famine (McNeely 2001a) and substantial economic losses. For example, Oliveira (2001) documented that the impact of introduced whiteflies on melon and sesame crops in Mexico costs farmers up to US $33 million annually.

In addition to impacting terrestrial agriculture, IAS can also have important repercussions for aquatic food production and other services (Lovell *et al.* 2006). For example, the introduction of Water Hyacinth into Lake Victoria has reduced the production and quality of fish, obstructed waterways and boat movement, damaged water supply intakes, contributed to the spread of waterborne diseases, and increased water loss through evapotranspiration. Because water hyacinth has little economic value for livestock food or paper production, its economic costs far outweigh any benefits (Joffe and Cooke 1997; Kasulo 2000).

In contrast, the physical and economic impact of IAS fish in African lakes depends on the original state of the invaded lake. Invasions have been less costly in lakes with low fish diversity and production and more costly where fish diversity and production was high (Kasulo 2000). This is particularly true in the case of Lake Victoria where the introduction of Nile Perch caused the collapse of the native cichlid fishery and has led socially unfavorable redistribution of wealth. Similarly, Comb Jelly (*Mnemiopsis leidyi*) have had a devastating impact on anchovy fisheries in the Black Sea (Shiganova *et al.* 2001), which Knowler and Barbier (2001) calculate as amounting to hundreds of millions of US dollars over several decades. Finally, Naylor (1996) demonstrated that the introduced Golden Apple Snail has resulted in a US $12.5–17.8 million loss to rice production in the Philippines.

The nature of the impact of IAS on food, fiber, and fuel is almost always a matter of scale and perspective. Serbesoff-King (2003) have shown that an invasive tree in Florida (*Melaleuca quinquenervia*) has a positive impact on honey production that is worth over $15 million annually, but removing this species would provide $168.6 million in tourism benefits (Diamond *et al.* 1991). The introduction of Brush-tailed Possums to New Zealand has resulted in massive defoliation, but is highly profitable for the fur industry. In both of these cases, the costs and benefits are distributed differently—those who benefit do not pay the costs and thus who lose are not compensated (McNeely 2001b).

Such non-crop IAS can have complex and sometimes beneficial impacts on subsistence communities in particular (Kaufmann 2004) (Table 12.1). For example, invasive *Acacia* and *Pinus* species have resulted in reduced stream flow and increased fire intensity (see Subsection 12.3.1.2; Richardson and van Wilgen 2004). However, these species are also an important resource for thatching, timber, medicine, charcoal, and firewood in South Africa (McNeely 2001b; de Neergaard *et al.* 2005); the economic value of the firewood alone is $2.8 million (Turpie *et al.* 2003). Because introduced species are often incorporated into local livelihoods, it is not possible to assume that harmful impacts on biodiversity or other ecosystem goods and services automatically translate into negative effects on human well-being (Shackleton *et al.* 2006) (Table 12.1).

Summary: In several well-documented cases, impacts of IAS on food, fiber, and fuel have been calculated in monetary terms and considered

Table 12.1 The impact of woody alien invasive plants on ecosystem services in the Cape Floristic Region, South Africa

Services Impacted	Mechanisms	Positive/ Negative	$-value (cost/benefit)	Source
Food	Reduced grazing area; less freshwater for fishing	+/−		Richardson and van Wilgen 2004; van Wilgen and Richardson Chapter 13 *this volume*
Fiber	Timber; flowers and thatching reed	+	$300 million/yr from forestry; $1.6 billion/yr in value-added wood products; $18 million/yr in lost flower and grass earnings	Cowling and Richardson 1995; Turpie and Heydenrych 2000; Le Maitre et al. 2002; Turpie et al. 2003; de Neergaard et al. 2005; van Wilgen and Richardson Chapter 13 *this volume*
Fuel	Firewood	+	$2.8 million	Turpie et al. 2003; de Neergaard et al. 2005; Shackleton et al. 2006
Fresh water	Uses more water than native species	−	$1.4 billion in water lost to transpiration; up to 30% of water supply; R1140/ha	Le Maitre et al. 1996; van Wilgen et al. 1996; Turpie et al. 2003; Le Maitre et al. 2002; De Wit et al. 2001
Medicine	Displace fynbos plants used for drugs and tea; loss of option value (undiscovered medicinal plants)	−	Rooibos tea exports worth $2.1 million (1993)	van Wilgen et al. 1996: Turpie and Heydenrych 2000
Pollination	Eucalyptus increase honey production; displaced flowers—loss of native nectar	+/−	R500/ha in lost pollination services	Johannsmeier and Mostert 1995; Turpie and Heydenrych 2000
Climate regulation	More carbon sequestration	+		
Erosion Control	More intense fires result in soil loss with rainwater runoff	−		Scott and van Wyk 1990; Scott and Schulze 1992; Scott et al. 1998; van Wilgen and Scott 2001
Natural hazards regulation	Increased biomass/fuel load; increased runoff following erosion causes flooding	−		van Wilgen and Richardson 1985; van Wilgen and Scott 2001; van Wilgen and Richardson Chapter 13 *this volume*
Aesthetic value	Ornamentals, shade trees; loss of fynbos wildflowers	+/−		van Wilgen and Richardson Chapter 13 *this volume*
Recreation and Tourism	Invasion of dunes has lead to loss of beaches; damage to fynbos ecotourism	−		Lubke 1985; Le Maitre et al. 1996; van Wilgen et al. 1996
Cultural Heritage	Displaces native flora for flower harvesting; disturbs sacred pools; wood used for ceremonies	+/−		Richardson and van Wilgen 2004; Shackleton et al. 2006

in decision-making (Born *et al.* 2005). Yet data on the economic impacts of most IAS on agriculture is lacking even for developed nations, and is rarely integrated with impacts on other services. In addition, the usual focus of science and assessments to-date in regards to food, fiber, and fuel is on economic impacts at national or regional scales which is not always meaningful at local scales and to the livelihoods of rural people, particularly in developing nations.

12.3.1.2 FRESH WATER

With several important exceptions (Zavaleta 2000; Gorgens and van Wilgen 2004; Holmes and Rice 1996), relatively few studies have documented the impacts of IAS on hydrologic services (Levine *et al.* 2003). A few well-documented examples, however, demonstrate that some IAS plants can fundamentally change the flow of water for drinking and irrigation if they have at least one of the following characteristics: 1) deeper roots, 2) higher evapotranspiration rates, or 3) greater biomass than native species (Mooney 2005).

Salt Cedar (*Tamarix ramosissima*), a widespread invasive alien tree along streams in the southwestern US, consumes more water than native riparian species by maintaining high transpiration and leaf area (Sala *et al.* 1996; but see Glenn *et al.* 1998). Because tamarisk uses an additional 1.4 billion cubic meters of water each year, $26.3–67.8 million of water is lost annually that would otherwise be available for irrigation, municipal drinking water, or hydropower (Zavaleta 2000). Similarly, Yellow Star Thistle (*Centaurea solstitialis*), an invasive late season annual in the Central Valley, depletes soil moisture, costing between 16–75 million dollars a year in lost water to the Sacramento watershed (Gerlach 2004).

Melaleuca quinquenervia, an invasive alien tree in Florida and Australia, and several *Eucalyptus* species, which are IAS in California, both use large amounts of water relative to their host native plant communities because of their deep tap roots (Schmitz *et al.* 1997). In contrast, in the midwestern US, invasive alien grasses have shallow roots and therefore may use less water than the native perennial grasses which they displace (Cline *et al.* 1977; Rickard and Vaughan 1998; Holmes and Rice 1996).

Gorgens and van Wilgen (2004) have documented that many invasive alien woody plants in South Africa (*Melia azedarach*, *Pinus* spp., *Acacia mearnsii*, *Prosopis* spp., *Lantana camara*) which have high evapotranspiration rates, have a negative impact on hydrologic services by decreasing the amount of surface water and the magnitude of stream flow (Table 12.1). Finally, invasive alien plants are not the only taxa that can affect hydrologic services; by building dams, introduced beavers can divert streams and slow flow (Naiman *et al.* 1998). The rapid spread of the North American Beaver (*Castor Canadensis*) in Tierra del Fuego has modified ecosystem hydrology by expanding wetlands, elevating the water table and altering nutrient cycling in riparian areas (Lizarralde 1993).

Summary: Water is increasingly limited yet indispensable. With the exception of several excellent examples, few studies document the economic impacts of IAS on water availability. Understanding how IAS affects hydrology, calculating costs, and making both transparent, are all part of gathering important information and support for protecting water resources.

12.3.1.3 MEDICINE
Summary: IAS may compromise current and potential medicinal resources (option value) provided by ecosystems by driving native plants to extinction. Research is needed to specifically address the link between IAS and the loss of medicinal plants.

12.3.2 Regulating Services

The economic impacts of IAS on regulating services remain largely undetermined. These impacts, however, may dwarf impacts on the far better understood provisioning services discussed above. As a result, invasive species could have hidden cascading impacts on regulating services such as pollination, water purification, pest control, natural hazards, and climate mitigation. These services are both the cornerstone of fisheries, agriculture, and forestry and fundamental to human well-being (Colautti *et al.* 2006).

12.3.2.1 POLLINATION

Non-native European honeybees (*Apis mellifera*) are widely used to pollinate crops (Gross 2001), providing indispensable services for farmers, particularly in areas where native pollinators are scarce. These pollination services are worth an estimated $14.8 billion annually in the US (Morse and Calderone 2000). In some cases, however, honeybees act as IAS by disrupting mutualisms (Kenta et al. 2007) and displacing native bees which may be better pollinators (Spira 2001). Non-native bees also have the potential to allow range expansion in pollinator-limited "sleeper weeds" (Stokes *et al.* 2006).

For these reasons, and because of the recent onset of colony collapse disorder, a multifactorial syndrome causing mass honeybee mortality (Oldroyd 2007), and due to the spread of dangerous Africanized bees (Williamson 1996)—an invasive and often lethal subspecies of the European honeybee (Schumacher and Egen 1995)—there is increasing interest in reducing dependence on honey bees. Native bees could provide important insurance against ongoing honey bee losses (Winfree *et al.* 2007). If sufficient habitat is restored to support native bees, both crops and wildlands will benefit from the "free" pollination services they provide (Kremen *et al.* 2002, 2004; Ricketts *et al.* 2004).

Invasive alien plants and pathogens also have an impact on pollination services. Pathogens can lead to the extinction of naïve bird or insect pollinators and IAS plants can distract both native and non-native pollinators from native species (Chittka and Schürkens 2001). Invasive plant species can also displace native plants that provide important nectar and pollen resources. For example, in South Africa the introduction of Acacia and Eucalyptus species have resulted in fewer native flowering shrubs and a $27 million loss to the bee keeping industry (Johannsmeier and Mostert 1995; Turpie and Heydenrych 2000) (Table 12.1).

Summary: Much research is now underway on the relative role of native and non-native arthropods in pollination which has repercussions for a variety of ecosystem services such as food, fiber, aesthetic, and cultural values. Little science exists on the impact of other alien invasive taxa (plants, parasites, pathogens) on pollination services. Understanding the ecological dimensions and economic impacts of pollination is crucial for food security and the maintenance of agricultural and natural plant communities.

12.3.2.2 CLIMATE REGULATION

When IAS replace native species, differences in carbon storage capacity could affect the amount of carbon dioxide released into the atmosphere. For example, non-native annual grasses have largely replaced native sagebrush in the US Great Basin region. This net loss of carbon sequestration over a large land area could contribute to climate warming (Prater *et al.* 2006). Carbon storage capacity has also been lost from the Brazilian Amazon as non-native pasture grasses, which have far less biomass, have steadily replaced rainforest (Kaufman *et al.* 1995, 1998).

In contrast, more carbon may be sequestered when non-native woody species replace native grassland (Robles and Chapin 1995). Hughes *et al.* (2006) found that the encroachment of *Prosopis glandulosa* (a native invasive species) in the Southern Great Plains of the US significantly increased above-ground carbon storage but had no impact on soil carbon pools. Non-native timber plantations can also make important contributions to carbon sequestration, depending on the site and the efficiency of the harvest (Marland and Schlamadinger 1997; Shan *et al.* 2001).

Summary: Although there is much focus on how climate change will affect the spread of invasive species, the extent and implications of the impact of NIS on climate change is yet to be determined.

12.3.2.3 WATER PURIFICATION

IAS in aquatic ecosystems have had mixed consequences for water purification. For example, by altering the food web, the Golden Apple Snail (*Pomacea canaliculata*) has transformed Southeast Asia's wetlands from a clear water purification system to a turbid, algae-dominated state (Carlsson *et al.* 2004). In addition to dramatic impacts on water quality, the Golden Apple Snail has also had serious economic repercussions for rice production (see Subsection 12.3.1.1 above; Naylor 1996). Introduced fish such as the Common Carp (*Cyprinus carpio*) in Spain can also degrade water quality by increasing nutrient concentrations (Angeler *et al.* 2002). In contrast, on occasion a non-native species can actually increase water filtration and purification, but often not without impacts on other important services. The Zebra Mussel (*Dreissena polymorpha*) is the poster child for a very effective biological filtration machine (Nalepa and Schloesser 1993) that also has done serious damage to the ecological and economic value of an entire region. In the Great Lakes (US) Zebra Mussels coat boats and beaches and clog water intakes of municipal water supplies and hydroelectric companies (Kovalak *et al.* 1993) (Table 12.2).

Summary: Wetlands are widely recognized as providing free water purification services that would otherwise be costly and chemical-intensive. Apple Snails and Zebra Mussels suggest that keeping both wetland and other aquatic areas free of IAS could save millions of dollars in replaced services.

12.3.2.4 EROSION REGULATION

Erosion is a natural process that shapes landscapes. If the rate of erosion is exacerbated by human impacts or IAS, however, erosion can negatively impact water quality, agricultural production, and compromise the stability of land under homes. IAS can change the nature and magnitude of erosion in an ecosystem through multiple mechanisms: 1) IAS can alter soil properties, 2) root structure of IAS can change soil-stabilization capacity, and 3) vertebrate IAS may eat plant biomass, including roots, causing increased erosion (Table 12.3).

By burrowing into coastal banks, *Sphaeroma quoyanum*, an invasive isopod, compromises bank stability and causes increased erosion into San Francisco Bay (Cohen and Carlton 1995). The extent to which this has detrimental impacts on fishing and navigation is not well-documented.

The introduction of Australian Pines (*Casuarinas equisetifolia*) has created forests on Florida's otherwise naturally tree-less coastlines causing increased erosion by excluding native soil stabilizers (Schmitz *et al.* 1997). When non-native forbs, such as *Centaurea maculosa*, replace native bunchgrasses in the western US rangelands, erosion increases because the root structure of bunchgrasses is better equipped to stabilize the soil (Lacey *et al.* 1989). The loss of topsoil and the grasses that hold it in place leads to lower quality rangelands.

Introduced ungulates, particularly on islands, have caused severe erosion by eating many plants that evolved in the absence of herbivory, which hold hillsides in place (Laughrin *et al.* 1994; North *et al.* 1994; Mack and D'Antonio 1998) (Table 12.3).

In several cases, IAS have been introduced deliberately for their ability to limit erosion but these introductions frequently have unintended consequences for other ecosystem services. For instance, kudzu (*Pueraria lobata*) was introduced to the southeastern US for erosion control (Bailey 1939) but is now an economic liability, smothering trees, homes, and telephone poles as well as impacting on air quality (see Subsection 12.3.2.7 below). Similarly, Pampas grass (*Cortaderia jubata*) is excellent at controlling erosion along roadsides, but also interferes with natural movement of dunes and aggressively displaces native plants (Lambrinos 2000).

Summary: Evidence exists that IAS impact erosion through several mechanisms. These impacts, however, are rarely translated into economic terms.

12.3.2.5 PEST AND DISEASE REGULATION

Invasive alien plants can serve as novel habitat for vectors, increasing incidence of disease. For example, the invasion of dense stands of *Lantana camara* in East Africa has provided new habitat for tsetse fly which carry sleeping sickness, leading to a higher occurrence of the disease (Greathead 1968). Similarly, by collecting water, non-native tank bromeliads provide new breeding habitat for potentially disease-carrying mosquitoes (O'meara *et al.* 1995). Water Hyacinth (*Eichhornia crassipes*), which was introduced to China to provide food for livestock and to control pollution by absorbing heavy metals, now smothers lakes in thick vegetation that provides habitat for disease-carrying mosquitoes and flies (Jianqing *et al.* 1995).

IAS can also themselves be vectors for dangerous diseases. Invasive alien snails serve as intermediate hosts of the Rat Lungworm (*Angiostrongylus cantonensis*) which causes the fatal disease *eosinophilic meningoencephalitis* in humans (Lo Re and Gluckman 2003). The Australian Brushtail Possum (*Trichosurus vulpecula*) transmits

Table 12.2 The multiple impacts of the Zebra Mussel (*Dreissena polymorpha*) on ecosystem services in the Great Lakes

Services Impacted	Mechanisms	Positive/ Negative	$-value (cost/benefit)	Sources
Food	Changes light environment and competes with fish for zooplankton prey (macrozooplankton); prey on microzooplankton; source of food for some fishes, crayfish, and diving waterfowl (adult mussels)	+/−	$32.3 million/yr in net costs to aquaculture	Mayer et al. 2002; MacIsaac et al. 2002; Beekey et al. 2004; Colautti et al. 2006; Lumb et al. 2007
Freshwater	Clogs intake pipes: increase in stranded macrophytes on water intake screens and in water treatment plants impairs flow	−	339 water dependent facilities reported total Zebra Mussel-related expenses of $69,070,780 from 1989–95; control costs of average large water user: $400,000–$460,000 annually	Kovalak et al. 1993; LePage 1993; O'Neil 1997; MacIsaac 1996; Lovell et al. 2006
Disease regulation	Accumulate mercury and lead (in fish we eat); avian botulism	−		Hogan et al. 2007
Water Purification	Efficient filter feeder; imparts odor in drinking water due to release of geosmin; changes nutrient fluxes, resulting in phytoplankton and cyanobacterial blooms	+/−	Windsor, Ontario spent $400,000 CDN/yr to eliminate taste and odor problems	Reeders and Bij de Vaate 1990; Holland 1993; MacIsaac 1996; MacIsaac et al. 2002; Conroy et al. 2005; Zhu et al. 2006; Knoll et al. 2008
Aesthetic value	Covers beaches and boats	−		
Recreation and Tourism	Cover beaches, boats, docks and piers; cyanobacteria blooms; increase in organochlorine and heavy metals in some recreational fish and ducks that prey on them	−	Threatens $4 billion sports fishery; costs boat owners $660 annually in upkeep	Badzinki and Petie 2006; Mazak et al. 1997; Fernald et al. 2007; Raikow et al. 2004; Lovell et al. 2006
Cultural Heritage	Shipwrecks spotted more easily; fouls shipwrecks; concentrates heavy metals dangerous to divers; can cut bathers feet	+/−		LaValle et al. 1999

Table 12.3 The impacts of feral pigs (*Sus scrofa*) on ecosystem services in Hawaii

Services Impacted	Mechanisms	Positive or Negative	$-value impact	Source
Food	damages crops; provides subsistence food	+/−	50% of some nut crops lost	Maguire et al. 1997; Burrows et al. 2007; Hadway pers. comm.
Medicine	Damages medicinal plants	−		Maguire et al. 1997
Water purification	Deposits fecal matter in waterways	−		
Erosion Control	Eat roots, create wallows and trample soils	−		Ralph and Maxwell 1984; Vitousek 1986; Stone and Loope 1987
Disease regulation	Creates breeding habitat for disease-carrying mosquitoes; transmits brucellosis/toxoplasmosis; spreads plant pathogens	−		Kliejunas and Ko 1976; Cuddihy and Stone 1990; Atkinson 1995
Natural hazards regulation	Probably increases risk of flooding through erosion	−		
Cultural Heritage	damages cultural plants; used for cultural events; has spiritual/religious value	+/−		Maguire et al. 1997; Maguire 2004; Burrows et al. 2007
Recreation and Tourism	Very important for hunting over last 150 yrs; damages trails and forests	+/−	$450,000/yr in national parks	Ralph and Maxwell 1984; Loope and Scowcroft 1985; Loope et al. 1991; Loh and Tunison 1999; Ikuma et al. 2002; Burrows et al. 2007

bovine tuberculosis to cattle and deer in New Zealand, posing a large economic threat that has led to millions of dollars in control costs (Clout 1999).

Invasive species can also change food web dynamics, making systems more vulnerable to disease. For example a native but aggressively invasive bird in Australia, the Noisy Miner (*Manorina melanocephala*), excludes native insectivorous birds in woodlands which leads to insect outbreaks and reduced tree health (Grey 1998). The predatory red imported fire ant (*Solenopsis invicta*) has negative impacts on native biological control agents in soybeans, leading to increased herbivory and economic loss (Eubanks 2001).

Introduced arthropods such as mosquitoes and Gypsy Moths (*Lymantria dispar*) can impair human health and happiness. Gypsy Moths defoliate forests, degrading the aesthetic and recreational quality of ecosystems. Invasive mosquitoes have exacerbated the spread of yellow fever and dengue fever, both deadly to humans if untreated, in the Americas and Asia (Juliano and Lounibos 2005). The spread of feral pigs (*Sus scrofa*) in Hawaii has catalyzed the spread of avian malaria (*Plasmodium relictum*), a devastating introduced disease that has played a major role in the extinction of Hawaii's avian pollinators and dispersers. Pigs dig hollows where water can collect and serve as breeding grounds for mosquitoes which carry avian malaria (Cuddihy and Stone 1990) (Table 12.3).

Summary: Pests and pathogens affect human health and well-being directly and cost society millions of dollars in control costs. Determining the role of IAS in disease pathways could help inform and motivate control efforts that would benefit multiple ecosystem services and save human lives.

12.3.2.6 NATURAL HAZARDS REGULATION: FIRES AND FLOODING

The alteration of fire regimes by IAS has been very well-studied (D'Antonio 2000). IAS can change fuel properties and ultimately alter the frequency, intensity, extent, type, and seasonality of fire (Brooks et al. 2004; D'Antonio and Vitousek 1992). Examples include the large-scale invasion of an annual grass, *Bromus tectorum*, into the North American shrub-steppe community. *Bromus tectorum* is fire-adapted and has increased the frequency of fires in this system (Billings 1990; Chambers et al. 2007). Because the native shrub-steppe plant community cannot regenerate in the face of this heightened fire frequency, the introduction of *B. tectorum* has permanently changed the plant community and increased the likelihood of fire with all of its potentially negative impacts on human livelihoods and communities (Whisenaut 1990), a problem that may be exacerbated by global increases in atmospheric CO_2 (Ziska *et al*. 2005). A similar transformation has taken place in Hawaii after the invasion of exotic grasses (D'Antonio and Vitousek 1992).

These altered fire regimes can come with substantial social and economic costs. An introduced tree, *Melaleuca quinqeunervia*, has increased fuel load and fire frequency in Australia and Florida (Schmitz *et al*. 1997). In Florida alone this species will cause $250 million in fire damages by 2010 (Serbesoff-King 2003). Another IAS has spread rapidly in riparian areas, *Arundo donax*. The fire tolerant *A. donax* increases fuel load in riparian areas and has serious negative consequences for plant communities that are generally not fire-adapted yet play important roles in erosion control and water purification (Jackson 1993; Scott 1993)

IAS can increase flood-risk by narrowing stream channels and decreasing holding capacity, causing millions in damages (Zavaleta 2000). *Tamarisk* (see Subsection 12.3.1.2 *freshwater* section above) is the best illustration of this impact on flooding. The floods that have resulted from the introduction of *Tamarisk* cost US $52 million annually in damages. The introduction of beavers into novel riparian areas may also increase flood risk to some communities, as well as altering other hydrologic services (Lizarralde 1993; see Subsection 12.3.1.2 above). Removing invasive alien aquatic plants from Florida's lakes and waterways results in $10 million annually in avoided flood damages to residential structures (Thunberg *et al*. 1992) and $6,345 per acre in avoided flood damage to citrus crops (Thunberg and Pearson 1993).

Summary: By increasing the intensity or frequency of fires or floods, IAS exclude native species that may provide key ecosystem service, and increase risk to nearby human communities. The "fire and flood-prevention" services that some native ecosystems supply are generally under-appreciated. These services should be accounted for in controlling IAS and paying for ecosystem services.

12.3.2.7 AIR QUALITY REGULATION
Summary: Native forests help maintain air quality by intercepting particulates. As described above, *Bromus tectorum* and other invasive alien species can increase fire frequency, releasing more particulates into the air. This impact is believed to be small, however, relative to other pollutants (D'Antonio and Vitousek 1992). Some evidence suggests that both kudzu (*Pueraria lobata*), an invasive vine that has smothered many parts of the southeastern US, and introduced *Eucalyptus* have been instrumental in releasing more isoprene into the atmosphere, which can enhance the production of air pollutants (Evans *et al*. 1982; Wolfertz *et al*. 2003). Additional research on invasive alien plants, which may degrade air quality with consequences for ecosystems and human health, is warranted.

12.3.3 Cultural Services

Impacts of IAS on cultural services, defined as those attributes of an ecosystem that are non-consumptive (i.e. hold value for recreation, tourism, culture, history, education, science, spirituality, and aesthetics) are difficult to assess because they are based on personal and local value systems. IAS nearly always alter cultural services, either negatively or positively and sometimes in opposition to impacts on other services. For example, Blue Gum Eucalyptus (*Eucalyptus globulus*), originally introduced for firewood, is now both adored and reviled in California. Blue gums are praised for their evocative beauty, their part in history and their contribution to a "sense of place" and for providing habitat for monarch butterflies. They are also reviled for being non-native, supporting far fewer species then native forest or shrubland, blocking views, and for their fragile branches which can be a threat to property and human lives (Miller 2007). Some cultural benefits and costs can easily be assigned monetary value (i.e. from recreation and tourism or from the sale of honey and ornamentals; Thompson *et al*. 1987) but impacts on other cultural services are more difficult to quantify.

12.3.3.1 RECREATION AND TOURISM

Land and water-based recreation are both strongly impacted by IAS. Water-based recreation alone in Lake Tahoe is worth $30–45 million/year. Even a one per cent loss in recreation revenue from the potential introduction of Eurasian Water Milfoil (*Myriophyllum spicatum*) into Lake Tahoe would cost up to $500,000/yr (Eisworth *et al.* 2000, 2005). Algal blooms on the coast of Maui have reduced property values and occupancy rates at hotels by fouling recreational beaches. Over a 16.1 km coastline, these blooms have resulted in $21.8 million in potential lost revenue (Cesar *et al.* 2002).

Revenue and enjoyment from sports fishing can also suffer as a result of aquatic invasive species. For example, halting control of sea lampreys, which have a heavy impact on angling in the US and Canada, would cost $675 million annually in lost fishing opportunities (OTA 1993). Leigh (1998) estimates that the cost of controlling introduced Ruffe fish in the Great Lakes ($13.6 million cumulative)—an aggressive species that competes with economically-valuable native fish for food and habitat—is far outweighed by benefits derived from fishing ($119 million to $1.05 billion cumulative).

Plant and animal invaders can be just as costly in terrestrial systems; feral pigs in Florida cost $5,331 to $43,257/ha in lost recreation value (Engeman *et al.* 2003) and introduced weedy plants in golf courses may cost up to $1 billion/yr to eradicate (Pimentel *et al.* 2000). By lacerating hikers and poisoning cattle, Star Thistle (*Centaurea solstitialis*) has also decreased the recreation value of large areas of the western US (Dudley 2000), as well as costing millions per year in lost livestock forage value (Eagle *et al.* 2007). Similarly, the red imported fire ant has created billions of dollars in costs across many southern US states. If introduced to Hawaii, this species is projected to result in $134 million in forgone outdoor opportunities to locals and tourists (Gutrich *et al.* 2007).

Combining ecological data with economic impacts can help determine whether slowing the spread of an IAS that impacts recreation could be economical beneficial. Keller *et al.* (2008) constructed a model which predicts which Wisconsin lakes are likely to be invaded by Rusty Crayfish (*Orconectes rusticus*), a species that reduces sport fish populations. They conclude that although optimum expenditures for control are high ($4.3 million in one scenario), protecting lakes would produce net economic benefits of at least $6 million based on market benefits alone.

12.3.3.2 AESTHETIC VALUE AND CULTURAL HERITAGE

Invasive alien species have transformed landscapes for better or worse depending on one's perspective (Mack *et al.* 2000). IAS such as Lantana (*Lantana* spp.) in Australia (Parsons and Cuthbertson 2001) or Scotch Broom (*Cytisus scoparius*) in North America (Bossard 2000) were imported because of their value as ornamentals: species that add beauty to the landscape. Prohibiting sales of such exotic plant species has social costs in the form of lost consumer benefits and profits for nurseries. Whether or not it is socially optimal to prevent the sale and use of such species depends on the level of invasion risk and where it is being planted (Knowler and Barbier 2005).

In many cases, people admire these plants as ornamentals, but for others their aesthetic and cultural value has changed with time and as a result of their invasiveness. For instance, in the 1890s a resident of New York City decided every bird mentioned by Shakespeare should be introduced into the US (Hunter 1996). Although the Starling was in Shakespeare's works just once, the Starling has thrived and now numbers more than 200 million and has spread across the continent to Alaska. Although initially attractive, Starlings are now often considered pests because they occur in large noisy groups and because they have a detrimental effect on native cavity-nesting birds such as Eastern Blue Birds (*Sialia sialis*) (Cabe 1993). Similarly, the noise and droppings from roosts of up to 4000 introduced Common Crows and Mynas in Singapore are quite a nuisance to nearby residents (Lim *et al.* 2003).

Both accidental and deliberate introductions such as Gypsy Moths (Hollenhorst *et al.* 1993) and pigeons (*Columba livia*) can have similarly negative impacts on quality of life. Gypsy Moths cause mass defoliation and tree mortality in neighborhoods (Campbell and Sloan 1977). Pigeons, which some people enjoy watching and feeding, also deface buildings, statues, and sidewalks with guano, costing millions of dollars annually to clean up (Pimentel et al. 2000).

IAS can have severe impacts on the audio-scape as well as the view. The Coqui Frog (*Eleutherodactylus coqui*), native to Puerto Rico, was introduced to Hawaii in the late 1980s with nursery plants. This tiny frog emits very loud (80–90 dBA at 0.5 m) mating calls and in Hawaii reaches densities of 55,000–133,000 frogs/ha, more than twice as high as Coqui densities in the frog's native Puerto Rico (Beard and Pitt 2005). Kaiser and Burnett (2006) found that property values of homes within 500 m of Coqui populations declined significantly more than other homes. If all houses in Hawaii County were exposed to Coqui, this would result in a loss of $7.6 million in direct property values.

In addition to aesthetics, IAS may be valued or reviled for their role in inspiration, spirituality, religion, ceremony, and tradition. The impacts of IAS on these culturally important elements of ecosystems remain virtually unstudied, complex and difficult to quantify (McNeely 2001a).

Summary: Impacts of IAS on recreation and tourism are far more likely to be quantified and incorporated into decision-making compared with cultural services such as aesthetic value and cultural heritage for which it is difficult to assign monetary value. Finding ways to account for impacts on these services could help gain broader support for managing invasive species.

12.4 Case studies

In the following tables, we introduce the many pathways through which IAS influence the provision of ecosystem services. Many non-native species have been introduced for economic purposes and can produce excellent sport fishing, food, fodder, timber, and biological control. However, these benefits can also come with costs:reducing crop production, water availability, and the provision of other services (McNeely 2001b; Jackson *et al.* 2005). Here, we choose three geographic regions with very different ecosystems and focus on one invasive alien species or functional group in each location to illustrate that synergies and tradeoffs exist, and to expose important gaps in knowledge. In each table we identify the IAS, the ecosystem services it impacts, the mechanisms by which it has an impact, whether the impact is positive or negative, the $-value of the impact (if available), and the source of the information.

12.5 Research and policy recommendations

1. Include all relevant ecosystem services in IAS impact assessments.

Many studies hint at impacts to ecosystem services but do not directly address them (Duncan *et al.* 2004). Doing so is critical because impact assessments are not complete without considering the full suite of "externalities"—or unaccounted for costs and benefits to society (McNeely 2001b; Born *et al.* 2005). Including these costs often makes prevention (Leung *et al.* 2002; Gutrich *et al.* 2007) or eradication (Simberloff 2003) worthwhile. For example, the Office of Technology Assessment (1993) calculated cost-benefit rations for major IAS in the US and found that the benefits of removal nearly always outweighed the costs of control.

Calculating costs and benefits does not automatically determine a decision because of the role of values and distributional issues (who pays and who benefits), but well-founded bio-economic risk analysis is likely result in better informed decision-making (Shogren 2000). For instance, the spread of *Miconia calvescens* in Hawaii threatens watersheds (reduces groundwater recharge), and recreational opportunities and aesthetic value for locals and tourists. A recent analysis incorporates all of these impacts into an optimal control model to weigh alternative policy options (Kaiser 2006; Burnett *et al.* 2007).

2. Focus research on fresh water and regulating and cultural services.

Assessing the affects of IAS on ecosystem services has not been an explicit focus of studies of invasion ecology (Richardson and van Wilgen 2004). This needs to change if ecologists are to inform public policy. Some ecosystem services are better understood than others. For example, the impacts of IAS on provisioning services (food, fiber, and fuel) are often well quantified. Impacts on other life-supporting services such as fresh water and most regulating services: pollination, disease and pest regulation, and flood and fire control are rarely calculated, but are likely to be substantial. Finally, of all the services, the interaction between IAS and culture is perhaps most complex and under-addressed. Yet these types of services resonate widely in society. There is a strong need for scientific research to help resolve the uncertainty surrounding the impacts of IAS on human well-being (Hoagland and Jin 2006).

3. Develop accurate predictions for which IAS groups are likely to have the greatest impact on important ecosystem services.

Much invasions research thus far has focused on predicting invisibility, comparing invader and native traits, and assessing environmental impacts, particularly on biodiversity. Do species with the greatest ecological impacts also have the greatest impacts on ecosystem services? Given that it is easier to prevent an introduction than to control an invasion, we must be capable of making good predictions regarding which species or groups of species will impact ecosystem services by understanding the underlying mechanisms. For example, are differences in impact due to functional traits (nitrogen fixers), or biomass (Dukes and Mooney 2004)? This is not an easy task. Karieva *et al*. (1996) found that models and short-term experiments were not very accurate predictors of invasions. The best approach may be intensive study and long-term monitoring of previous invasions of the same or similar species (Williamson 1999).

4. Make the impacts of IAS on ecosystem services and human well-being explicit and use economic instruments to develop tools for compliance.

Global trade and travel is likely to exacerbate the problem of invasions and continue to compromise vital ecosystem services (Perrings *et al*. 2002; Keller *et al*. 2007). More effective inspection systems are critical at international borders in order to identify and cut off pathways of introduction (Lodge *et al*. 2006). Because losses from IAS are not always transparent and are spread across many stakeholders, there is little incentive to cooperate with border inspections and no strong group has emerged to pressure government. Using economic incentives and disincentives such as taxes, fines, and grants could result in greater compliance for those in danger of introducing IAS (McNeely 2001b) and are working well to control established IAS in many places such as South Africa (van Wilgen and Richardson, Chapter 13 this volume). Investing in education in tandem with economic incentives could also lead to better bottom-up enforcement (Le Maitre *et al*. 2004) and more public support for affective top down programs for species pre-screening (Keller *et al*. 2007).

5. The problems arising from biological invasions deserve a global response.

Much of invasion biology focuses on ecological impacts, predicting spread and developing control methods rather than documenting economic and social damage to society from impacts on ecosystem services. More of the latter research is required to place this problem in the proper context relative to other environmental stressors. Science and policy should reflect the fact that IAS have taken their place next to human-caused global warming as a driver of global environmental change (Vitousek *et al*. 1997).

6. Measure the impacts of IAS using broader criteria than biodiversity.

Because ecosystems provide life-support services to all of human society, when using the ecosystem service framework for prevention and control of IAS it is particularly important to incorporate cultural and subsistence values into policy decisions.

Acknowledgements

We thank the organizers and participants of the October 2007 Diversitas Meeting on the economics and ecology of invasive species for stimulating discussion on this topic. The case studies greatly benefited from review by H. MacIsaac, B. Van Wilgen, J. Jeffrey, and L. Hadway. We thank M. Williamson and C. Perrings for comments on the manuscript.

References

Angeler, D.G., M. Álvarez-Cobelas, S. Sánchez-Carrillo and M.A. Rodrigo. (2002). Assessment of exotic fish impacts on water quality and zooplankton in a degraded semi-arid floodplain wetland. *Aquatic Sciences* 64:76–86.

Atkinson, C. (1995). Wildlife disease and conservation in Hawaii: pathogenicity of avian malaria (*Plasmodium relictum*) in experimentally infected Iiwi (*Vestiaria coccinea*). *Parasitology* 111:S59–S69.

Badzinski, S.S. and S.A. Petrie. (2006). Diets of lesser and great scaup during autumn and spring on the lower Great Lakes. *Wildlife Society Bulletin* 34:664–74.

Bailey, R.Y. (1939). Kudzu for erosion control in the Southeast. USDA Farmer's Bulletin, Washington, D.C.

Beard, K.H. and W.C. Pitt. (2005). Potential consequences of the coqui frog invasion in Hawaii. *Diversity and Distributions* 11:427–33.

Beekey, M.A., D.J. McCabe, and J.E. Marsden. (2004). Zebra mussels affect benthic predator foraging success and habitat choice on soft sediments. Oecologia 141:164–170.

Billings, W.D. (1990). *Bromus tectorum*, a biotic cause of ecosystem impoverishment in the Great Basin. In G.M. Woodwell (ed.) *The Earth in Transition*, Cambridge University Press, New York.

Born, W., F. Rauschmayer and I. Brauer. (2005). Economic evaluation of biological invasions – a survey. *Ecological Economics* 55:321–36.

Bossard, C. (2000). "*Cytisus scoparius*." In Bossard, C., Randall, M. and Hoshovsky, M. *Invasive Plants of California's Wildlands*. Los Angeles: University of California Press, Ltd, pp. 145–49.

Brooks, M.L., C.M. D'Antonio, D.M. Richardson, J.B. Grace, J.E. Keeley, J.M. D'Tomaso, R.J. Hobbs, M. Pellant, and D. Pyke. (2004). Effects of invasive alien plants on fire regimes. *BioScience* 54:677–88.

Burnett, K., B. Kaiser, and J. Roumasset. (2007). Economic lessons from control efforts for an invasive species: *Miconia calvescens* in Hawaii. *Journal of Forest Economics*, 13:151–67.

Burrows, C.P.M., C.L. Isaacs and K. Maly. (2007). Pua'a (pigs) in Hawaii, from traditional to modern. www: [http://www.nature.org/wherewework/northamerica/states/hawaii/files/puaa_fact_sheet.pdf]

Cabe, P.R. (1993). European Starling (*Sturnus vulgaris*). In: Poole, A. and Gill, F. (eds) *The Birds of North America* no. 48. The Academy of Natural Sciences, Philadelphia; The American Ornithologist's Union, Washington, DC.

Campbell, R.W. and R.J. Sloan. (1977). Forest stand responses to defoliation by the Gypsy Moth. *Forest Science* 19:1–34.

Carlsson, N.O.L., C. Bronmark, and L. Hansson. (2004). Invading herbivory: the golden apple snail alters ecosystem functioning in Asian wetlands. *Ecology* 85:1575–80.

Cesar, H., P. Vanbeukering and S. Prince. (2002). An economic valuation of Hawaii's coral reefs. Report prepared for the Hawaii Coral Reef Initiative Research Program, Honolulu, Hawaii.

Chambers, J.C., B.A. Roundy, R.R. Blank, S.E. Meyer and A. Whittaker. (2007). What makes Great Basin sagebrush ecosystems invasible by *Bromus tectorum*? *Ecological Monographs* 77:117–45.

Charles, H. and J. S. Dukes. (2007). Impacts of invasive species on ecosystem services. In W. Nentwig (Ed.) *Biological Invasions*. Ecological Studies Vol. 193 Springer-Verlag Berlin, Germany.

Chittka, L. and Schürkens S. (2001). Successful invasion of a floral market. *Nature* 411, 653–53.

Cline, J.F., D.W. Uresk and W.H. Rickard. (1977). Comparison of soil water used by a sagebrush bunchgrass and a cheatgrass community. *Journal of Range Management* 30: 199–201.

Clout, M.N. (1999). Biodiversity conservation and the management of invasive animals in New Zealand. In: O.T. Sandlund, P.J. Schei and A. Viken (eds.), *Invasive Species and Biodiversity Management*, Kluwer Press, London, pp. 349–359.

Cohen, A.N. and J. T. Carlton. (1995). Biological Study. Nonindigenous Aquatic Species in a United States Estuary: A Case Study of the Biological Invasions of the San Francisco Bay and Delta. A Report for the United States Fish and Wildlife Service, Washington, D.C., and The National Sea Grant College Program, Connecticut Sea Grant, NTIS Report Number PB96-166525, 246 pp.

Colautti, R.I., S.A. Bailey, C.D.A. van Overdijk, K. Amundsen and H.J. MacIsaac. (2006). Characterised and projected costs of nonindigenous species in Canada. *Biological Invasions* 8:45–59.

Conroy, J.D., W.J. Edwards, R.A. Pontius, D.D. Kane, H.Y. Zhang, J.F. Shea, J.N. Richey and D.A. Culver. (2005). Soluble nitrogen and phosphorus excretion of exotic freshwater mussels (*Dreissena* spp.): potential impacts for nutrient remineralisation in western Lake Erie. *Freshwater Biology* 50:1146–62.

Convention on Biological diversity (CBD). (2008). Article 8(h). http://www.cbd.int/convention/articles.shtml?a=cbd-08

Coombs, E.M., H. Radtke, D.L. Isaacson, and S. Snyder. (1996). Economic and regional benefits from biological control of tansy ragwort, *Senecio jacobaea* in Oregon. In: *International Symposium of Biological Control of Weeds*. V.C. Moran and J.H. Hoffman (eds.) University of Cape Town, Stellenbosch, South Africa, pp. 489–94.

Cordy, D.R. (1978). *Centaurea* species and equine nigropallidal encephalomalacia. In: R.F. Keeler, K.R. Van Kampen, and L.F. James (eds.). *Effects of Poisonous Plants on Livestock*. Academic Press, New York, pp. 327–36.

Costello, C. and C. McAusland. (2003). Protectionism, trade and measures of damage from exotic species introductions. *American Journal of Agricultural Economics* 85:964–75.

Cowling, R.M. and D.M. Richardson. (1995). Fynbos: South Africa's unique flora kingdom. Fernwood Press, Cape Town South Africa.

Crooks, J.A. (2002). Characterizing ecosystem-level consequences of biological invasions the role of ecosystem engineers. *Oikos* 97:153–66.

Cuddihy, L.W. and C.P. Stone. (1990). *Alteration of Native Hawaiian Vegetation: Effects of Humans, Their Activities and Introductions*. University of Hawaii Press, Honolulu.

Daily, G.C., S. Alexander, P.R. Ehrlich, L. Goulder, J. Lubchenco, P. Matson, H. Mooney, S. Postel, S. Schneider, D. Tilman, and G. Woodwell. (1997). Ecosystem services: benefits applied to human societies by natural ecosystems. *Issues in Ecology* 2:1–16.

D'Antonio, C.M. and P.M. Vitousek. (1992). Biological Invasions by Exotic Grasses, the Grass/Fire Cycle, and Global Change. *Annual Review of Ecology and Systematics* 23:63–87.

D'Antonio, C.M. (2000). Fire, plant invasions, and global changes. In: Mooney H.A., Hobbs R.J. (eds.) *Invasive Species in a Changing World*. Island Press, Washington DC, pp. 65–93.

D'Antonio, C.M., N.E. Jackson, C.C. Horvitz, and R. Hedberg. (2004). Invasive plants in wildland ecosystems: merging the study of invasion processes with management needs. *Frontiers in Ecology and the Environment* 2:513–521.

De Neergaard, A., C. Saarnak, T. Hill, M. Khanyile, A.M. Berzosa, and T. Birch-Thomsen. (2005). Australian wattle species in the Drakensberg region of South Africa—An invasive alien or a natural resource? *Agricultural Systems* 85:216–233.

De Wit, M.P., D.J. Crooks and B.W. Van Wilgen. (2001). Conflicts of interest in environmental management: estimating the costs and benefits of a tree invasion. *Biological Invasions* 3:167–178.

Diamond, C., D. Davis and D.C. Schmitz. (1991). Economic impact statement: the addition of *Melaleuca quinqeunervia* to the Florida Prohibited Aquatic Plant List. *Proceedings of the Symposium on Exotic Pest Plants*. US Department of the Interior, Washington, D.C.

Dudley, D.R. (2000). Wicked weed of the west. California Wild 2000:32–35.

Dukes, J. S and H. A. Mooney. (2004). Disruption of ecosystem processes in western North America by invasive species. Revista Chilena de Historia Natural 77:411–437.

Duncan, C.A., J.J. Jachetta, M.L. Brown, V.F. Carrithers, J.K. Clark, J.M. DiTomaso, R.G. Lym, K.C. McDaniel, M.J. Renz, and P.M. Rice. (2004). Assessing the economic, environmental, and societal losses from invasive plants on rangelands and wildlands. *Weed Technology* 18:1411–16.

Eagle, A.J., M.E. Eiswerth, W.S. Johnson, S.E. Schoenig, and G.C. van Kooten. (2007). Costs and losses imposed on California ranchers by Yellow Starthistle. *Rangeland Ecology and Management* 60:369–77.

Eiswerth, M.E., S.G. Donaldson and W.S. Johnson. (2000). Potential environmental impacts and economic damages of Eurasian water milfoil (*Myriophyllum spicatum*) in Western Nevada and Northeastern California. *Weed Technology* 14:511–18.

Eiswerth, M.E. and G. C. van Kooten. (2002). Uncertainty, economics, and the spread of an invasive plant species. *American Journal of Agricultural Economics* 84: 1317–22.

Eiswerth, M.E., T.D. Darden, W.S. Johnson, J. Agapoff, and T.R. Harris. (2005). Input -output modeling, outdoor recreation, and the economic impact of weeds. *Weed Science* 53:130–37.

Engeman, R. M., H. T. Smith, S. A. Shwiff, B. U. Constantin, M. Nelson, D. Griffin, and J. Woolard. (2003). Estimating the prevalence and economic value of feral swine damage to native habitats in three Florida state parks. *Environmental Conservation* 30:319–24.

Eubanks, M. D. (2001). Estimates of the direct and indirect effects of red imported fire ants on biological control in field crops. *Biological Control* 21:35–43.

Evans, R.C., D.T. Tingey, M.L. Gumpertz and W.F. Burns. (1982). Estimates of isoprene and monoterpene emission rates in plants. *Botany Gazette* 143:304–10.

Fernald, S.H., N.F. Caraco and J.J. Cole. (2007). Changes in cyanobacterial dominance following the invasion of the zebra mussel *Dreissena polymorpha*: long-term results from the Hudson River Estuary. *Estuaries and Coasts* 30:163–70.

Gerlach, D. (2004). The impacts of serial land-use changes and biological invasions on soil water resources in California, USA. *Journal of Arid Environments* 57:365.

Glenn, E., R. Tanner, S. Mendez, T. Kehret, D. Moore, J. Garcia and C. Valdes. (1998). Growth rates, salt tolerance and water use characteristics of native and invasive riparian plants from the delta of the Colorado River, Mexico. *Journal of Arid Environments* 40:281–94.

Gorgens, A. and B.W. van Wilgen. (2004). Invasive alien plants and water resources in South Africa: Current understanding, predictive ability and research challenges. *South African Journal of Science* 100:27–33.

Greathead, D.J. (1968). Biological control of Lantana—a review and discussion of recent in East Africa. *PANS* 14:167–175.

Grey, M.J., M.F. Clarke and R.H. Loyn. (1998). Influence of the Noisy Miner Manorin melanocephala on avian diversity and abundance in remnant Grey Box woodland. *Pacific Conservation Biology* 4:55–69.

Greathead, D.J. (1968). Biological control of Lantana—a review and discussion of recent developments in East Africa. *PANS* 14:167–75.

Grey, M.J., M.F. Clarke and R.H. Loyn. (1998). Influence of the Noisy Miner *Manorina melanocephala* on avian diversity and abundance in remnant Grey Box woodland. *Pacific Conservation Biology* 4:55–69.

Gross, C.L. (2001). The effect of introduced honeybees on native bee visitation and fruit-set in *Dillwynia juniperina* (Fabaceae) in a fragmented ecosystem. *Biological Conservation* 102:89–95.

Gutrich, J.J., E. VanGelder and L. Loope. (2007). Potential economic impact of introduction and spread of the red imported fire ant, *Solenopsis invicta*, in Hawaii. *Environmental Science and Policy* 10:685–696.

Hirsch, S.A. and J.A. Leitch. (1996). The impact of knapweed on Montana's economy. *North Dakota State University Agricultural Economics Report* 355:1–23.

Hoagland, P. and D. Jin. (2006). Science and economics in the management of an invasive species. *Bioscience* 56:931–935.

Hogan, L.S., E. Marschall, C. Folt, and R.A. Stein. (2007). How non-native species in Lake Erie influence trophic transfer of mercury and lead to top predators. *Journal of Great Lakes Research* 33:46–61.

Holland, R. (1993). Changes in planktonic diatoms and water transparency in Hatchery Bay, Bass Island area, western Lake Erie since the establishment of the zebra mussel. *Journal of Great Lakes Research* 19:617–24.

Hollenhorst, S.J., S.M Brock, W.A. Freimund and M.J Twery. (1993). Predicting the effects of gypsy moths on near-view aesthetic preferences and recreation appeal. *Forest Science* 39:28–40.

Holmes T.H. and K. J. Rice. (1996). Patterns of growth and soil-water utilization in some exotic annuals and native perennial bunchgrasses of california. *Annuals of Botany* 78:233–43.

Hughes, R.F., S.R. Archer, G.P. Asner, C.A. Wessman, C. McMurty, J. Nelson, and R.J. Ansley. (2006). Changes in aboveground primary production and carbon and nitrogen pools accompanying woody plant encroachment in a temperate savanna. *Global Change Biology* 12:1733–47.

Hunter, M. L., Jr. (1996). *Fundamentals of Conservation Biology*. Blackwell Science. Cambridge, Massachusetts.

Ikuma, E.K., D. Sugano, and J.K. Mardfin. (2002). *Filling the Gaps in the Fight against Invasive Species*. Legislative Reference Bureau, Honolulu, Hawaii, January 2002.

Jackson, R.B., E.G. Jobbagy, R. Avissar, S.B. Roy, D.J. Barrett, C.W. Cook, K.A. Farley, D.C. le Maitre, B.A. McCarl, and B.C. Murray. (2005). Trading water for carbon with biological carbon sequestration. *Science* 310:1944–47.

Jetter, K.M., J.M. DiTomaso, D.J. Drake, K.M. Klonsky, M.J. Pitcairn and D. A. Sumner. (2003). Biological control of yellow starthistle. In: *Exotic Pests and Diseases: Biology and Economics for Biosecurity*. D.A. Sumner (ed.). Iowa State University Press, Ames, Iowa, pp. 225–41.

Jianqing, D., R. Wang, and Z. Fan. (1995). Distribution and infestation of water hyacinth and the control strategy in China. *Journal of Weed Science* 9:49–51.

Joffe, S. and S. Cooke. (1997). *Management of the Water Hyacinth and other Invasive Aquatic Weeds: Issues for the World Bank*, Global IPM Facility, CAB Bioscience.

Johannsmeier, M.F. and A.J.N. Mostert. (1995). South African nectar and pollen flora. In *Beekeeping in South Africa*, 3rd edn, ed. M.F. Johannesmeier, pp. 127–148. Plant Protection Research Institute Handbook No. 14, Agricultural Research Council, Pretoria.

Johnson, P., A. Gerbolini, D. Ethridge, C. Britton, and D. Ueckert. (1999). The economics of redberry juniper control in the Texas Rolling Plains. *Journal of Range Management* 52:569–74.

Juliano, S.A. and L.P. Lounibos. (2005). Ecology of invasive mosquitoes: effects on resident species and on human health. *Ecology Letters* 8: 558–74.

Kaiser, B.A. (2006). Economic impacts of non-indigenous species: Miconia and the Hawaiian economy. *Euphytica* 148:135–150.

Kaiser, B. and K. Burnett. (2006). Economic impacts of *E. coqui* frogs in Hawaii. Interdisciplinary *Environmental Review* 8:1–11.

Kareiva, P., I.M. Parker, and M. Pascual. (1996). Can we use experiments and models in predicting the invasiveness of genetically engineered organisms? *Ecology* 77: 1670–75.

Kasulo, V. (2000). The impact of invasive species in African lakes. In: *The Economics of Biological Invasions*. C. Perrings, M. Williamson and S. Dalmazzone (eds.), Edward Elgar, Cheltenham.

Kaufmann, J.B., D.L. Cummings, D.E. Ward, and R. Babbitt. (1995). Fire in the Brazilian Amazon 1: biomass, nutrient pools and losses in slashed primary forests. *Oecologia* 104:397–408.

Kaufmann, J.B., D.L Cummings and D.E. Ward. (1998). Fire in the Brazilian Amazon: 2. biomass, nutrient pools, and losses in cattle pastures. *Oecologia* 113:415–27.

Kaufman, J.C. (2004). Prickly pear cactus and pastoralism in southwest Madagascar. *Ethnology* 43:345–61.

Keller, R.P., D.M. Lodge and D.C. Finnoff. (2007). Risk assessment for invasive species net bioeconomic benefits. PNAS 104:203–207.

Keller, R.P., K. Frang and D.M. Lodge. (2008). Preventing the spread of invasive species economic benefits of intervention guided by ecological predictions. *Conservation Biology* 22:80–88.

Kenta, T., N. Inari, T. Nagamitsu, K. Goka, and T. Hiura. (2007). Commercialized European bumblebee can cause pollination disturbance: an experiment on seven native plant species in Japan. *Biological Conservation* 134:298–309.

Kliejunas, J.T. and W.H. Ko. (1976). Dispersal of *Phytophthora cinnamomi* on Island of Hawaii. *Phytopathology* 66:457–60.

Knoll, L.B., O. Sarnelle, S.K. Hamilton, C.E.H. Kissman, A.E. Wilson, J.B. Rose and M.R. Morgan. (2008). Invasive zebra mussels (*Dreissena polymorpha*) increase cyanobacterial toxin concentrations in low-nutrient lakes. *Canadian Journal of Fisheries and Aquatic Science* 65:448–55.

Knowler, D. and E.B. Barbier. (2000). The economics of an invading species: a theoretical model and case study application. In: *The Economics of Biological Invasions*. C. Perrings, M. Williamson, and S. Dalmazzone (eds.) Edward Elgar, Cheltenham, pp. 70–93.

Knowler, D. and E. Barbier. (2005). Importing exotic plants and the risk of invasion: are market-based instruments adequate? *Ecological Economics* 52:341–54.

Kovalak, W., G. Longton and R. Smithee. (1993). Infestation of power plant water system by the zebra mussel (*Dreissena polymorpha*). In: *Zebra Mussels: Biology Impacts, and Control*, ed. T.F. Nalepa and D.W. Schloesser, pp. 359–80. Lewis Publishers, Boca Raton.

Kremen, C., N.M. Williams and R.W. Thorp. (2002). Crop pollination from native bees a risk from agricultural intensification. *PNAS* 99:16812–16.

Kremen, C., N.M. Williams, R.L. Bugg, J.P. Fay and R.W. Thorp. (2004). The area requirements of an ecosystem service: crop pollination by native bee communities in California. *Ecology Letters* 7:1109–19.

Lacey, J.R., C.B. Marlow, and J.R. Lane. (1989). Influence of spotted knapweed on surface runoff and sediment yield. *Weed Technology* 3:627–63.

Lambrinos, J.G. (2000). The impact of the invasive alien grass *Cortaderia jubata* (Lemoine) Stapf on an endangered mediterranean-type shrubland in California. *Diversity and Distributions* 6:217–231.

Laughrin, L., M. Carroll, A. Bromfield, and J. Carroll. (1994). Trends in vegetation changes with the removal of feral animals grazing pressures on Santa Catalina Island. In Halvorson, W. L. and Maender, G. J. (eds.). *The Fourth California Islands Symposium: Update on the Status of Resources*, pp. 523–530. Santa Barbara Museum of Natural History. Santa Barbara, CA.

LaValle, P.D., A. Brooks, and V.C. Lakhan. (1999). Zebra mussel wastes and concentrations of heavy metals on shipwrecks in Western Lake Erie. *International Association of Great Lakes Research* 25:330–38.

Leigh, P. (1998). Benefits and costs of the ruffe control program for the Great Lake Fishery. *Journal of Great Lakes Research* 24:351–60.

Leitch, J.A., F.L. Leistritz, and D.A. Bangsund. (1994). Economic effect of leafy spurge in the upper Great Plains: methods, models, and results. *North Dakota State University Agricultural Economics Report* 316:1–8.

Le Maitre, D.C., B.W. van Wilgen, R.A. Chapman, and D.H. McKelly. (1996). Invasive plants and water resources in the western Cape Province, South Africa: modeling the consequences of a lack of management. *Journal of Applied Ecology* 33:161–72.

Le Maitre, D.C., B.W. van Wilgen, C.M. Gelderblom, C. Bailey, R.A. Chapman, and J.A. Nel. (2002). Invasive alien trees and water resources in South Africa: case studies of the costs and benefits of management. *Forest Ecology and Management* 160:143–59.

Le Maitre, D.C., D.M. Richardson, and R.A. Chapman. (2004). Alien plant invasions in South Africa: driving forces and the human dimension. *South African Journal of Science* 100:103–112.

LePage, W. (1993). The impact of *Dreissena polymorpha* on water works operations at Monroe, Michigan: a case study. In: *Zebra Mussels, Biology, Impacts, and Control*. Ed. T.F. Nalepa and D.W. Schloesser, Lewis Publishers, Boca Raton, pp. 333–58.

Leung, B., D.M. Lodge, D. Finnoff, J.F. Shogren, M.A. Lewis, and G. Lamberti. (2002). An ounce of prevention or a pound of cure: bioeconomic risk analysis of invasive species. *Proceedings of the Royal Society London B* 269:2407–13.

Levine, J.M., M. Vila, C.M. D'Antonio, J.S. Dukes, K. Grigulis, and S. Lavorel. (2003). Mechanisms underlying the impacts of exotic plant invasions. *Proc. R. Soc. Lond. B* 270:775–81.

Lim, H.C., N.S. Sodhi, B.W. Brook, and M.C.K. Soh. (2003). Undesirable aliens: factors determining the distribution of three invasive bird species in Singapore. *Journal of Tropical Ecology* 19:685–95.

Lizarralde, M.S. (1993). Current status of the introduced Beaver (*Castor Canadensis* population in Tierra-del-Fuego, Argentina. *Ambio* 22:351–58.

Lodge, D.M., S. Williams, H.J. MacIsaac, K.R. Hayes, B. Leung, S. Reichard, R.N. Mack, P.B. Moyle, M. Smith, D.A. Androw, J.T. Carlton, and A. McMichael. (2006). Biological invasions: recommendations for US policy and management. *Ecological Applications* 16:2035–54.

Loh, R. K., and J. T. Tunison. (1999). Vegetation recovery following pig removal in 'Ola'akoa rainforest unit, Hawaii Volcanoes National Park. Technical Report 123. University of Hawaii Cooperative National Park Resources Studies Unit, Honolulu.

Long, J.L. (1981). *Introduced Birds of The World*. David and Charles, London.

Lonsdale, W.D. (1999). Global patterns of plant invasions and the concept of invasibility. *Ecology* 80:1522–36.

Loope, L.L. and P.G. Scowcroft. (1985). Vegetation response within exclosures in Hawaii: A review. In *Hawaii's Terrestrial Ecosystems: Preservation and Management*, C. P. Stone, and J. M. Scott, eds. University of Hawaii Cooperative National Park Resources Studies Unit, Honolulu.

Loope, L.L., A.C. Medeiros and B.H. Gagné. (1991). Recovery of vegetation of a montane bog in Haleakala National Park following protection from feral pig rooting. Technical Report 77. University of Hawaii Cooperative National Park Resources Studies Unit, Honolulu.

Lo Re, V. III. and S.J. Gluckman. (2003). Eosinophilic meningitis. *The American Journal of Medicine* 114:217–223.

Lovell, S.J., S.F. Stone and L. Fernandez. (2006). The economic impacts of aquatic invasive species: a review of the literature. *Agricultural and Resource Economics Review* 35:195–208.

Lubke, R.A. (1985). Erosion of the breach at St. Francis Bay, Eastern Cape, South Africa. *Biological Conservation* 32:99–127.

Lumb, C.E., T.B. Johnson, H.A. Cook, and J.A. Hoye. (2007). Comparison of lake whitefish (*Coregonus clupeaformis*) growth, condition and energy density between lakes Erie and Ontario. *Journal of Great Lakes Research* 33:314–325.

MacIsaac, H.J. (1996). Potential abiotic and biotic impacts of zebra mussels on the inland waters of North America. *American Zoologist* 36:287–99.

MacIsaac, H.J., S.A. Bandoni, R.I. Colautti, C.D.A. van Overdijk, and K. Admundsen. (2002). Economic impacts of invasive nonindigenous species in Canada: a case study approach. A report to the Office of the Auditor General of Canada, pp. 1–185.

Mack, M.C. and C.M. D'Antonio. (1998). Impacts of biological invasions on disturbance regimes. *Trends in Ecology and Evolution* 13:195–198.

Mack, R.N., D. Simberloff, W. M. Lonsdale, H. Evans, M. Clout, and F.A. Bazzaz. (2000). Biotic invasions: causes, epidemiology, global consequences, and control. *Ecological Applications* 10:689–710.

Maguire, L.A., P. Jenkins, and G. Nugent. (1997). Research as a route to consensus? Feral ungulate control in Hawaii. *Transactions of the 62nd North American Wildlife and Natural Resources Conference*, pp. 135–45.

Maguire, L.A. (2004). What can decision analysis do for invasive species management? *Risk Analysis* 24:859–80.

Marland, G. and B. Schlamadinger. (1997). Forests for carbon sequestration or fossil fuel substitution? A sensitivity analysis. *Biomass and Bioenergy* 13:389–97.

Mayar, C.M., R.A. Keats, L.G. Rudstam and E.L. Mills. (2002). Scale-dependent effects of zebra mussels on benthic invertebrates in a large eutrophic lake. *Journal of the North American Benthological Society* 21:616–33.

Mazak, E.J., H.J. MacIsaac, M.R. Servos, and R. Hesslein. (1997). Influence of feeding habits on organochlorine contaminant accumulation in waterfowl on the Great Lakes. *Ecological Applications* 7:1133–43.

McNeely, J.A. (2001a). *The Great Reshuffling: Human Dimensions of Invasive Alien Species*. IUCN, Gland, Switzerland.

McNeely, J.A. (2001b). Invasive species: a costly catastrophe for native biodiversity. *Land Use and Water Resource Research* 1:1–10.

Millennium Ecosystem Assessment. (2005). *Ecosystems and Human Well-Being: Synthesis*. Island Press, Washington, D.C.

Miller, J.C. (2007). A lovely nuisance: ethics and eucalyptus in California. A Master in Liberal Arts Thesis, Stanford University.

Mooney, H.A. and R.J. Hobbs. (2000). *Invasive Species in a Changing World*. Island Press, Washington D.C.

Mooney, H.A. (2005). Invasive alien species: The nature of the problem. In: *Invasive Alien Species*, H.A. Mooney, R.N. Mack, J.A. McNeely, L.E. Neville, P.J. Sheia, and J.K. Waage (eds.), Island Press, Washington, DC. pp. 1–15.

Morse, R.A. and N.W. Calderone. (2000). The value of honey bees as pollinators of U.S.crops in (2000). *Bee Culture* 128:1–14.

Naiman, R.J., S.R. Elliott, J.M. Helfield, and T.C. O'Keefe. (1998). Biophysical interactions and the structure and dynamics of riverine ecosystems: the importance of biotic feedbacks. *Hydrobiologica* 410:79–86.

Nalepa, T.F. and D.W. Schloesser. (1993). *Zebra Mussels: Biology, Impacts and Control*. Lewis Publishers, Boca Raton, U.S.A., pp. 1–810.

Naylor, R.L. (1996). Invasions in agriculture: assessing the cost of the Golden Apple Snail in Asia. *AMBIO* 25:443–448.

North, S.G., D.J. Bullock and M.E. Dulloo. (1994). Changes in the vegetation and reptile populations on Round Island Mauritius, following eradication of rabbits. *Biological Conservation* 67:21–28.

Office of Technology Assessment. (1993). *Harmful Non-Indigenous Species in the United States*. US Government Printing Office, Washington, D.C.

Oldroyd, B.P. (2007). What's killing American honey bees? *PLOS Biology* 5:1195–99.

Oliveira, M.R.V., T.J. Henneberry, and P. Anderson. (2001). History, current status and collaborative research projects for *Bemisia tabaci*. Crop Protection 20:709–23.

O'meara, G. F., L. F. Evans, A. D. Gettman, and A. W. Patteson. (1995). Exotic tank bromeliads harboring immature *Aedes albopictus* and *Aedes bahamensis* (Diptera: Culicidae) in Florida. *Journal of Vector Ecology* 20:216–24.

O'Neill, C.R. (1997). Economic Impact of Zebra Mussels—Results of the 1995 National Zebra Mussel Information Clearinghouse Study. *Great Lakes Research Review*, 1–8.

Orr, S.P., J.A. Rudgers, K. Clay. (2005). Invasive plants can inhibit native tree seedlings: testing potential allelopathic mechanisms. *Plant Ecology* 181:153–65.

Parker, I.M., D. Simberloff, W.M. Lonsdale, K. Goodell, M. Wonham, P.M. Kareiva, M.H. Williamson, B. Von Holle, P.B. Moyle, J.E. Byers, and L. Goldwasser. (1999). Impact: towards a framework for understanding the ecological effects of invaders. *Biological Invasions* 1:3–19.

Parsons, W.T. and E.G. Cuthbertson. (2001). *Noxious Weeds of Australia*, pp. 627–33.

Perrings, C.M., M. Williamson and S. Dalmazzone (eds.). (2000). *The Economics of Biological Invasions*. Edward Elgar, Cheltenham.

Perrings, C., M. Williamson, E.B. Barbier, D. Delfino, S. Dalmazzone, J. Shogren, P. Simmons, and A. Watkinson. (2002). Biological invasion risks and the public good: an economic perspective. *Conservation Ecology* 6:1. [online] URL: http://www.consecol.org/vol6/iss1/art1.

Pimentel, D., L. Lach, R. Zuniga, and D. Morrison. (2000). Environmental and economic costs of nonindigenous species in the United States. *BioScience* 50:53–65.

Pimentel, D., R. Zuniga, and D. Morrison. (2005). Update of the environmental and economic costs associated with alien-invasive species in the United States. *Ecological Economics* 52:273–288.

Prater, M.R., D. Obrist, J.A. Arnone III and E.H. Delucia. (2006). Net carbon exchange and evapotranspiration in postfire and intact sagebrush communities in the Great Basin. *Oecologia* 146:595–607.

Raikow, D.F., O. Sarnelle, A.E. Wilson, and S.K. Hamilton. (2004). Dominance of the noxious cyanobacterium *Microcystis aeruginosa* in low-nutrient lakes is associated with exotic zebra mussels. *Limnology and Oceanography* 49:482–87.

Ralph, C.J. and B.D. Maxwell. (1984). Relative effects of human and feral hog disturbance on a wet forest in Hawaii. *Biological Conservation* 30:291–303

Reeders, H.H. and A. Bij de Vaate. (1990). Zebra mussels (*Dreissena polymorpha*): a new perspective for water quality management. *Hydrobiologia* 200/201: 437–50.

Richardson, D.M. and B.W. van Wilgen. (2004). Invasive alien plants in South Africa: how well do we understand the ecological impacts? *South African Journal of Science* 100:45–52.

Rickard, W. H. and B. E. Vaughan. (1988). Plant community characteristics and responses. In W. H. Rickard, L. E. Rogers, B. E. Vaughan, and S. F. Liebetrau, eds. *Shrubsteppe: Balance and Change in a Semi-Arid Terrestrial Ecosystem*. Elsevier, Amsterdam, pp.109–179.

Ricketts, T.H, G.C. Daily, P.R. Ehrlich, and C.D. Michener. (2004). Economic value of tropical forest to coffee production. *PNAS* 101:12579–12582.

Robles M and F.S. Chapin, III. (1995). Comparison of the influence of two exotic species on ecosystem processes in the Berkeley Hills. *Madrono* 42:349–357.

Sakai, A., F.W. Allendorf, J.S. Holt, D.M. Lodge, J. Molofsky, K.A. With, S. Baughman, R.J. Cabin, J.E. Cohen, N.C. Ellstrand, D.E. McCauley, P. O'Neil, I.M. Parker, J.N. Sala, A., S.D. Smith, and D.A. Devitt. (1996). Water use by *Tamarix ramosissima* and associated phreatophytes in a Mojave desert floodplain. *Ecological Applications* 6:888–98.

Schmitz, D.C., D. Simberloff, R. Hoffstetter, W. Haller, and D. Sutton. (1997). The ecological impact of nonindigenious plants. In: Simberloff, D. D.C. Schmitz and T.C. Brown (edis), *Strangers in Paradise*. Island Press, pp. 39–61.

Schumacher, M.J. and N.B. Egen. (1995). Significance of Africanized bees for public-health—a review. *Archives of Internal Medicine* 155:2038–2043.

Scott, D.F. and D.B. van Wyk. (1990). The effects of wildfire on soil wettability and hydrological behaviour of an afforested catchment. *Journal of Hydrology* 121:239–256.

Scott, D.F. and R.E. Schulze. (1992). The hydrological effects of a wildfire in a Eucalypt afforested catchment. *South African Forestry Journal* 160:67–74.

Scott, D.F. (1993). The hydrological effects of fire in South African mountain catchments. *Journal of Hydrology* 150:409–32.

Serbesoff-King, K. (2003). Melaleuca in Florida: A literature review on the taxonomy, distribution, biology, ecology, economic importance and control measures. *Journal of Aquatic Plant Management* 41:98–112.

Shackleton, C.M., D. McGarry, S. Fourie, J. Gambiza, S.E. Shackleton, and C. Fabricius. (2006). Assessing the effects of invasive alien species on rural livelihoods: case examples and a framework from South Africa. *Human Ecology* 35:113–27.

Shan, J.S., L.A. Morris, and R.L. Hendrick. (2001). The effects of management on soil and plant carbon sequestration in slash pine plantations. *Journal of Applied Ecology* 38:932–41.

Shiganova, T., Z. Mirzoyan, E. Studenikina, S. Volovik, I. Siokou-Frangou, S. Zervoudaki, E. Christou, A. Skirta, H. Dumont. (2001). Population development of the invader ctenophore *Mnemiopsis leidyi*, in the Black Sea and in other seas of the Mediterranean basin. *Marine Biology* 139:431–45.

Shogren, J.F. (2000). Risk reduction strategies against the 'explosive invader'. In: *The Economics of Biological Invaders*. C. Perrings, M. Williamson, and S. Dalmazzone (eds.) Edward Elgar, Cheltenham, pp. 56–69.

Simberloff, D. (2003). Eradication-preventing invasions at the outset. *Weed Science* 51:247–253.

Somers, C.M and R.D. Morris. (2002). Birds and wine grapes: foraging activity causes small-scale damage patterns in single vineyards. *Journal of Applied Ecology* 39:511–523.

Spira, T.P. (2001). Plant-pollinator interactions: A threatened mutualism with implications for the ecology and management of rare plants. *Northwest Areas Journal* 21:78–88.

Stokes, K.E., Y.M. Buckley, and A.W. Sheppard. (2006). A modeling approach to estimate the effect of exotic pollinators on exotic weed population dynamics: bumblebees and broom in Australia. *Diversity and Distributions* 12:593–600.

Stone, C.P. and L.L. Loope. (1987). Reducing negative effects of introduced animals on native biotas in Hawaii: what is being done, what needs doing, and the role of national parks. *Environmental Conservation* 14:245–58.

Texas Soil and Water Conservation Board. (1991). A comprehensive study of Texas watersheds and their impacts on water quality and water quantity. *Texas Soil and Water Conservation Board*. Temple, Texas.

Thompson, D.Q., R.L. Stuckey, and E.B. Thompson. (1987). Spread, impact and control of Purple Loosestrife (Lythrum salicaria) in North American Wetlands. Fish and Wildlife Research 2. USDA Fish and Wildlife Service, Washington DC 1–55.

Thunberg, E.M., C.N. Pearson, and J.W. Milon. (1992). Residential flood control benefits of aquatic plant control. *Journal of Aquatic Plant Management* 30:66–70.

Thunberg, E.M. and C.N. Pearson. (1993). Flood control benefits of aquatic plant control in Florida's flatwoods citrus groves. *Journal of Aquatic Plant Management* 31:248–54.

Traveset, A. and D.M. Richardson. (2006). Biological invasions as disruptors of plant reproductive mutualisms. *Trends in Ecology and Evolution* 21:208–16.

Turpie, J. and B. Heydenrych. (2000). Economic consequences of alien infestation of the Cape Floral Kingdom's Fynbos vegetation. In: *The Economics of Biological Invasions*. C. Perrings, M. Williamson and S. Dalmazzone (eds.). Edward Elgar, Cheltenham.

Turpie, J.K., B.J. Heydenrych, and S.J. Lamberth. (2003). Economic value of terrestrial and marine biodiversity in the Cape Floristic Region: implications for defining effective and socially optimal conservation strategies. *Biological Conservation* 112:233–251.

van Wilgen, B.W. and D.M. Richardson. (1985). The effects of alien shrub invasions on vegetation structure and fire behaviour in South African fynbos shrublands: a simulation study. *Journal of Applied Ecology* 22:955–966.

van Wilgen, B.W., R.M. Cowling, and C.J. Burgers. (1996). Valuation of ecosystem services: a case study from the fynbos, South Africa. *Bioscience* 46:184–89.

van Wilgen, B.W. and D.F. Scott. (2001). Managing fires on the Cape Peninsula, South Africa: dealing with the inevitable. *J. Med. Ecology* 2:197–208.

Vitousek, P.M. (1986). Biological invasions and ecosystem processes: can species make a difference? In *Biological invasions in North America and Hawaii*, H. A. Mooney and J. Drake, (eds). Spinger-Verlag, New York, pp.163–76.

Vitousek, P.M. (1990). Biological invasions and ecosystem processes: towards an integration of population biology and ecosystem studies. *Oikos* 57:7–13.

Vitousek, P.M, C.M. D'Antonio, L.L. Loope, M. Rejmanek and R. Westbrooks. (1997). Introduced species:

A significant component of human-caused global change. *New Zealand Journal of Ecology* 21:1–16.

Vivrette, N.J. and C.H. Muller. (1977). Mechanism of invasion and dominance of coastal grassland by *Mesembryanthemum crystallinum*. *Ecological Mongraphs* 47: 301–18.

Watkinson, A.R., R.P. Freckleton, and P.M. Dowling. (2000). Weed invasions of Australian farming systems: from ecology to economics. In: *The Economics of Biological Invasions*. C. Perrings, M. Williamson and S. Dalmazzone (eds.), Edward Elgar, Cheltenham, pp. 94–114.

Whisenant, S. (1990). Changing fire frequencies on Idaho's Snake River plains: Ecological and management implications. In *Proceedings from the Symposium on Cheatgrass Invasion, Shrub Dieoff, and Other Aspects of Shrub Biology and Management*. U.S. Forest Service General Technical Report INT-276, pp. 4–10.

Wilcove, D.S., D. Rothstein, J. Dubow, A. Phillips, and E. Losos. (1998). Quantifying threats to imperiled species in the United States. *BioScience* 48:607–15.

Williamson, M. (1996). *Biological Invasions*. Population and Community Biology Series:15. Chapman and Hall, London, pp. 156–160.

Williamson, M. (1999). Invasions. *Ecography* 22:5–12.

Wilson, G.R. (1995). The economic impact of rabbits on agricultural production in Australia—a preliminary assessment, ACIL Economics and Policy Pty Ltd, International Report to International Wool Secretariat, Melbourne, Australia.

Winfree, R., N.M. Williams, J. Dushoff, and C. Kremen. (2007). Native bees provide insurance against ongoing honey bee losses. *Ecology Letters* 10: 1105–1113.

Wolfertz, M., T.D. Sharkey, W. Boland, F. Kuhnemann, S. Yeh, and S.E. Weise. (2003). Biochemical regulation of isoprene emission. *Plant Cell and Environment* 26: 1357–64.

Xu, H.G., H. Ding, M.Y. Li, S. Qiang, J.Y. Guo, Z.M. Han, Z.G. Huang, H.Y. Sun, S.P. He, H.R. Wu and F.H. Wan. (2006). The distribution and economic losses of alien species invasion to China. *Biological Invasions* 8: 1495–1500.

Zavaleta, E. (2000). The economic value of controlling an invasive shrub. *Ambio* 29:462–67.

Zhu, B., D.G. Fitzgerald, C.M. Mayer, L.G. Rudstam, and E.L. Mills. (2006). Alteration of ecosystem function by zebra mussels in Oneida Lake: Impacts on submerged macrophytes. *Ecosystems* 9:1017–28.

Ziska, L.H., J.B. Reeves, and B. Blank. (2005). The impact of recent increases in atmospheric CO_2 on biomass production and vegetative retention of Cheatgrass (*Bromus tectorum*): implications for fire disturbance. *Global Change Biology* 11:1325–32.

CHAPTER 13

Current and Future Consequences of Invasion by Alien Species: A Case Study from South Africa

B.W. van Wilgen and D.M. Richardson

13.1 Introduction

The invasion of ecosystems by alien species is an increasingly important aspect of global change. Up until the late nineteenth century, mountains, large rivers, deserts, and oceans provided formidable barriers to the movement and migration of species. As a result, ecosystems evolved in relative isolation. Early human migration saw the first intentional introductions of alien species as our ancestors attempted to satisfy their physical and social needs by domesticating wild animals and planting crops.

However, the magnitude and frequency of these introductions were minor compared to those associated with today's global agriculture and vast volumes of trade and passenger movements. The ongoing and increasing human redistribution of species to support agriculture, forestry, horticulture, and recreation supplies a continuous pool of species from which invasive aliens are recruited. We use the term "invasive" to denote alien organisms that have spread substantially from sites of human-mediated introduction to a region outside their natural distribution range. Unless specifically stated, the term carries no connotation of impact (Pyšek *et al.* 2004). Invasive alien species are also a by-product of accidental introductions, and include disease organisms, agricultural weeds, and insect pests.

The problem is growing in severity and geographic extent as global trade and travel accelerate, and as human-mediated disturbance, changes in the world's climate, and biogeochemical cycling make ecosystems more susceptible to invasion by alien species. As a result, all human communities and natural ecosystems are under siege from a growing number of destructive invasive alien species that erode natural capital, compromise ecosystem stability, and threaten economic productivity.

Invasive alien species cause damage estimated at billions of dollars worldwide (Pimentel 2002; Mc Neely 2001). The situation is of major concern in South Africa, especially with regard to invasive alien plants. Invasive plants affect almost 10 million hectares (over 8 per cent) of the country, (Versfeld *et al.* 1998), and have a range of negative impacts on ecosystem services and agricultural practices. Perhaps best known in South Africa are the impacts that invading tree species have on water resources. However, there are many others. They include the exacerbation of problems with wildfires, the reduction of the grazing potential of the land by replacing palatable plants with unpalatable or poisonous plants, the reduction of biodiversity and the potential extinction of many endemic species, and the degradation of water bodies by invading aquatic weeds.

South Africa has several features that make it very useful as a case study area. These include:

- A relatively good understanding of the problems of invasions exists in the country. South Africa has played an important role in the development of invasion ecology as a science. This began with the establishment of the Scientific Committee on Problems in the Environment's (SCOPE) program on biological invasions (Macdonald *et al.* 1986; Drake *et al.* 1989). This program was conceptualized at the 3rd International Conference on Mediterranean Ecosystems, held in South Africa in 1980.

- South Africa is a relatively large (1.2 million km^2), topographically and climatically varied country with a wide range of ecosystems. These include savannas, grasslands, arid shrublands, mediterranean shrublands, deserts and forests (Mucina and Rutherford 2006), perennial, seasonal and ephemeral rivers, estuaries, coastal and marine environments in the Indian and South Atlantic Oceans (Payne and Crawford 1989), and offshore islands. In line with patterns elsewhere in the world, all of these ecosystems now harbor well-established populations of invasive species.
- The country has a history of concerted effort in dealing with problematic invasive species, and has implemented a highly successful program for managing invasive alien plants at a national scale (van Wilgen *et al.* 2002).
- South Africa is a microcosm of the world. The ratio of rich to poor approximates the global average (which is rare in a single country), and accentuates the problem of finding equitable solutions that will address the needs of developed and developing sectors of society.

This chapter reviews what is known about those invasive alien species that impact on the terrestrial, freshwater, and marine ecosystems of the country. We briefly review the pathways by which these species have arrived, how these pathways have changed over time, and how many have become invasive. The current levels of understanding with respect to impacts, both ecological and economic, are reviewed. This is followed by an account of the prognosis for the growth in these impacts should current trends continue.

13.2 Invasive species in South Africa

This section reviews the current understanding with respect to alien species that have been introduced to, and have become invasive in, the country. We have dealt with these species within major taxonomic groupings for terrestrial species, and for all species in freshwater and marine ecosystems (Table 13.1).

13.2.1 Plants

About 750 tree species and around 8000 shrubby and herbaceous species have been introduced into South Africa (Henderson 1998; von Breitenbach 1989). About a decade ago, 161 introduced species, (38 herbaceous, 13 succulent, and 110 woody) were regarded as invasive (Henderson 1995), and the number grows all the time. Of these, 44 species have been legally declared or proposed for declaration as noxious weeds (i.e. their removal is required by law), while 31 have been declared or proposed for declaration as invaders (i.e. their spread has to be controlled). Invading alien plants have been introduced into South Africa from South and Central and North America, Australia, Europe, and Asia. Plants introduced from Australia are particularly invasive—almost half (45 per cent) of species introduced from this continent have become important pests, compared to only 12 per cent of those from South America and 3 per cent of those from Europe and Asia (Richardson *et al.* 1997). A more recent analysis of 233 invasive species (Henderson 2006) found that almost as many originated from temperate regions (112) as from the tropics (121). About 33 per cent of these species were regarded as "transformers" (plants that form monospecific stands, thereby altering the structure, integrity, and functioning of ecosystems).

13.2.2 Mammals

Many alien mammal species have been introduced to South Africa, but few have become invaders (Skinner and Smithers 1990). Richardson *et al.* (2003) mention 6 mammal species that have established feral populations in South Africa. Two of these (the House Rat *Rattus rattus* and the Brown Rat *Rattus norvegicus*) are commensal with humans, while the American Grey Squirrel (*Scurius carolinensis*) is dependent on introduced oaks and pines,

Table 13.1 The total numbers of alien and alien-invasive species in different taxonomic or ecological groups in South Africa, and estimates of current and potential future levels of impact

Group	Number of species introduced	Number of invasive species	Major pathways of introduction	Current levels of impact	Future levels of impact
Plants in terrestrial ecosystems	8000?	~250	Horticulture Forestry Pasture improvement Accidental	Reduction of surface water resources (3.3 billion m^3) Other impacts on grazing, erosion, and biodiversity	Impacts set to increase dramatically if current trends continue (8 × greater for water resources, for example)
Mammals	No estimates available	6	Deliberate introductions for agriculture and pet trade; some accidental	Negligible/minor (Himalyan Thar—now eradicated—is an exception)	Not known
Birds	48	7	Deliberate introductions, and one range expansion	Negligible/minor	Not known
Reptiles and amphibians	280	1	Mainly through the pet trade; some accidental introductions	Negligible	Not known
Freshwater biota	>60 (fish and invertebrates) 23 (aquatic plants)	At least 58 species regarded as invasive 13 plants (including 5 serious invaders)	Introduced to support recreational fishing; inter-basin water transfers; aquaculture	Most fish have detrimental impacts on biodiversity, and some on water quality. Three considered beneficial Aquatic plants can have serious negative impacts on biodiversity, fishing, water extraction and navigation	Severe impacts on the biodiversity of freshwater ecosystems predicted if current trends continue
Terrestrial invertebrates	No estimates available	10–100?	All accidental	Many species have negative impacts on crops and livestock Some have impacts on natural ecosystems, including predation of pollinators and disruption of seed dispersal mechanisms	Not known
Marine biota	10 ?	1	All accidental, via seaborne trade	The invasive species is considered beneficial	Not known

and does not invade areas with natural vegetation alone. Fallow Deer (*Cervus dama*) have established widespread populations, and are the most successful alien mammal species in the region to date. The Hymalayan Thar (*Hemitragus jemlahicus*) established a successful population on the Cape Peninsula, but a recent eradication campaign in this isolated area of suitable habitat has apparently been successful. Domestic Pigs (*Sus scrofa*) have also established a number of viable feral populations in several areas.

13.2.3 Birds

Some 48 alien bird species have been introduced into, or have invaded, southern Africa, but only 7 of these species have established viable feral populations, and the populations of a further 13 species are geographically restricted or decreasing (Dean 2000). All 7 species that have viable populations are commensal with humans, and 4 (the Rock Pigeon *Columbia livia*, the Common Myna *Acridotheres tristis*, the Common Starling *Sturnis vulgaris*, and the House Sparrow *Passer domesticus*) are widespread. With the exception of the house crow *Corvus splendens* (which was "self-introduced") all of these species were deliberate introductions (Richardson *et al*. 2003).

13.2.4 Reptiles and amphibians

Comprehensive information on the number of alien species in these groups in South Africa has yet to be compiled. CITES trade records and permit data indicate that over 280 taxa (species or subspecies) of alien reptiles and amphibians have been introduced into South Africa, but most are in captivity (van Wilgen *et al*. 2008). There are apparently no aggressively invasive species in terrestrial ecosystems, although some (for example the Tropical House Gecko *Hemidactylus mabouia* and the Flowerpot Snake *Ramphotyphlops braminus*) have established feral populations (Brooke *et al*. 1986). In freshwater ecosystems, the Red-eared Slider (a freshwater turtle, *Trachemys scripta*) has established several feral populations.

13.2.5 Terrestrial invertebrates

There is no detailed account of invertebrate invasive species in South Africa, and the impacts of such species on natural ecosystems is either absent, escaping notice, or remains uninvestigated (Lach *et al*. 2002). Those species that attack crops have received some attention, because of their direct economic impacts. Of the top 40 crop pests in the country, 42 per cent are non-native (Dennill and Moran 1989). These include the Red Scale (*Aonidiella aurantii*), Mediterranean Fruit Fly (*Ceratitus capitata*), and Coddling Moth (*Cydia pomonella*) (Annecke and Moran 1982). Several other species are known to cause adverse effects, and some examples are summarized below.

Tribe and Richardson (1994) reported that, 20 years after the first record of the European Wasp, *Vespula germanica*, in southern Africa, it was still confined to the Cape Peninsula, an area of approximately 470 km^2. Its spread in southern Africa has been remarkably slow compared with other regions in the southern hemisphere due to several ecological factors. Tribe and Richardson urged that efforts should be made to exterminate the wasp before it moves beyond the Cape Peninsula where it will encounter more favorable conditions and may well spread rapidly.

The Argentine Ant has invaded large parts of the fynbos (mediterranean shrubland) biome, where it displaces native ants and disrupts seed dispersal. This change in a vital ecological process (seed dispersal by native ants) has been identified as a major potential threat to the biodiversity of the species-rich fynbos vegetation (Bond and Slingsby 1984), although the ant only invades human-modified ecosystems.

The Wood Wasp (*Sirex noctilio*), a native of Europe, Asia, and northern Africa, has been found in South Africa, where it poses a serious threat to plantations of introduced pine trees, causing up to 80 per cent tree mortality (Tribe 1995). The importance of this species is increasing rapidly as it spreads northwards.

The ectoparasitic mite, *Varroa destructor*, has caused widespread collapse of honeybee colonies throughout the world. The Varroa Mite is relatively harmless on its natural host, the Eastern honeybee,

Apis cerana, but has crossed onto the Western honeybee, *Apis mellifera*, and spread from its Asian origins throughout most of the world. In regions of the world where the Varroa Mite is well established, such as Europe and the USA, wild honeybee populations have all but disappeared. In 1997 the Varroa Mite was found only in the Western Cape, but as expected the mite has spread rapidly throughout South Africa, almost entirely as a result of migratory beekeeping activities, and is now present in commercial honeybee colonies in all provinces (Allsopp 2007).

13.2.6 Freshwater biota

All major southern African river systems are inhabited by alien animal species, including introduced alien animals as well as translocated indigenous species. De Moor and Bruton (1988) listed 11 alien invertebrates, including 4 parasites of fishes or bivalves, and 7 free-living invertebrates (including snails and other molluscs, and 1 crab). Alien fish species, introduced for a range of reasons, make up the bulk of the species though. Of the 58 alien and translocated species listed by De Moor and Bruton (1988), 37 were considered to be detrimental, 3 to be beneficial, and 18 to be equivocal, that is beneficial under some circumstances and detrimental under others. Of the 55 detrimental and equivocal species, 18 were considered to have a major detrimental impact on indigenous species or ecosystems.

A more recent study reported that ten species of freshwater gastropod have been introduced into South Africa, mostly through the aquarium trade. Two of these, *Lymnaea columella* (Lymnaeidae) and *Physa acuta* (Physidae), have been invasive in river systems across the country for many years, probably since the 1940s or 1950s, and two recent arrivals, *Tarebia granifera* (Thiaridae) and *Aplexa marmorata* (Physidae), are spreading (Appleton 2003).

The recent increase in aquaculture has resulted in the introduction, and subsequent escape, of four alien species of freshwater crayfish (*Cherax tenuimanus*, *C. destructor*, *C. quadricarinatus* and *Procambarus clarkii*) in South Africa. Although feral populations are small and localized at present, the potential impact of these species on South African freshwater ecosystems could be severe, and De Moor (2002) recommended that importation permits should not be granted for the further import and cultivation of any of these species.

In addition to animals, freshwater ecosystems are invaded by a range of alien aquatic plant species. Henderson and Cilliers (2002) list 21 important species of which 13 have been declared as weeds or invader plants, and as such are subject to legally-binding control requirements. These plants include what Henderson and Cilliers call "the big five": Red Water Fern (*Azolla filiculoides*), Water Hyacinth (*Eichhornia crassipes*), Water Lettuce (*Pistia stratiodes*), Parrots Feather (*Myriophyllum aquaticum*), and Salvinia (*Salvinia molesta*). These species are widespread in South Africa, and have significant economic impacts.

13.2.7 Marine organisms

A recent review of the status and impacts of marine alien species in South Africa (Robinson *et al.* 2005) identified 10 alien species that had established viable populations in South Africa. Most of these were restricted in distribution to sheltered bays, estuaries, and harbors. Only one species, the Mediterranean Mussel (*Mytilus galloprovincialis*) has spread extensively along the coast, with significant ecological and economic effects. The species displaces indigenous species, and results in a dramatic increase in intertidal biomass. These changes have been beneficial, as far as can be seen. The invasions have increased the available habitat for many infaunal species and resulted in enhanced food supply for intertidal predators. Considerable economic benefits have also resulted from this invasion, as the Mediterranean Mussel forms the basis of the South African mussel culture industry.

13.2.8 Pathways of invasion

The number of invasive species in the different taxonomic groups and their current extent of spread has been influenced by numerous factors, including biological traits of the species themselves, features of the environment, and the time since they arrived (Wilson *et al.* 2007). Fundamentally important in all cases has been the combination of socio-economic factors that have defined the pathways for their introduction to, and dissemination within, the country. The dimensions of these pathways have changed markedly over time, continue to change, and will change in different ways in the future. Some examples illustrate the complexity of the phenomenon.

For plants, many of the most widespread invaders were introduced and disseminated for specific purposes in the forestry and horticultural sectors. Such purposes were dictated by the socio-political forces of the day. Richardson *et al.* (2003) show, for example, how radical changes in policies relating to forestry using alien trees in reasonably discrete phases between 1652 and the present shaped the way that alien tree species were introduced and planted, and how problems that arose when some species became invasive were perceived and addressed in policy and management.

For birds and mammals, there have been marked changes in pathways of introduction and spread. For example, many alien species were introduced for agricultural purposes, for commercial fur trade, and for pets, with fewer species being imported for biological control, hunting, and through self-introductions. The use of alien taxa in these groups for agriculture, the fur trade, and for biological control has declined markedly, while introductions for the pet trade have remained important, and self-introductions have increased. For these groups, no radical qualitative changes to the magnitude of introductions via different pathways are predicted in the near future (Richardson *et al.* 2003). The legal and illegal trade in alien amphibians and reptiles has, as in other parts of the world, escalated dramatically (van Wilgen *et al.* 2008b).

Very marked changes have occurred in the relative importance of known pathways for the introduction of alien freshwater fishes into open waters. Whereas introductions by nature conservation agencies to stock water bodies has decreased, many other pathways (e.g. introductions for angling and for the aquarium trade) have remained fairly constant, while introductions via engineered inter-basin transfers has increased substantially. For marine organisms, the most profound changes have involved the decline in the importance of dry ballast and the marked increase in the magnitude and volume of ballast water discharge. Further details regarding the dimensions of the abovementioned changes in pathways of introduction and dissemination for different groups are discussed by Richardson *et al.* (2003). The drivers of such changes are detailed by Le Maitre *et al.* (2004). Strategic measures to deal with invasive species in the long term must address these socio-political drivers.

13.3 Current impacts of invasive species in South Africa

This section reviews the current levels of understanding of the impacts of invasive alien species. The focus is of necessity on those impacts where information is available. These included, for terrestrial ecosystems, the key services of water provision and grazing, ecosystem processes such as fire and erosion, as well as on biodiversity and human health. We also briefly review several studies that have sought to quantify the economic consequences of these impacts.

13.3.1 Impacts on water resources

The effects of invasive alien trees and shrubs on surface water resources has been identified as a major impact in South Africa. While not all alien plant infestations use more water than the natural vegetation that they replace, trees tend to use more water than grasses or shrubs (Bosch and Hewlett 1982; Dye 1988; Dye 1996; Smith and Scott 1992; Dye and Versfeld 2007). The greatest impacts occur when seasonally dormant vegetation is replaced by evergreen plants (Dye *et al.* 1995). Thus, where grasslands or shrublands are invaded by alien trees, the overall water use by the vegetation increases, leaving less water for the streams. Most of the evidence for this comes from catchment experiments in South Africa and elsewhere. These experiments have been established in high-rainfall areas where shrublands or grasslands were afforested with pines or eucalypts. These afforestation experiments can be likened to invasion by alien trees where the end result is the same—the replacement of a low shrubland or grassland with a tall woodland or forest. The results of catchment experiments have been used, together with crude estimates of the extent of invasion, to make a preliminary estimate of the size and significance of the problem in South Africa (Versfeld et al. 1998). The current estimate is that invading aliens cover 10 million ha, and use 3.3 billion m^3 of water in excess of that used by native vegetation every year (almost 7 per cent of the runoff of the country). The estimate is based on models which make a number of assumptions (Le Maitre et al. 1996), and can therefore be regarded as only preliminary. Nonetheless, the estimate indicates that the reductions are already significant.

13.3.2 Impacts on grazing resources

Grazing lands in South Africa are invaded by a number of alien plants. Of these, Mesquite (*Prosopis* spp.), Jointed Cactus (*Opuntia aurantiaca*), and Nasella Tussock Grass (*Stipa trichotoma*) are currently the most important, although several other species (for example Triffid Weed *Chromolaena odorata*, Satansbos *Solanum elaeagnifolium* and Pompom Weed *Campuloclinium macrocephalum*) are increasing in importance. In arid parts of the country, invasive Mesquite trees form dense stands of stunted trees that exclude livestock and reduce herbaceous ground cover (Zimmermann 1991), although the economic consequences of these invasions have yet to be quantified. The Jointed Cactus currently occupies about 1.9 million ha in South Africa, where its impacts amount to about R280/ha/yr (van Wilgen et al. 2004), indicating potential annual impacts of around R530 million (1 US$ = R7). Tussock-Grass invasions became evident in summer rainfall montane grassland pastures in the Eastern Cape, Free State, and Mpumalanga provinces in the 1970s (Milton 2004). These grasses reduce the forage value of natural pastures, and control by a combination of herbicide treatment, manual removal, and improved grazing practice is costly. The potential of this South American grass to transform large areas of natural pasture had been demonstrated in similar habitats in New Zealand in the 1940s.

13.3.3 Impacts on biodiversity in terrestrial ecosystems

Biodiversity, in its broadest sense, includes the composition, structure, and functioning of ecosystems at all scales (Noss 1990). The impacts of invasive species on biodiversity in South Africa are at best poorly understood. In a recent review, Richardson and van Wilgen (2004) showed that most South African research on alien-plant impacts has focused at small spatial scales (plots or communities), and much of this work has been in the fynbos (Mediterranean-climate shrubland) biome. This research has shown that dense stands of alien trees and shrubs in fynbos can rapidly reduce abundance and diversity of native plants at the scale of small plots. Such invasions also dramatically increase biomass, and change litterfall dynamics and nutrient cycling. Tree and shrub invasions in fynbos change many aspects of faunal communities. Studies have documented altered abundance and composition in native ant communities, with implications for seed dispersal functions of native plants. Changed feeding behavior of native generalist birds that disperse seeds, with likely detrimental effects on native species, has also been described. Richardson and van Wilgen's review concluded that studies from other biomes have produced only scattered information on impacts. In arid savannas, the widespread replacement of native *Acacia*-dominated communities by alien *Prosopis* species has been shown to impact on biodiversity. These invasions radically change bird habitats, leading to reduced species richness and diversity (Dean et al. 2002). These changes include the elimination of raptors, and reductions in frugivores and insectivores. *Prosopis* invasions also reduce the numbers of dung beetle species, with the most marked declines being found among large species, and rare species (Steenkamp and Chown 1996). In mesic savannas, *Chromolaena odorata* invasions in riparian situations increase shading on riverbanks, leading to altered sex ratios of native Nile Crocodiles due to reduced soil temperatures in nests (Leslie and Spotila 2001). The Argentine Ant (*Linepithema humile*) displaces native ants in the Cape fynbos shrublands, and disrupts the ecologically important process of seed dispersal and underground protection of seeds from fires (Bond and Slingsby 1984). This could potentially have severe implications for the structure and diversity of these ecosystems, although it has yet to be shown that this is happening on a large scale.

13.3.4 Impacts on fire regimes and erosion

Invasion of grasslands and shrublands by tall trees and shrubs increases the amount of plant material (fuel load) that can burn. Typical fuel loads in grass and shrublands are around 0.3–4 tonnes per hectare (van Wilgen and Scholes 1997), while invaded sites have up to 10 times more fuel (10–25 tonnes per hectare, van Wilgen and Richardson 1985). While ecosystems in South Africa are normally quite resilient to regular burning, these increased fuel loads lead to higher intensity fires and a range of detrimental effects. Physical damage to the soil can occur, resulting in increased erosion after fire. For example, 6 tonnes of soil per hectare was lost following fires in pine stands compared to 0.1 tonnes per hectare following fire in adjacent fynbos in the Western Cape (Scott *et al*. 1998); and 37 tonnes per hectare was lost following fires in pine stands compared to 1.8 tonnes per hectare in adjacent grassland in the KwaZulu-Natal Drakensberg (Scott and van Wyk 1990; Scott and Schulze 1992; van Wyk 1985). The viability of indigenous seeds can be reduced, causing poor regeneration from soil-stored seed banks. The seeds of some alien species such as *Acacia saligna* can tolerate the high soil temperatures associated with intense fires, while the seeds of some indigenous fynbos species, such as *Podalyria calyptrata*, are killed after brief exposure to this temperature (Cocks and Stock 1997; Jeffrey *et al*. 1988). Intense fires will therefore favor recruitment for some alien species by promoting their germination over that of indigenous species. The physical damage caused by fires is also increased. Fire intensities in normal grassland or fynbos fires range from 200 to 5000 kW m^{-1}; indications are that fire intensities in stands of invading plants can be as high as 50 000 kW m^{-1} (van Wilgen and Richardson 1985). Damage to physical structures, and mortality in plants, is directly proportional to fire intensity.

13.3.5 Impacts on human health and safety

Very little information is available regarding the impacts of invasive alien species on human health. However, negative effects can be significant, as illustrated by the case of Parthenium Weed (*Parthenium hysterophorus*). This species has invaded several countries such as Australia, India, and Ethiopia (Medhin, 1992) with serious levels of impact. Consequently, its spread through various African countries including Ethiopia, Mozambique, Swaziland, Kenya, and now South Africa (Parsons and Cuthbertson, 1992) is a matter of great concern. The species impacts on stock and crop production, and is also a serious threat to environmental and human health as a result of its ability to produce chemicals that cause severe dermatitis, allergy, and toxicity in humans (with corresponding reductions in quality of life and productivity) and animals. Parthenium is highly toxic to domestic animals, and if eaten, the meat gets tainted and this causes direct economic losses. Crop losses are caused primarily through allelopathic effects and direct competition with crops. Wise *et al*. (2007) examined the potential impact of this species on small-scale and commercial farmers in South Africa's Mpumalanga Province. The analysis showed that if Parthenium Weed were allowed to spread without control, returns to small-scale farmers would decline between 26 and 41 per cent, while commercial farmer's annual total economic returns would decline by between US$38,818 and US$60,957.

13.3.6 Impacts on freshwater aquatic ecosystems

Invasive alien fish have a significant impact on indigenous fish faunas. Of the 30 IUCN red-listed fish species in South Africa, more than half (16) have invasive alien species (almost exclusively predatory fish) listed as the primary threat, and a further 3 have aliens as a secondary threat

(Skelton 2002). Alien fish species now almost outnumber indigenous species, and are now often the dominant fish in many river systems. For example, of the 35 fish species found in freshwater ecosystems in the Cape Floristic Region, 16 are alien (Impson *et al*. 2002). Predation on fish is not the only impact alien fish have on indigenous biota; they impact on other biotic components such as aquatic macro invertebrates and can alter habitat structure, which in turn affects ecosystem functioning (Anon. 2007).

Invasive alien trees along riparian habitats can lead to the local extinction of shade-intolerant dragonflies and damselflies; this is of special concern where the species are endemic. Invasive alien plants, especially Australian *Acacia* trees along watercourses, are by far the most important threat to these endemic species in South Africa (Samways and Taylor 2004). Removal of the invasive alien trees has led to the recovery of several species that had previously been thought to be extinct (Samways *et al*. 2005).

The impacts of freshwater invertebrate invasives have been poorly studied. Freshwater snails will almost certainly compete with native biota for both food and physical space. They may also act as intermediate hosts for liver flukes that infect livestock (as they do where they are invasive in other parts of the world), but there is no direct evidence for this yet (Appleton 2003).

Many factors impact on the quality of freshwater ecosystems, and invasive alien plants is but one of them. Many freshwater ecosystems are under threat from nutrient enrichment, chemical pollution, water extraction, erosion, and impoundments. Invasion by alien species can serve to exacerbate this, but identifying the costs associated with each of the driving factors is problematic. One study was able to show that the Red Water Fern (*Azolla filiculoides*) formed dense mats (5–20 cm thick), on dams of up to 10 ha and on slow-moving water bodies in South Africa (McConnachie *et al*. 2003). It seriously affected the biodiversity of aquatic ecosystems and had severe implications for all aspects of water utilization. These effects were considered to be most severe in the agricultural sector, where the weed increased siltation of dams and rivers, reduced the quality of water for agricultural and domestic use, clogged irrigation canals and pumps, and caused drowning of livestock that were unable to differentiate between pasture land and a weed covered dam.

13.3.7 Economic assessments of impact

Attempts to quantify the economic impacts of invasive species began in the mid-1990s, with a focus on the effects of reductions in water resources (van Wilgen *et al*. 1992; Le Maitre *et al*. 1996; van Wilgen *et al*. 1996). These studies were pivotal in persuading government to take the issue very seriously (van Wilgen *et al*. 2002), while at the same time providing the impetus for further studies on the economics of plant invasions. The studies showed that invasive alien plants may be using as much as 6.7 per cent of the country's runoff (Versfeld *et al*. 1998); that clearing the invasive plants is a good investment simply to prevent water losses (van Wilgen *et al*. 1997; Hosking and du Preez 1999); and that failure to clear stands of invading trees will result in exponential increases in the costs of clearing as catchment areas become further invaded (Le Maitre *et al*. 2002). The economic consequences of "lost" water would be even more important. As water is a limiting resource in South Africa, losses of water will restrict the potential for economic growth. At least one study (De Wit *et al*. 2001) has sought to carefully quantify the impacts of lost water for urban, agricultural, and industrial use, and the results showed that the cost of clearing programs can easily be justified in terms of the economic benefits derived from preventing water losses or restoring them to pristine levels. Some examples are provided below.

One of the few detailed studies calculated the value of a hypothetical $4\,km^2$ (4000 ha) mountain fynbos ecosystem at between US$ 3 million (with no management of alien plants), and US$ 50 million (with effective management of alien plants), based on six components: water production, wildflower harvest, hiker visitation, ecotourist visitation, endemic species, and genetic storage (Higgins *et al*. 1997b).

Turpie and Heydenrych (2000) estimated a range of economic impacts associated with the invasion of natural ecosystems in the Cape Floristic

Region. Their estimates included harvested products (~R80/ha), non-consumptive use (R25/ha from protected areas), the value of reductions in water runoff (R1140/ha) and pollination services (R500/ha) (R7 = 1US$). Given the large areas over which these benefits are, or could be, generated, this impact could be substantial.

In a study on Black Wattle (*Acacia mearnsii*) invasions, De Wit, Crookes and van Wilgen (2001) calculated the economic value of streamflow lost to invasions of Black Wattle in South Africa using the opportunity-cost approach. First, the value added by water over the different demand sectors (irrigation, domestic and urban use, mining and industry, the environment, and afforestation) was calculated. Secondly, the value added by additional water where Black Wattles were eradicated was estimated. These estimates were adjusted to allow for evaporation and spillage of flood water (33 per cent of additional water was assumed to be unusable), changes in the numbers of downstream water users over the next 20 years, and the degree to which water would contribute to the economic value added in each sector (assumed to be 10 per cent of predicted growth in economic value added). This study revealed a "net present cost" of US$ 1.4 billion attributed to Black Wattle invasions (it should be noted that this study considered only Black Wattles and not the many other invasive trees in the country).

Van Wilgen *et al*. (2004) estimated the costs and benefits of the biocontrol of six invasive alien plant species in South Africa. They estimated the costs of biological control research that was done on the species, the rate at which each of these species spreads in the absence of biological control, and the degree to which spread has been arrested or reversed by biological control. This, in turn, was used to estimate the extent to which the species would have spread had biological control not been introduced. Estimates of the impacts associated with uncontrolled spread were expressed in the form of three categories of benefits associated with the prevention of invasion: the loss of water due to excessive transpiration by invasive plants; reductions in the values of land that became invaded; and reductions in value added by biodiversity to ecosystem services. These benefits were compared to the costs of biological control research in order to derive cost: benefit ratios. The economic benefits of preventing invasion ranged from $R300\,ha^{-1}\,yr^{-1}$ for Jointed Cactus (*Opuntia aurantica*) to $R3600\,ha^{-1}\,yr^{-1}$ for Golden Wattle (*Acacia longifolia*) (values are discounted to the year 2000). The economic value of water accounted for 70 per cent of the combined benefits. Benefit: cost ratios for the historic analysis (from the release of the biocontrol agent to the year 2000) ranged from 8:1 for Lantana (*Lantana camara*) to 709:1 for Jointed Cactus. When future estimates of benefits were considered, benefit: cost ratios were greater, and ranged from 34:1 for lantana to 4333:1 for Golden Wattle. These large differences were attributed to the length of time that the biocontrol agents have been released (this ranged from 13 to 65 years for different weed species) as well as to the 30-fold differences in the potential area that different weed species would eventually invade.

The above negative impacts should not be presented in a one-sided manner, however. Almost all the important crops in South Africa are harvested from alien plants, and the point needs to be made that a relatively small percentage of these alien plants become invasive. In addition, some invasive alien species have considerable value, despite their negative impacts, and these need to be taken into account when assessing the costs resulting from invasions. Conflicts of interest arise from time to time in cases where important commercial species become invasive and spread beyond the areas where they are cultivated. These include plantation forestry (*Pinus* species; Richardson 1998); where alien plants provide firewood (many *Acacia* species; Higgins *et al*., 1997a), food (*Opuntia* species; e.g. Brutsch and Zimmermann 1993), fodder (*Prosopis* species), or nectar for bees (*Eucalyptus* species; Johannesmeier 1985); and where they have aesthetic or utilitarian value (ornamentals, shade trees, or windbreaks).

A good example is provided by plantation forestry, which is an important part of the South African economy, contributing US$300 million, or 2 per cent, to the GDP and employing over 100,000 people. Downstream industries, based on forestry, produce timber products worth a further US$1.6 billion, much of which is exported, earning valuable foreign exchange. Clearly, these activities are significant. However, a large proportion (38 per cent) of the area invaded by woody alien plants in South Africa is occupied by species used in commercial forestry (especially *Pinus* and *Acacia* species). It is thus clear that forestry has been one of the country's major sources of alien infestation (Richardson 1998; Richardson *et al.* 2003).

13.4 Future impacts

13.4.1 Estimates of magnitude of future impacts

Indications are also that the already serious problem will get worse. New invasive species continue to arrive, and many potential invasive species are already established—but not yet invading. For example, a recent study has quantified the degree to which invasive species that are already a problem will increase in South Africa's major terrestrial biomes (Fig. 13.1). Many serious invasions have exhibited a "lag period" in which the introduced species may occur at very low population levels for several decades before becoming invasive, sometimes suddenly. This could be the result of exponential population growth, a period of selection of genotypes suited to the newly invaded environment, or the occurrence of a change in environmental conditions that constrain invasions. With the rapid growth in the rate of introduction of new species, most introductions of alien species have occurred recently. It is therefore likely that a large number of new potential invasions are currently in their "lag period", and the rate of new invasive species problems will increase dramatically in future. Climate change may also worsen the situation, by bringing about conditions more favourable for invasions.

A recent study (van Wilgen *et al.* 2008) reported on an assessment of the current and potential impacts of invasive alien plants on selected ecosystem services in South Africa. The study used data on the current and potential future distribution of 56 invasive alien plant species to estimate their impact on four services (surface water runoff, groundwater recharge, livestock production, and biodiversity) in five terrestrial biomes. The estimated reductions in surface water runoff as a result of current invasions were >3000 million m^3 (about 7 per cent of the national total), most of which is from the fynbos (shrubland) and grassland biomes; the potential reductions would be more than eight times greater if invasive alien plants were to occupy the full extent of their potential range (Table 13.2). Impacts on groundwater recharge would be less severe, potentially amounting to approximately 1.5 per cent of the estimated maximum reductions in surface water runoff. Reductions in grazing capacity as a result of current levels of invasion amounted to just over 1 per cent of the potential number of livestock that could be supported. However, future impacts could increase to 71 per cent (Table 13.2). A "biodiversity intactness index" (the remaining proportion of pre-modern populations) ranged from 89 per cent to 71 per cent for the five biomes (Table 13.2). With the exception of the fynbos biome, current invasions have almost no impact on biodiversity intactness. Under future levels of invasion, however, these intactness values decrease to around 30 per cent for the savanna, fynbos, and grassland biomes, but to even lower values (13 per cent and 4 per cent) for the two karoo biomes. Thus, while the current impacts of invasive alien plants are relatively low (with the exception of those on surface water runoff) the future impacts could be very high. While the errors in these estimates are likely to be substantial, the predicted impacts are sufficiently large to suggest that there is serious cause for concern.

13.4.2 Motivating for control on the basis of expected benefits

Invasions normally proceed through distinct stages, from introduction through to ecosystem transformation. Each stage will have an associated cost of control, which rises as the infestation proceeds to the next stage (Fig. 13.2). Generally speaking, the ratio of cost of control to damage avoided will increase as invasions proceed to more advanced stages (Fig. 13.2). At later stages of invasion, the costs of control may well outweigh the value of

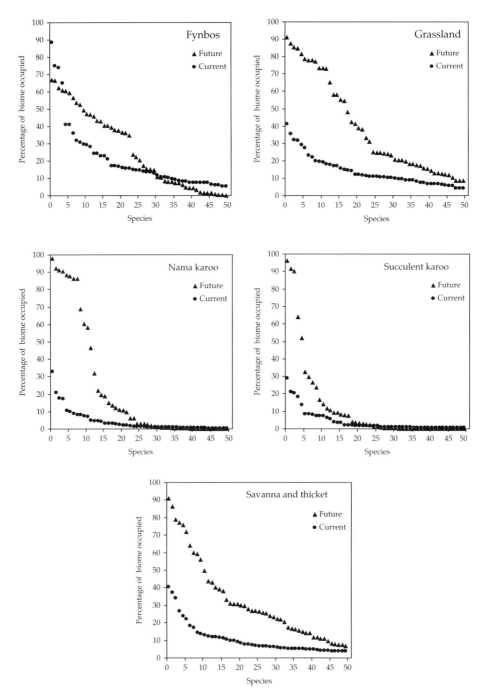

Figure 13.1 Percentage of the biome occupied by the 50 most abundant invasive alien plant species in five biomes in South Africa. The current distribution was estimated from records of presence in quarter-degree squares, and the future distribution from climatic suitability modeling (Figure reproduced from van Wilgen et al. 2008a, with permission).

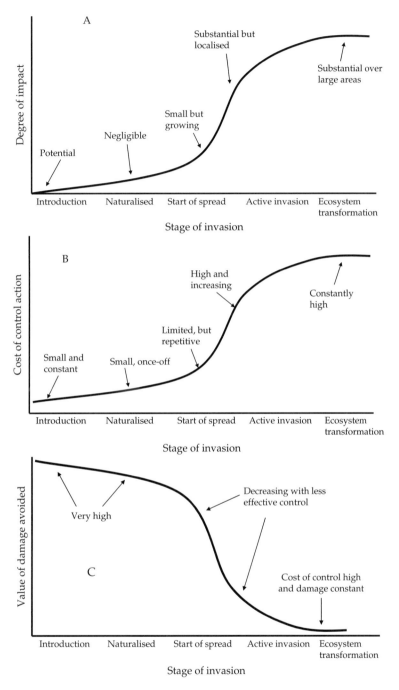

Figure 13.2 Degree of impact (A), cost of control (B), and value of damage avoided if control is successful (C), associated with different stages of invasion by alien plants.

Table 13.2 Estimates of the current and potential impacts of invasive alien plants on three ecosystem services in five terrestrial biomes in South Africa (from van Wilgen et al. 2007). The services are: water resources measured as surface water runoff; grazing measured as the number of live stock units that could be supported; and biodiversity, expressed as a "biodiversity intactness index" (Scholes and Biggs 2005), which is the estimated proportion of remaining populations of vertebrates and vascular plants

Predicted impact	Degree of invasion	Biome				
		Fynbos	Grassland	Succulent karoo	Nama karoo	Savanna and thicket
Mean annual surface water runoff (millions of m^3)	No invasion	6765	26,615	366	2764	14,306
	Current levels of invasion	5701	25,777	268	2621	13,981
	Future levels of invasion	4271	13,188	145	1951	6562
Number of large stock units (millions)	No invasion	0.436	4.80	0.45	2.96	3.66
	Current levels of invasion	0.362	4.78	0.45	2.96	3.63
	Future levels of invasion	0.143	1.3	0.28	0.82	1.01
Biodiversity intactness (%)	No invasion	73	71	87	89	82
	Current levels of invasion	70	71	87	89	82
	Future levels of invasion	31	32	13	4	32

damages avoided, while at earlier stages, the ratio of costs to damage avoided can be very attractive. However, it is often difficult to motivate for expenditure at early stages of invasion, when the future costs have yet to manifest themselves clearly. In this section, we briefly review some aspects of this problem, using examples from South Africa.

The first stage, preventing the introduction of potentially invasive species into new areas, would presumably offer large returns on investment if detrimental species could be detected and blocked from entry. However, this aspect of invasive species management has been neglected in South Africa. The reasons for this include the paucity of capacity to deal with the problem, and until recently the low levels of awareness of the problem. The most advanced countries with regard to dealing with this problem are island nations (such as Australia and New Zealand), where awareness is higher, and such controls are arguably easier to enforce and more likely to succeed. This aspect of control remains weak in South Africa, and deserves attention.

The second stage includes species that have become naturalized, and are currently in a "lag" phase prior to potentially becoming invasive. Australian trees in the genus *Acacia* have become serious invaders in South Africa, as reviewed above. The genus is large, and some species are known to occur in naturalized stands in very small numbers in South Africa at this stage. One example is *Acacia paradoxa*, which is limited to a single population on the slopes of Table Mountain. Its potential to do damage is amply illustrated by comparison to its congeners, and this has been sufficient to persuade the landowners (in this case the national parks authority) to spend limited funds on a one-off eradication effort, rather than ongoing control.

The third stage includes species that are aggressively invasive, but still have a limited distribution range. The localized but rapidly spreading Pompom Weed (*Campuloclinium macrocephalum*) provides an example of a species at the start of its spreading stage. This species was introduced from South America to South Africa. It invades grasslands and can potentially have very large impacts on grazing resources. This species, along with four other "emerging" weeds, has been selected as a candidate for biological control research, based on the prediction that it will have severe impacts if it should be allowed to spread further (Olckers 2004). The approach of targeting "emerging" species as candidates for biological control, before their impacts are fully manifested, is (as far as we know) unprecedented elsewhere. In this case, the control was justified with reference to cost-benefit studies that demonstrated the enormous benefits arising from relatively small investments in biological control at early stages (van Wilgen *et al.* 2004).

Triffid Weed (*Chromolaena odorata*) provides an example of a weed species that is currently at the active invasion stage. Triffid Weed is a scrambling shrub introduced to Africa from central America and the Caribbean. It impacts negatively on agricultural practices and on biodiversity. The species was first recorded in the Hluhluwe/Imfolozi Game Reserve in the 1970s, and it has recently become dominant over much of the reserve's northern sector. The reserve is a major tourist attraction in the region, and is home to Africa's densest populations of both black and white rhinoceros (both severely endangered species). Simple models were used to show that this weed could cover the entire reserve if left uncontrolled, with serious consequences for both conservation and tourism. This was sufficient to persuade the provincial government to dedicate R100 million (approximately US$ 14 million) annually to control operations (Working for Water 2004).

In the case of species that have already reached the stage where they are dominant and can transform ecosystems, motivations for very costly control interventions have to be made based on large economic benefits. An example of this is provided by the tree and shrub species in the genera *Hakea*, *Acacia*, and *Pinus* that decrease water resources in South Africa (Le Maitre *et al.* 1996, van Wilgen *et al.* 1996). South Africa is a dry country, and like many others the demand for water resources often exceeds the capacity of ecosystems to provide them. While the country as a whole still has a water surplus, recent studies have shown that demand already exceeds supply in more than half of the 87 water management areas in South Africa (van Wilgen *et al.* 2007). Thus any further reduction in water supplies as a result of watershed areas becoming invaded by trees and shrubs will seriously retard the prospects for economic growth. Similar statements could be made with regard to livestock production. Although estimates are difficult to make at a national level, livestock production from natural ecosystems (including both domestic stock and "game" farming with indigenous species) generates in the order of R1.25 billion annually in South Africa (Department of Agriculture 2005). A significant proportion of this economic benefit may well be lost as a result of invasion of rangelands by trees, shrubs, succulents, and unpalatable grasses.

13.5 Conclusions

The above review reflects the current state-of-the-art with regard to our understanding of the ecology of invasive species, and their impacts and economic consequences in South Africa. The review highlights a number of points that can serve to inform a future research agenda that would improve our understanding, and our ability to deal with the problem. These points are discussed briefly below.

- Although we have a fair idea of what species have arrived in the country, there is no proper catalogue of alien species. Thus, although South Africa is better off than most other African countries in this regard, there is a need for such a catalogue to be established and maintained as a basis for assessment and management.
- Terrestrial plants and freshwater aquatic organisms have the biggest impacts. Mammals, birds, reptiles, amphibians, and marine organisms have not had any major impacts to date. Not enough information is available on invasive alien invertebrates to be able to assess their impact at this stage.
- The species that have had major impact can be characterized in many cases by the fact that they have been extensively propagated, widely distributed, are predisposed to our circumstances, and have been here for a considerable length of time. This suggests strongly that the number of invasive species, and therefore the level of impacts, is set to grow.
- We don't know nearly enough about the actual impacts, as thorough ecological studies are rare. The translation of ecological impacts into economic terms is in its infancy. Both of these aspects require attention.
- We do not know the rate at which new species are arriving, and we pay little attention to this, despite rising awareness of the problem. Many are discovered only once they start to become a problem.

- We do not have robust models that can predict the rate of expansion and time to reach maximum impacts for species that are here, let alone new ones that will arrive.
- Problems are emerging with regard to developing pragmatic, practical legislation to deal with the problem, and in finding capacity to implement this legislation.

The paucity of data on the impacts of invasive alien species, combined with an imperfect understanding of the ecological processes that will govern future rates and extents of invasion, means that estimates of future impacts have to make many assumptions. In addition, many such estimates do not allow for the calculation of error associated with predicted impacts. While the errors in these estimates could be large, the predicted impacts are of sufficient magnitude to suggest that, even with significant overestimates, there is cause for serious concern; for example, even if the levels of impact are one tenth of those predicted, they would result in significant losses of benefit.

References

Anon. (2007). *Western Cape State of Biodiversity 2007*. CapeNature Scientific Services, Cape Town.

Allsopp, M. (2007). Honeybee pests and diseases. http://www.arc.agric.za. [Online: Accessed 26 April 2008.]

Annecke, D.P. and Moran, V.C. (1982). *Insects and Mites of Cultivated Plants in South Africa*. Butterworths, Durban.

Appleton, C.C. (2003). Alien and invasive freshwater Gastropoda in South Africa. *African Journal of Aquatic Science*, 28, 69–81.

Bond, W. and Slingsby, P. (1984). Collapse of an ant-plant mutualism: The Argentine ant (Iridomyrmex humilis). and myrmechchorous Proteaceae. *Ecology*, 65, 1031–1037.

Bosch, J.M. and Hewlett, J.D. (1982). A review of catchment experiments to determine the effect of vegetation changes on water yield and evaporations. *Journal of Hydrology*, 55, 3–23.

Brooke, R.K., Lloyd, P.H. and De Villiers, A.L. (1986). Alien and translocated terrestrial vertebrates in South Africa. In IAW Macdonald, FJ Kruger and AA Ferrar, eds *The Ecology and Management of Biological Invasions in Southern Africa*. Oxford University Press, Cape Town.

Brutsch, M.O. and Zimmermann, H.G. (1993). The prickly pear (*Opuntia ficus-indica* [Cactaceae]) in South Africa: Utilization of the naturalized weed, and of the cultivated plants. *Economic Botany*, 47, 154–162.

Cocks, M.P. and Stock, W.D. (1997). Heat-stimulated germination in relation to seed characteristics in fynbos legumes of the Western Cape Province, South Africa. *South African Journal of Botany*, 63, 129–132.

Dean, W.R.J. (2000). Alien birds in southern Africa: what factors determine success? *South African Journal of Science*, 96, 9–14.

Dean, W.R.J., Anderson, M.D., Milton, S.J., and Anderson, T.A. (2002). Avian assemblages in native Acacia and alien Prosopis drainage line woodland in the Kalahari, South Africa. *Journal of Arid Environments*, 51, 1–19.

De Moor, I.J. (2002). Potential impacts of alien freshwater crayfish in South Africa. *African Journal of Aquatic Science*, 27, 125–139.

De Moor, I.J. and Bruton, M.N. (1988). *Atlas of Alien and Translocated Indigenous Aquatic Animals in Southern Africa*. South African National Scientific Programmes report No. 144, CSIR, Pretoria.

Dennill, G.B. and Moran, V.C. (1989). On insect-plant associations in agriculture and the selection of agents for weed biocontrol. *Annals of Applied Biology*, 114, 157–66.

Department of Agriculture, 2005. Census of commercial agriculture 2002. http://www.nda.agric.za/ [Online: Accessed 26 April 2008.]

De Wit, M, Crookes, D, and van Wilgen, BW. (2001). Conflicts of interest in environmental management: Estimating the costs and benefits of a tree invasion. *Biological Invasions*, 3, 167–78.

Drake. J.A., Mooney, H.A., di Castri, F., *et al.* (1989). *Biological Invasions: A Global Perspective*. Wiley and Sons, Chichester.

Dye. P.J. (1988). How plants make water flow uphill. *African Wildlife*, 42, 7–8.

Dye, P.J. (1996). Climate, forest and streamflow relationships in South African afforested catchments. *Commonwealth Forestry Review*, 75, 31–8.

Dye, P.J., Olbrich, B.W. and Everson, C.S. (1995). The water use of plantation forests and montane grassland in summer-rainfall forestry regions of South Africa. In *Proceedings of the Seventh South African National Hydrological Symposium*. Institute for Water Research, Grahamstown.

Dye, P.J. and Versfeld, D.B. (2007). Managing the hydrological impacts of South African plantation forests: An overview. *Forest Ecology and Management*, 251, 121–28.

Henderson, L. (1998). South African plant invaders atlas (SAPIA). *Applied Plant Sciences*, 12, 31.

Henderson, L. (2006). Comparisons of invasive plants in southern Africa originating from southern temperate, northern temperate and tropical regions. *Bothalia*, 36, 201–22.

Henderson, L. and Cilliers, C.J. (2002). *Invasive Aquatic Plants: A Guide to the Identification of the Most Important and Potentially Dangerous Invasive Aquatic and Wetland Plants in South Africa*. Agricultural Research Council, Pretoria.

Higgins, S.I., Azorin, E.J., Cowling, R.M., and Morris, M.J. (1997a). A dynamic ecological-economic model as a tool for conflict resolution in an invasive alien, biological control and native-plant scenario. *Ecological Economics* 22, 141–54.

Higgins, S.I., Turpie, J.K., Costanza, R. *et al.* (1997b). An ecological economic simulation model of mountain fynbos ecosystems: Dynamics, valuation and management. *Ecological Economics* 22, 155–69.

Hosking, S.G. and Du Preez, M. (1999). A cost-benefit analysis of removing alien trees in the Tsitsikamma mountain catchment. *South African Journal of Science* 95, 442–48.

Impson N.D., Bills R. and Cambray J.A. (2002) A conservation plan for the unique and highly threatened freshwater fishes of the Cape Floristic Kingdom. In Collares-Periera M.J., Cowx I.G. and Coelho M.M., eds *Conservation of Freshwater Fishes: Options for the Future*, pp. 432–440. Fishing News Books, Blackwell Science, London.

Jeffery, D.J., Holmes, P.M., and Rebelo, A.G. (1988). Effects of dry heat on seed germination in selected indigenous and alien legume species in South Africa. *South African Journal of Botany* 54, 28–34.

Johannesmeier, M.F. 1985, *Beeplants of the South-Western Cape*. Department of Agriculture, Pretoria.

Lach, L., Picker, M.D., Colville, J.F., Allsopp, M.H., and Griffiths, C.L. (2002). Alien invertebrate animals in South Africa. In D Pimental ed. *Biological Invasions: Economic and Environmental Costs of Alien Plant, Animal and Microbe species*, CRC Press, Boca Raton, pp. 267–82.

Le Maitre, D.C., Richardson, D.M., and Chapman, R.A. (2004). Alien plant invasions in South Africa: driving forces and the human dimension. *South African Journal of Science* 100, 103–112.

Le Maitre, D.C., van Wilgen, B.W., Chapman, R.A., and McKelly, D.H. (1996). Invasive plants and water resources in the western Cape Province, South Africa: modelling the consequences of a lack of management. *Journal of Applied Ecology* 33, 161–72.

Le Maitre, D.C., van Wilgen, B.W., Gelderblom, C.M., Bailey, C., Chapman, R.A., and Nel, J.A. (2002). Invasive alien trees and water resources in South Africa: Case studies of the costs and benefits of management. *Forest Ecology and Management* 160, 143–59.

Leslie, A.J. and Spotila, J.R. (2001). Alien plant threatens Nile crocodile (*Crocodylus niloticus*) breeding in Lake St. Lucia, South Africa. *Biological Conservation* 98, 347–55.

Macdonald, I.A.W., Kruger, F.J., and Ferrar, A.A. (1986). *The Ecology and Management of Biological Invasions in Southern Africa*. Oxford University Press, Cape Town.

McConnachie, A.J., de Wit, M.P., Hill, M.P. and Byrne, M.J. (2003). Economic evaluation of the successful biological control of *Azolla filiculoides* in South Africa. *Biological Control* 28, 25–32.

McNeely, J. (2001). Invasive species: a costly catastrophe for native biodiversity. *Land Use and Water Research* 1, 1–10.

Medhin, B.G., 1992. *Parthenium hysterophorus*, a new weed problem in Ethiopia. FAO Plant Protection Bulletin number 40, Forestry and Agricultural Organization, Rome.

Milton, S.J. (2004). Grasses as invasive alien plants in South Africa. *South African Journal of Science* 100, 69–75.

Mucina, L. and Rutherford, M.C. (2006). *The vegetation of South Africa, Lesotho and Swaziland*. South African National Biodiversity Institute, Pretoria.

Noss, R.F. (1990). Indicators for monitoring biodiversity: A hierarchical approach. *Conservation Biology* 4, 355–64.

Olckers, T. (2004). Targeting emerging weeds for biological control in South Africa: the benefits of halting the spread of alien plants at an early stage of their invasion. *South African Journal of Science* 100, 64–8.

Parsons, W.T. and Cuthbertson, E.G., 1992. *Noxious Weeds in Australia*. Inkata Press, Melbourne.

Payne, A.I.L. and Crawford, R.J.M. (1989). *Oceans of Life off Southern Africa*. Vlaeberg Publishers, Johannesburg.

Pimentel, D. (2002). *Biological Invasions: Economic and Environmental Costs of Alien Plant, Animal and Microbe Species*. CRC Press, Boca Raton.

Pyšek, P., Richardson, D.M., Rejmánek, M., Webster, G.L., Williamson, M., and Kirschner, J. (2004). Alien plants in checklists and floras: towards better communication between taxonomists and ecologists. *Taxon* 53, 131–43.

Richardson, D.M. (1998). Forestry trees as invasive aliens. *Conservation Biology* 12, 18–26.

Richardson, D.M., Cambray, J.A., Chapman, R.A. *et al.* (2003). Vectors and pathways of biological invasions in South Africa—past, present and future. In G.M. Ruiz and J.T. Carlton, eds. *Invasive Species: Vectors and Management Strategies*. Island Press, Washington.

Richardson, D.M., Macdonald, I.A.W., Hoffmann, J.H. and Henderson, L. (1997). Alien plant invasions. In R.M. Cowling, D.M. Richardson, and S.M. Pierce, eds. *The Vegetation of Southern Africa*, Cambridge University Press, Cambridge, pp. 535–70.

Richardson, D.M. and van Wilgen, B.W. (2004). Invasive alien plants in South Africa: how well do we understand the ecological impacts? *South African Journal of Science* 100, 45–52.

Robinson, T.B., Griffiths, C.L., McQuaid, C.D. and Rius, M. (2005). Marine alien species in South Africa—status and impacts. *African Journal of Marine Science* 27, 297–306.

Samways, M.J., Taylor, S. and Tarboton, W. (2005). Extinction reprieve following alien removal. *Conservation Biology* 19, 1329–30.

Samways, M.J. and Taylor, S. (2004). Impacts of invasive alien plants on Red-Listed South African dragonflies (Odonata). *South African Journal of Science* 100, 78–80.

Scott, D.F. and Schulze, R.D. (1992). The hydrological effects of a wildfire in a eucalypt afforested catchment. *South African Forestry Journal* 160, 67–74.

Scott, D.F. and van Wyk, D.B. (1990). The effects of wildfire on soil wettability and hydrological behaviour of an afforested catchment. *Journal of Hydrology*, 121, 239–56.

Scholes, R.J., and Biggs, R. (2005). A biodiversity intactness index. *Nature* 434, 45–49.

Scott, D.F., Versfeld, D.B. and Lesch, W. (1998). Erosion and sediment yield in relation to afforestation and fire in the mountains of the Western Cape Province, South Africa. *South African Geographical Journal* 80, 52–59.

Skelton, PH (2002) An overview of the challenges of conserving freshwater fishes in South Africa. In M.J. Collares-Pereira, M.M. Coelho, and I.G. Cowx, eds. *Conservation of Freshwater Fishes: Options for the Future*. Fishing News Books, Blackwell Science, London.

Skinner, J. and Smithers, R.H.N. (1990). *The Mammals of the Southern African Subregion*. University of Pretoria, Pretoria.

Smith, R.E. and Scott, D.F. (1992). *Simple empirical models to predict reductions in annual and low flows resulting from afforestation*. Report FOR-DEA 465. CSIR Division of Forest Science and Technology, Pretoria.

Steenkamp, H.E. and Chown, S.L. (1996). Influence of dense stands of an exotic tree, *Prosopis glandulosa* Benson, on a savanna dung beetle (Coleoptera: Scarabaeinae) assemblage in southern Africa. *Biological Conservation* 78, 305–11.

Taye, T. and Tanner, D.G. (1998). Determination of economic threshold densities for major weed species competing with bread wheat in Ethiopia. *African Crop Science Journal* 5, 371–84.

Tribe, G.D. (1995). The woodwasp *Sirex noctilio* Fabricius (Hymenoptera: Siricidae), a pest of *Pinus* species, now established in South Africa. *African Entomology* 3, 215–217.

Tribe, G.D. and Richardson, D.M. (1994). The European wasp, *Vespula germanica* (Fabricius) (Hymenoptera: Vespidae), in southern Africa and its potential distribution as predicted by ecoclimatic modelling. *African Entomology* 2, 1–6.

Turpie, J. and Heydenrych, B. (2000). Economic consequences of alien infestation of the Cape Floral Kingdom's Fynbos vegetation. In C. Perrings, M. Williamson, and S, Dalmazzone, eds. *The Economics of Biological Invasions*, Edward Elgar, Cheltenham, pp. 152–82.

van Wilgen, B.W., Bond, W.J. and Richardson, D.M. (1992). Ecosystem Management. In R.M. Cowling, ed. *The Ecology of Fynbos; Nutrients, Fire and Diversity*, Oxford University Press, Cape Town, pp. 345–71.

van Wilgen, B.W., Cowling, R.M., and Burgers, C.J. (1996). Valuation of ecosystem services—A case study from South African fynbos ecosystems. *BioScience* 46, 184–89.

van Wilgen, B.W., de Wit, M.P., Anderson, H.J. et al. (2004). Costs and benefits of biological control of invasive alien plants: case studies from South Africa. *South African Journal of Science* 100, 113–22.

van Wilgen, B.W., Little, P.R., Chapman, R.A., Görgens, A.H.M., Willems, T., and Marais, C. (1997). The sustainable development of water resources: History, financial costs, and benefits of alien plant control programmes. *South African Journal of Science* 93, 404–11.

van Wilgen, B.W., Marais, C., Magadlela, D., Jezile, N., and Stevens, D. (2002). Win-win-win: South Africa's Working for Water programme. In S.M. Pierce, R.M. Cowling, T. Sandwith and K. MacKinnon, eds. *Mainstreaming Biodiversity in Development: Case Studies from South Africa*, pp. 5–20. The World Bank, Washington D.C.

van Wilgen, B.W., Nel, J.L. and Rouget, M. (2007). Invasive alien plants and South African rivers: A proposed approach to the prioritization of control operations. *Freshwater Biology* 52, 711–23.

van Wilgen, B.W., Reyers, B., Le Maitre, D.C., Richardson, D.M., and Schonegevel, L. (2008a). A biome-scale assessment of the impact of invasive alien plants on ecosystem services in South Africa. *Journal of Environmental Management*, 89, 336–349.

van Wilgen, B.W. and Richardson, D.M. (1985). The effects of alien shrub invasions on vegetation structure and fire behaviour in South African fynbos shrublands: A simulation study. *Journal of Applied Ecology*, 22, 955–66.

van Wilgen, B.W. and Scholes, R.J. (1997). The vegetation and fire regimes of southern hemisphere Africa. In B.W. van Wilgen, M.O. Andreae, J.A. Lindesay and J.G. Goldammer, eds. *Fire in Southern African Savannas—Ecological and Atmospheric Perspectives*, Witwatersrand University Press, South Africa, pp. 27–46.

van Wilgen N.J., Richardson, D.M., and Baard, E.H.W. (2008b). Alien reptiles and amphibians in South Africa:

Towards a pragmatic management strategy. *South African Journal of Science*, 104, 13–20.

van Wyk, D.B. (1985). The effects of catchment management on sediment and nutrient exports in the Natal Drakensberg. In R.E. Schulze, ed. *Proceedings of the 2nd South African National Hydrological Symposium*, University of Natal, Pietermaritzburg, pp. 266–74.

Versfeld, D.B., le Maitre, D.C., and Chapman, R.A. (1998). *Alien Invading Plants and Water Resources in South Africa: A Preliminary Assessment*. Report TT99/98, Water Research Commission, Pretoria.

Von Breitenbach, F. (1989). *National List of Introduced Trees*. Dendrological Foundation, Pretoria.

Wilson, J.R.U., Richardson, D.M., Rouget, M. *et al.* (2007). Residence time and potential range: crucial considerations in modelling plant invasions. *Diversity and Distributions*, 13, 11–22

Wise, R.M., van Wilgen, B.W., Hill, M.P. *et al.* (2007). The economic impact and appropriate management of selected invasive alien species on the African continent. Report CSIR/NRE/ER/2007/0044/C, CSIR, Pretoria.

Working for Water (2004). *Annual Report 2003/04*. Department of Water Affairs and Forestry, Cape Town.

Zimmermann, H.G. (1991). Biological control of mesquite, *Prosopis* spp. (Fabaceae) in South Africa. *Agriculture, Ecosystems and Environment*, 37, 175–86.

CHAPTER 14

Invasive Plants in Tropical Human-Dominated Landscapes: Need for an Inclusive Management Strategy

R. Uma Shaanker, Gladwin Joseph, N.A. Aravind, Ramesh Kannan, and K.N. Ganeshaiah

14.1 Introduction

Invasive species have been regarded as one of the most important yet subtle threats to biological diversity (Cronk and Fuller 1995; Humphries *et al.* 1991; Luken and Thieret 1997; Schmitz *et al.* 1997; Simberloff 1998; Vitousek *et al.* 1997). In fact, according to the Convention on Biological Diversity (CBD), invasive species are defined as those that result in considerable damage to ecosystem processes (Secretariat of the Convention on Biological Diversity 2005). It is frequently claimed that the impact of invasive species on biological diversity is the second largest threat next only to that due to habitat destruction and climate change, especially in the Islands (D'Antonio and Kark 2002; Vitousek *et al.* 1997; Walker and Steffen 1997). Free from their native habitats, invasive species are no longer reined in by ecological forces that may have otherwise kept their population in check. Consequently, through their path of invasion, invasive species compete, marginalize, and frequently usurp native biological diversity (Mooney and Hobbs 2000). In extreme cases, invasive species can also lead to the extinction of native species (Atkinson *et al.* 1995, 2000; Gurevitch and Padilla 2004; Jenkins *et al.* 1989; Van Riper *et al.* 1986; Warner 1968; Wikelski *et al.* 2004; Williamson 1996 and references therein; Work *et al.* 2000).

Besides impacting on the native flora and fauna, invasive species are also reported to impair critical ecosystem services (Braithwaite and Lonsdale 1987; Braithwaite *et al.* 1989; Cronk and Fuller 1995; Hobbs and Mooney 1986; Vitousek and Walker 1989) such as disrupting pollinator and dispersal services to native species (Feinsinger 1987; Ghazoul 2001; see review in Levine *et al.* 2003) and altering the nutrient recycling and disturbance regime (see review in Levine *et al.* 2003). Parker *et al.* (1999) identified at least five distinct environmental consequences of invasive species ranging from their impacts on the phenotypic characteristics of native individuals to their impacts on the ecosystem services and processes.

While there are few estimates, it is well recognized that invasive species do have a significant economic impact (Perrings *et al.* 2000). It is speculated that the annual cost of invasive species in the US alone could be in the order of 137 billion US$, which amounts to nearly two per cent of the country's GDP (McNeely 2001; Pimentel *et al.* 2000; Pimentel *et al.* 2001). These costs are difficult to audit as often it is necessary to impute the indirect economic losses as well. Thus in South Africa, besides the direct cost of invasive species, the loss of agricultural productivity due to competition for scarce water by the invasive species is also being considered to evaluate the impact of invasive species (van Wilgen and van Wyk 1999).

Quite understandably, against the overarching negative effects of the invasive species on biodiversity and ecosystem services, the global agenda has been to prevent and contain the invasive species and thereby to mitigate their impacts on local biodiversity and the ecosystem function and human health (www.gisp.org). However, apart from a few prominent examples (such as the control of *Optunia* in Australia), management of invasive species has met with limited success (Cilliers 1983; see review in *Day et al.* 2003). The low degree of success is further compounded by the fact that preventing and controlling the spread of invasive species is often enormously expensive. Obviously, management of invasive species in low income countries of the tropics, which incidentally also happen to have some of the most invasive species ridden landscapes, is heavily constrained (Perrings 2005).

The management of invasive species is especially challenging in tropical human-dominated landscapes. First, in these habitats, invasive species can exacerbate biodiversity crises by reducing the population densities of indigenous species, many of which fulfil subsistence needs of the rural poor (Shackleton *et al.* 2007 and references therein; Uma Shaanker *et al.* 2004a). In many of these regions, non-timber and minor forest products (NTFPs) constitute an important source of livelihood for millions of people (Uma Shaanker *et al.* 2004a). In India alone, it is estimated that over 50 million people are dependent on NTFPs for their subsistence and cash income (National Centre for Human Settlements and Environment 1987; Hegde *et al.* 1996). Recent studies in India have shown that traditional income sources from forest based resources could be jeopardized by invasive species, both directly due to impact on resources and indirectly by rendering these resources less accessible for collection and harvest (Uma Shaanker *et al.* 2004a). Thus, under these circumstances, invasive species could potentially lead to further marginalization of the already impoverished livelihoods of the people in the tropical human dominated landscapes.

Second, in tropical human-dominated landscapes, invasive species have often been viewed as a resource that can accrue potential economic benefits that can aid rural livelihoods. Thus it is not uncommon that control of invasive species in these habitats may actually lead to loss of rural livelihoods (Perrings 2005). While the resolution of this interesting dilemma—to control or not—is admittedly a difficult one to address, it throws open serious challenges to the conventional wisdom and approach in managing invasive species, at least for the tropical landscapes.

Unfortunately, mainstream research on management of invasive species has scarcely addressed this question. In fact, as argued by Shackleton *et al.* (2007), the impact of invasive species on rural livelihoods has received little attention, despite the fact that rural land and waters are most affected by invasive species.

In summary, solutions to the management of invasive species need to be reworked to take into account the fact that invasive species can impact human livelihoods both negatively and positively. Besides the existing strategies for the management of invasive species, there is a need for alternative strategies in terms of the net benefit they yield, taking of course all benefits and costs into account.

Here we consider a specific case of control of invasive species in largely tropical landscape, with the attendant problems of human dependence on natural resources as well as lack of investment portfolios to control invasive species. We propose management strategies that promote use of the invasive as a way of minimizing the net costs of the invasive species.

While the above thesis seems to be at variance with conventional management strategies that always work to exclude the species, it could actually be viewed as one among the spectrum of management options, also trying to maximize the marginal returns of limiting the damage and control costs of the invasive species.

In the following section we first briefly review how the management of invasive species scales with the dynamics of the invasion; clearly management decisions do not necessarily conform to "one size fits all" and hence different strategies, including perhaps the exploitation of the invasive at some stage of its invasion, could be regarded as a rational approach to minimizing its net costs. Second, we review briefly three examples of invasive species, where exploitation of the species may help reduce the net costs imposed by the species. In the last section, we present a recent case study that has explored the use of *Lantana camara* as a substitute for bamboo and canes in India and discuss how such use could be viewed in the context of limiting the damage as well as control costs of the species and hence increasing the marginal benefits of managing the invasive species.

14.2 Dynamics of invasive species and management scenarios

Optimizing management strategies for invasive species requires an evaluation of the full range of costs and benefits of alternative control options on one hand and the reduction of the ecological impacts on the other. Accordingly, in managing invasive species, it might be useful to consider the following questions: a) how should management scale with the temporal dynamics of invasive species, b) what are the threshold levels of invasion that warrant management (when benefits due to investments in control exceed the damage costs) and, as a corollary, c) at what threshold (or when), should management effort be considered to have failed (that is, when benefits due to control are actually less than the damage costs)? Clearly, answers to these questions would not only put the current management strategies in perspective but also help raise interesting alternatives to existing management solutions. We address these questions briefly below.

14.2.1 How does management scale with the dynamics of invasive species?

The temporal dynamics of invasive species have been characteristically represented as a logistic function (Fig. 14.1; Hobbs and Humphries 1994; McGarry *et al.* 2005; http://tncweeds.ucdavis.edu) with 1) an initial lag phase where the invasion has set in and the species has been able to found single or multiple populations at the site of invasion; 2) a logistic growth phase where, upon successful establishment, the species reproduces rapidly and occupies a disproportionately large ecological space, and finally 3) a stable phase where the species is seen to have nearly saturated the available niche and possibly attained an equilibrium state

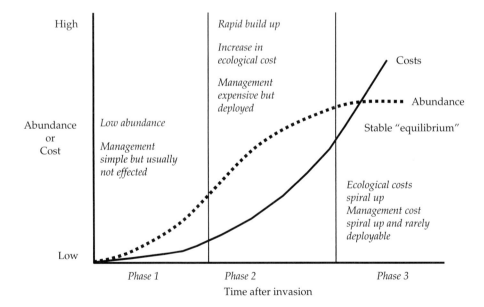

Figure 14.1 A schematic representation of the temporal dynamics of an invasive species and its associated ecological costs (please see text for explanation).

(McGarry et al. 2005; Williamson 1996). The time taken from invasion to attaining the stable phase varies with species, ranging from as low as 4 years to as high as 90 years (http://tncweeds.ucdavis.edu) and may be dependent upon a number of factors including the life history strategies of the species, extraneous drivers, and so on. Obviously, for invasive pathogens, the timescales would be even smaller (Jules et al. 2002; http://www.cbd.int/doc/submissions/ias/ias-diversitas-risk-2007-en.pdf).

Assuming that the ecological cost of an invasive species is a function of its spread and abundance, least impacts would be predicted when the species is still in its lag phase, but with higher impacts during the subsequent logistic and stable "equilibria" phases (Fig. 14.1). Thus across the time period of invasion, the ecological cost curve would tend to also increase at an increasing rate until perhaps the stable phase is attained. However, in the latter phase, the cost curve may still be expected to increase reflecting multiplier effects, such as the accumulated effects on local ecosystems including arrest of recruitment of native flora, allelopathic effects, and pollinator or seed dispersal disruptions (Achhireddy and Singh 1984; Dunbar and Facelli 1999; Feinsinger 1987; Fensham et al. 1994; Ghazoul 2001; Jain et al. 1989; Lamb 1991; Lyon and French 1991; Martin 1999; Mersie and Singh 1987; Sharma et al. 1988; Singh and Achhireddy 1987; Vivrette and Muller 1977; also see review in Levine et al. 2003).

It follows from the above, that the resources committed to control invasives would not be equal across the temporal dynamics of invasion. In the early stages of invasion, eradication is the generally preferred option, both because the marginal costs are low and the potential benefits are large. For example, when *Caulerapa*, a highly invasive algae from the Mediterranean, invaded California waters, immediate action was taken to eradicate the weed (Anderson 2003). Ironically, however, because the implied ecological costs are either too small or not yet perceived at this stage, often no management effort is initiated. In the logistic phase, with greater impacts and increased perception of the invasion, management investments are made, though they are expensive. Most management programs end up being deployed during this phase (logistic phase; Fig. 14.1).

In the last, stable phase, management interventions are not only going to be very expensive but tend to be least successful, and hence are rarely deployed. That is, in this stage, the marginal benefits due to control tend to be less than the marginal control costs. Thus for all practical purposes, management of invasive species in the third phase may be regarded to be a futile exercise. But should management options to control invasive species at this stage be abandoned? Could alternate management strategies be developed? We are particularly interested in addressing this question, because, at least in tropical landscapes, it is not uncommon to find invasive species that have swamped entire ecosystems (presumably reflecting the stable phase) and thus exhausted most management options that are economically viable.

While the answers to the questions raised above are as challenging as they are interesting, we believe that part of the problem lies in our current view of invasive species purely from the point of their negative impacts. We argue that incorporating a plurality of view, that invasive species across their temporal dynamics may also have some positive impacts and can probably be exploited for use, could drastically change the calculus of their management.

In summary, the management options would not be expected to be consistent across the dynamics of invasion, but would tend to be a function of the species, habitats occupied, naturalization and establishment, spread process, and so on. In extreme cases, where the net benefits of control are negative, it might even pay to encourage the invasive, not restricting it. Under this overarching view, management of invasive species may not preclude any options as long as the marginal benefits of control are positive.

Some introduced species have the potential to yield benefits that may outweigh any negative impacts they have on the flow of ecosystem services in the affected habitats. For example, a few introduced species have been shown to provide direct and indirect benefits to the ecosystem, including prevention of soil erosion, pollinator services, soil enrichment, non-timber wood requirement,

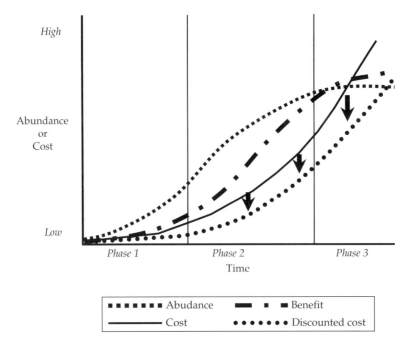

Figure 14.2 A schematic representation of the temporal dynamics of an invasive species showing the discounted ecological cost should the invasive accrue some benefit from its use (please see text for explanation).

bio-fuel, biofertilizers, and with the more recent recognition that they may contribute to carbon sequestration (Geesing *et al.* 2004; Kannan *et al. in press*). In fact, Zavaleta *et al.* (2001) argued that control of the invasive species during the third phase (stable phase) could be confounded by the fact that it may actually lead to loss of certain functions that the invasive species has been able to replace during the process of stabilization in its new habitat. Also, in several instances, invasive species could have both negative and positive impacts, and it might be necessary to adopt a holistic frame and not be eclectic in the choice of the impacts. For example, Wootton *et al.* (2005) reported that *Carex kobomugi*, an invasive species on the Florida coast has been recognized to be beneficial in stabilizing sand dunes while also imposing an ecological cost by displacing other species.

In tropical human-dominated landscapes, the introduced species can serve as an important resource to aid rural livelihoods. Siges *et al.* (2005) reported that the invasive shrub, *Piper aduncum*, could actually provide livelihoods to rural people in Papua New Guinea. As will be described in the next section, there are now a number of cases where the introduced species has been integrated into the livelihood requirement of local people. In an interesting treatment of biological invasions and poverty, Perrings (2005) acknowledged that in low income countries, because people might exploit invasive species for food, fiber, and fuel, there could be an ambivalent attitude to the control of invasive species. Indeed in these countries, control of invasive species may have to be weighed against the loss that may be incurred due to the loss of resources.

From a management perspective, the increasing ecological cost of invasive species should be weighed against any benefits that may accrue (Fig. 14.2). Accordingly, the strategy for management or otherwise will depend upon the boundary condition such as (a) opportunity cost of use of invasive and (b) the trade-off between potential ecological costs and benefits of use of invasive species.

Thus it appears that the management strategy, within the limits of temporal dynamics of the invasive species, would be a function of the relative marginal benefits and costs involved in controlling the invasive species. In the range of options available for management, one can allow for the entire spectrum from complete eradication to an adaptive management (wherein the species can actually be gainfully used; http://tncweeds.ucdavis.htm).

14.2.2 When control of invasive species fails: three examples

The history of management of invasive species, or perhaps the lack of successful management, has unleashed several notorious invasive species in the world, notable among them being *Eichhornia*, *Prosopis*, *Optunia*, and *Lantana* (See review in Day *et al*. 2003; Julien *et al*. 2001; Navarro and Phiri 2000; Matthews 2004). A combination of factors, including a rapid multiplication rate and the lack of successful control measures, has lead to a widespread colonization of those species across the face of earth, save those landscapes that are inherently unsuitable for the species. Control or management of these invasive species, especially in tropical human-dominated landscapes with traditionally poor economies, has been nearly absent or woefully unsuccessful. On the contrary, over their time of residence, at least some of these invasive species have blended into the local ecosystems and become integrated into the livelihoods of people (Shackleton *et al*. 2007).

One of the best illustrated examples of an invasive species is Water Hyacinth (*Eichhornia*). The plant was introduced from South America to many parts of the world between 1879 and 1890 for its ornamental value (www.gisp.org; http://www.invasivespeciesinfo.gov). A century and half later, the species has invaded nearly all parts of the tropical world (Navarro and Phiri 2000). The plant is known for its incredibly high rate of multiplication, producing about 3000 saplings in just 50 days and covering an area of 600 m^2 in a year. The plant is considered one of the worst aquatic invaders in the world (Holm *et al*. 1991). Efforts to control the plant, in the conventional sense of the term, have been not very successful (Navarro and Phiri 2000).

Confronted by the huge biomass of the plant, efforts have been made in many parts of the world to explore the possibility of using the invasive as a resource in a variety of scenarios. For example, in a number of countries including the Philippines, Indonesia, and India, Water Hyacinth is being used as a substrate in a small-scale cottage industry for making paper. In many African and South East Asian countries such as Kenya, Tanzania, Thailand, Philippines, Malaysia, and Bangladesh, Water Hyacinth is used to make furniture, carpets, mats, pillows, and ropes. Wicker items made from the stems have proved extremely popular in Germany and Japan. In Bangladesh, Water Hyacinth is being used to make fiber boards and paper (http://practicalaction.org/docs/technical_information_service/water_hyacinth_control.pdf). Water Hyacinth is also used to make low cost organic fertilizer for farms. In Sri Lanka, Water Hyacinth is mixed with organic municipal waste, ash, and soil, composted and sold to local farmers and market gardeners. In Malaysia, fresh water hyacinth is cooked with rice bran and fishmeal and mixed with copra meal as feed for pigs, ducks, and pond fish (Gopal 1987). Similar practices are used in Indonesia, the Philippines, and Thailand (National Academy of Sciences 1976). Water Hyacinth has also been used for bio-remediation especially in the removal or reduction of nutrients, heavy metals, organic compounds, and pathogens from water (Gopal 1987).

Another equally notable species is *Prosopis*. The species was introduced from south and central America to many parts of the world, to meet the fuel wood requirement and thus reduce the pressure on indigenous forests (Geesing *et al*. 2004; Mwangi and Swallow 2005). Unfortunately, and as has been the case with a number of such introduced species, *Prosopis* came to invade large swathes of landscape and be associated with a number of negative impacts. Control measures proved to be quite unsuccessful. However over time, with its naturalization in many parts of the world, *Prosopis* has come to be used as an important resource (Geesing *et al*. 2004; Mwangi and Swallow 2005). In Niger and Yemen, *Prosopis* has been exploited for fuelwood, free-grazing forage, construction materials, and pods are even used to make biscuits (Geesing *et al*. 2004).

The flowers of *Prosopis* species are regarded as a valuable source of bee forage, and honey has become the food product most often derived from *Prosopis* (Geesing *et al*. 2004). In Niger and Mauritania, *Prosopis* plantations have been established for sand dune stabilization (Jensen and Hajej 2001), restoration of degraded land in Cape Verde, and remediation of saline land in India. It has also been used as shelterbelts, with animal fodder and other uses as co-products (Geesing *et al*. 2004). It has been estimated that the annual income generated by selling *Prosopis* wood in the rural market would yield 3.25 million US dollars per annum (Boureima *et al*. 2001). In an interesting experiment, more than 500 women and farmers have been trained in making *Prosopis* flour for making food for human consumption and also to utilize this new bio-resource (Geesing *et al*. 2004).

Our final example is of *Optunia* cacti that were introduced into South Africa in 1700 (Larsson 2004). Over the next three centuries, *Optunia* invaded virtually all parts of South Africa with few effective control systems in place. Again, as in the case of Water Hyacinth and *Prosopis*, local ingenuity found that the fruits of the invasive *Opuntia ficus-indica*, could serve as an important food resource for humans, and the cladodes as fodder for the livestock (Larsson 2004). More recently, new products such as Prickly Pear wine and jam are being made from the fruit (Shackleton *et al*. 2007). However, in Australia, Opuntia was successfully contained by biological control agent *Cactoblastis cactorum* (Parsons and Cuthbertson 2001; Tu *et al*. 2001). At present, *Opuntia* spp. remains as isolated and scattered populations (Greenfield and Nicholson 2007).

The three examples presented above are all symptomatic of invasive species that represent the far end of the temporal dynamics curve (Fig. 14.2). Having spread far and wide, they have either attained what one might refer to as a stable equilibrium condition or at best a stable phase in their invasion. Management options at this stage, as mentioned earlier, are very limited, basically because of the impracticability of management. In fact in all of the three examples listed above, the costs of containment may not be justified by the damage avoided. It is under these conditions that the species in question lend themselves to alternate management streams including using them as resource. While the latter approach does not classify itself as conventional in so far as controlling invasive species goes, it can, as argued earlier, be viewed as an alternative nevertheless that aims to use the resource and thereby lead to a reduction in the net cost of the species.

In fact, a number of invasive species the world over have been incorporated into the daily livelihoods of people (Table 14.1). Why would communities use alien invasive species? Kannan *et al*. (*in press*) conjectured that among other factors the following may be the common drivers that compel communities to use resources that they hitherto had no knowledge of: a) availability of the invasive species in abundance—towards the end of their logistic expansion and stable phase the invasive species become quite abundant and, in a typical ecosystem setting, perhaps constitute the predominant vegetation community; b) zero investment resource—invasive species are frequently a zero-investment resource, freely available for harvest, in fact, forest managers and farmers alike, in most circumstances, would be very willing to have their forest or field cleared of the invasive; c) substitutability of invasive species with some locally available but rare or otherwise expensive resource—invasive species could offer a suitable if not a perfect substitute for an existing resource, that is either scarce or is expensive; d) opportunity cost of collection and utilization—because of their abundance and substitutability, the opportunity cost of collection and utilization of many invasive species is very low.

14.3 The specific case of *Lantana* in India

14.3.1 Invasion, spread, and ecosystem impacts

Lantana camara (hereafter referred to as *Lantana*) is one of the most notable alien invasive plant species with a pan-continental distribution. Native to Central and South America, the plant is now reportedly distributed and established in over 60 countries around the world (Parsons and Cuthbertson 2001; Day *et al*. 2003). The plant has been considered one of the worst weeds recorded in

Table 14.1 Some examples of use of invasive species

Sl no	Invasive species	Uses	References
1	Lantana camara	Basket, furniture, charcoal, medicine, toys	Joshi 2002; Kannan et al. In Press
2	Eichhornia crassipes	Pillows, furniture and carpets	http://www.water-hyacinth.com/crafts.html http://ecosyn.us/ecocity/Links/My_Links_Pages/water_hyacinth01.html
		High quality paper and cattle feed	Nolan and Kirmse, 1974 http://itdg.org/docs/technical_information_service/water_hyacinth_control.pdf
		Bioconversion of water-hyacinth hemicellulose acid hydrolysate to motor fuel ethanol by xylose–fermenting yeast	Nigam 2002
		Used for fiber board, fertilizer, cooking wood and cultivation of mushrooms, briquettes, sewage treatment, animal feed production and mushroom cultivation and composting	http://library.thinkquest.org/C0126023/uses.htm Haider 1989
3	Prosopis spp.	Fuel wood, fencing poles, furniture, food, honey and charcoal, wood flooring	http://www.gardenorganic.org.uk/pdfs/international_programme/Prosopis-PolicyBrief-1.pdf http://www.vetiver.org/OT_prosopis.htm Geesing et al. 2007
4	Opuntia ficus-indica	Fruits	Shackleton et al. 2007
5	Oreochromis mossambicus	Food for humans and Zoo animals	Dahanukar et al. 2005 (Cited in McGarry et al. 2005)
6	Mimosa pigra	Fuel wood	http://www.worldagroforestrycentre.org/Sea/Products/AFDbases/AF/asp/SpeciesInfo.asp?SpID=672

human history (Cronk and Fuller 1995). *Lantana* has usurped numerous native plants of their niches as well as unsettling farm lands and forest gaps. Forest managers and farmers alike are at their wits end to control the weed (Ganeshaiah and Uma Shaanker 2001).

Lantana was introduced into India at the National Botanical Gardens, Calcutta in 1807, as an ornamental plant by the British (Thakur *et al.* 1992). Since then, the plant has successfully invaded virtually all parts of the country (Fig. 14.3). In India, it is distributed from the sub-montane regions of the outer Himalayas to the southernmost part of India, occurring in every forest type from about 400 m above sea level (ASL) and above (Thakur *et al.* 1992). Millions of hectares of grazing land as well as agricultural lands are infested by the weed. It forms the major undergrowth in forestry plantations (Singh 1976) and is generally found to be associated with anthropogenic disturbances (Ganeshaiah and Uma Shaanker 2001). An ecological niche model of the species indicated that among the possible sites of invasion, two of the country's megadiversity hotspots, namely, the Western Ghats and the Eastern Himalayas are also very suitable and thus prone for invasion by the plant (Fig. 14.3).

Innumerable studies have been conducted to assess the impacts of *Lantana* on ecosystems and ecosystem services. While most of them demonstrate an adverse effect, a few studies have shown a positive effect of the plant. A brief summary of the impacts of *Lantana* is presented in Table 14.2.

Several methods for controlling *Lantana*, including chemical, mechanical, fire, and biological controls have been used but with limited success (see review in Day *et al.* 2003; Davis *et al.* 1992). Fire is one of the cheapest methods for controlling *Lantana* and is often used in grazing areas. However, mature *Lantana* is fire tolerant and re-growth from seeds and basal shoots is common. Extensive efforts have been made to find effective biocontrol agents for

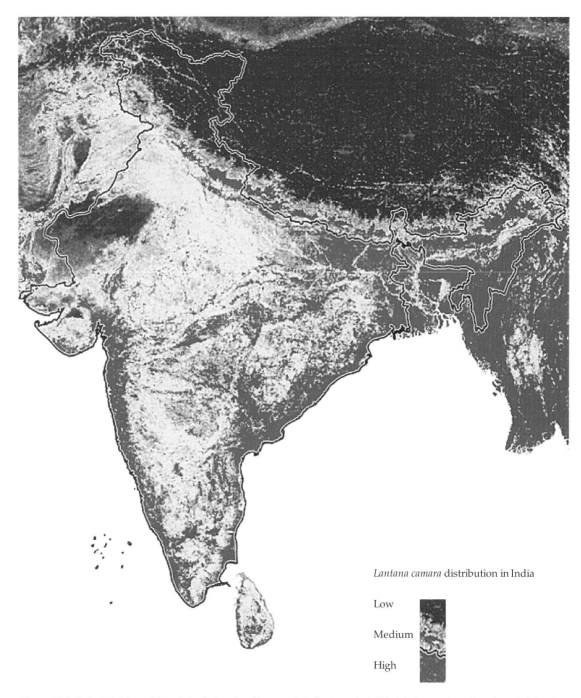

Figure 14.3 Ecological niche model prediction for invasion of *Lantana* in India. Areas shaded black indicate regions where the probability of invasions is low. Areas shaded solid grey indicate regions where the probability of invasions is high. Areas that include light grey shades indicate regions where the probability of invasions is intermediate. Note that both the biodiversity hotspots are highly prone for invasion. Inset: Cumulative increase in records of Lantana in published floras in India. (Courtsey: Map of the ecological niche model prediction from Mohammed Irfan, ATREE, Bangalore).

Table 14.2 Summary of the studies of impacts of *Lantana* on ecosystem and ecosystem services

Effect	References
Positive effects	
Increases regeneration of native species*	Murali and Setty 2001
Increased soil nutrient pools and nutrient mobility	Lamb 1982
Increases soil nutrients	Wilson 1968; Lamb 1982
Increases regeneration of non timber forest products	Ganesan *Pers. Comm*
Antifungal potential in soil	Shaukat and Siddiqui *et al.* 2001
Antimicrobial, fungicidal, insecticidal, and nematicidal activity, but not antiviral activity	Chavan and Nikam 1982; Sharma and Sharma 1989
Lantana pulp is used for writing and printing paper	Gujral and Vasudevan 1983
Used as a cover crop in deforested areas and also used to enrich the soil and protect against erosion	Anon. 1962; Greathead 1968; Willson 1968; Ghisalberti 2000
Negative effects	
Decrease in regeneration of native species*	Jain *et al.* 1989
Decrease in biodiversity	Lamb 1991; Lyon and French 1991
Decrease in species richness	Fensham *et al.* 1994
Contamination of gene pool of native *Lantana* species	Sanders 1987; Anon. 1999
Threat to native species of *Lantana* from competition	Sanders 1987, 2001
Reduces the pollinator loads of native plants	Feinsinger 1987
Extinction of the shrub *Linum cratericola*	Mauchamp *et al.* 1998
Alters fire regime	Humphries and Stanton 1992
Affects human health by harboring malarial mosquitoes and tsetse flies	Gujral and Vasudevan 1983; Greathead 1968; Katabazi 1983; Okoth and Kapaata 1987; Mbulamberi 1990
Allelopathic effects, resulting in either no growth or reduced growth	Achhireddy and Singh 1984; Achhireddy *et al.* 1985; Mersie and Singh 1987; Sharma *et al.* 1988; Jain *et al.* 1989; Singh and Achhireddy 1987
Decrease in community biomass and a proportional increase in the foliage component in the vegetation	Bhatt *et al.* 1994
Loss of pasture land	Culvenor 1985
Threat to agriculture	Holm *et al.* 1991
Poisoning of cattle, buffalo, sheep, goats horses and dogs, guinea pigs and captive red kangaroos	Sharma *et al.* 1988
Unripe fruit are mildly poisonous	Morton 1994; Sharma 1994

Lantana (Cilliers 1983; Greathead 1968; Harley 1973; Neser and Cilliers 1989; Perkins and Swezey 1924). Unfortunately most of these attempts have been unsuccessful in India and elsewhere in the world (Julien and Griffiths 1998).

The invasion and current spread of *Lantana* across the world represents a typical case of a successful invasive species through its temporal dynamics of initial establishment, logistic spread and, finally, successful colonization and establishment (the stable phase). Management options at this stage (as mentioned above) are few and even if deployed would be expensive and futile. Save for the discovery of a silver bullet, the species has come to stay in most places of its invasion. Managers have their backs to the wall, just as for the cases of Water Hyacinth, *Optunia*, and *Prosopis*. The cost of containing *Lantana* may not justify the damage avoided.

Under these circumstances, as we have argued earlier, one of the strategies could be to look at the possible utilization of the invasive and explore whether this can lead to an adaptive management of the invasive in a manner that would reduce the net cost of the species, by partially offsetting both control and damage costs. Here we present a case

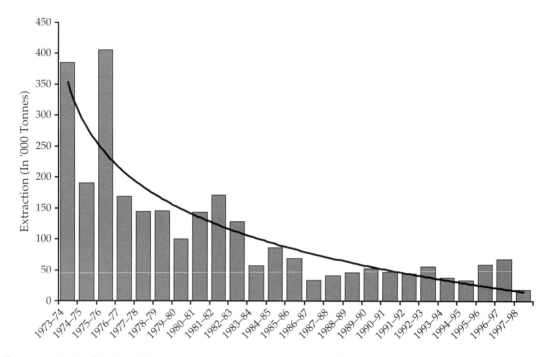

Figure 14.4 Decline of bamboo yields in Karnataka, India (From Uma Shaanker et al. 2004b).

study of a recent initiative that encouraged the use of *Lantana* among the marginalized communities in southern India.

14.3.2 When you cannot break it, at least bend it!

In India, a number of forest dwelling communities depend almost exclusively on forest resources for their livelihood (Murali et al. 1996). Among these communities are the *Medars* and *Koravas* and a number of scheduled castes and tribal communities, such as the *Soligas* (Uma Shaanker et al. 2004b). These traditional weaving communities are often hereditarily dependent on bamboo and cane resources, with most of them having no other means of livelihood (Uma Shaanker et al. 2004b). In recent years, indiscriminate extraction of bamboo and canes (the two materials most preferred for weaving) has severely depleted the natural stocks and in many places has directly threatened the livelihoods and further marginalized these communities (Uma Shaanker et al. 2004b; Fig. 14.4). Lack of alternative sources of income and land tenure has further aggravated these marginalized communities (Uma Shaanker et al. 2004b). Any effort that can offer an appropriate substitute for the declining wild bamboo and rattan resources could make a substantial difference to their livelihood.

Kannan et al. (2008) explored the possibility of using the locally abundant invasive species, *Lantana*, as a substitute for bamboo and canes such that a) it could maintain or even enhance the livelihoods of the traditional weaving communities dependent upon scarce bamboo resources and b) alleviate the stress on natural population of bamboo and canes and therefore help in conserving native biological diversity.

Lantana forms one of the most dominant plant communities in the open forest, roadsides, wastelands, and fallows. Typically with dry to moist deciduous vegetation, most forest sites have been invaded by *Lantana*. At one site, namely MM Hills, Karnataka State, nearly 80 per cent of the 290 km^2 of the forest has been invaded by *Lantana* (Ganeshaiah and Uma Shaanker 2001). The density of the weed ranged from 932 stems ha^{-1} in moist

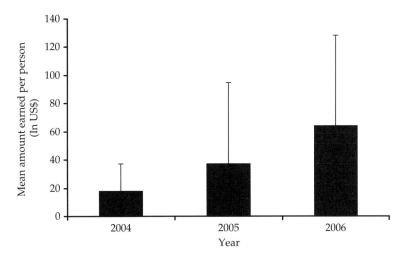

Figure 14.5 Mean per capita increase in cash income of artisans using *Lantana* as a substitute for bamboo and canes (From Kannan *et al.* In press).

deciduous forest to 1941 stems ha^{-1} in dry deciduous forest (Ganeshaiah and Uma Shaanker 2001). At none of the sites was *Lantana* ever used as a resource with cash value; its only use was as a hedge plant around fields to protect crops from cattle and wild animals.

Kannan *et al.* (2008) promoted the use of *Lantana* as a substitute for bamboo and canes and helped designed appropriate *Lantana* products for rural and urban markets. Over 350 men and women were trained in the use of Lantana at several field sites in south India. More than 50 different products, from baskets to furniture were developed. The average number of man-days employed in *Lantana* craft increased from about 30 in 2004 to above 80 man-days in 2006 (Kannan *et al. in press*). For the same period, the mean annual income per capita from *Lantana* increased from US$ 17.90 to US$ 63.93 (Fig. 14.5). There was a significant increase in the annual income of families after adoption of *Lantana* craft compared to that before adoption ($p < 0.035; df = 8$).

In summary, the use of *Lantana* as a substitute for bamboo and canes significantly enhanced the income profile of the forest-based communities at several field sites in south India. Having been sensitized to the income generating potential of a resource that they had not regarded as convertible to cash income, use of *Lantana* provides a safety net for the very poor and marginalized tribal communities.

While this initiative was prompted by the immediate need to alleviate the livelihoods of the forest dwelling communities, viewed in the context of managing invasive species it offers an unconventional approach to contributing to minimizing the net cost of the invasive species. For instance, the use of *Lantana* could allow for the regeneration and recruitment of native plant species and mitigate other ecosystem damages such as pollinator loss. The stems, when harvested, provide a window of time during which regeneration of at least some native plants might be facilitated. The extraction of the weed might also help in reducing the spread of forest fires, which otherwise are fanned by *Lantana* stems, and also offer greater accessibility for both animals as well as men.

In other words, the use of the weed by the local communities provides for both reducing the control as well as damage costs of the species. However, it is obvious that these mitigating effects would be dependent upon the scale at which *Lantana* is being used as also in the manifold ways through which the invasive species could be used. For instance, in a more recent initiative, there has been an attempt at biorefining abundant biomass (typical of invasive species) for chemical prospecting (Uma Shaanker and Ganeshaiah 2006).

Clearly such innovative exploitation of "uncontrollable invasive species" would tend to substantially reduce the control and damage costs and hence allow for a net minimization of the cost of invasive species.

The use of *Lantana* as a substitute for the dwindling wild bamboo and cane resources by poor rural communities in India provides a new perspective on the use of invasives (Shackleton *et al.* 2007). The approach is easily replicable elsewhere in the world and could have important implications for much of the human-dominated forested landscapes in the world. The idea has already found favor in Madagascar and Sri Lanka, both to address problems with local livelihoods and to prevent the spread of this invasive weed (Kannan *et al. in press*).

Careful management of the use of invasive species should lead to interesting outcomes in its ecological and livelihood impacts. Can the use of the invasive be sustainable (and should this be of interest?). What are the potential conflicts in the use of the invasive species on one hand (by the poor) and the control of the invasive (by the forest managers)? Can the managed use of the invasive lead to a win-win situation for both the ecological services and functions that are otherwise impaired by the invasive species and the livelihood gains it might accrue to the communities? Ralph Waldo Emerson (1803–82) once quipped *"What is a weed? A plant whose virtues have not been discovered"* (1878). The work reported on the use of *Lantana* by Kannan *et al.* (2008) quite well exemplifies this quip.

14.4 The management of uncontrollable invasive species: from exclusion to inclusion

The current views on managing invasive species stem from the fact that these species frequently have high ecological and economic costs. While the weight of evidence does justify this view, management, or the lack of it, has often resulted in invasive species occupying a predominant ecological space that is simply beyond any control measures. Under these circumstances, our currently held view has often constrained the development of alternate management regimes to address the issue of invasive species. One of these alternate strategies that we have argued for in this chapter is to promote the utilization of the invasive species as one of the means to manage it. This approach might appear unconventional and at times defeat the entire purpose of controlling the invasive species. However, as argued earlier, viewed from the point of minimizing the net cost of the invasive, exploitation of the invasive could represent one possible approach towards this end. Thus, when conventional management options are no longer cost effective and the invasive species has come to a state of stabilization in the ecosystem, it might be worthwhile to reach out to alternative strategies even if it means exploiting the species. Under these circumstances we argue that any action (including promoting the use of the invasive) that can reduce the ecological cost of the invasive should be a potentially useful strategy. In short, this calls for a shift in our view of managing invasive species, from one of exclusion to that of inclusion. This may not only be pragmatic but also realistic, especially in tropical human-dominated landscapes where low income and the abundance of the invasive species constrain effective control, while the lack of rural options encourages utilization of invasive species. In this context, invasive species, especially those that have escaped effective control, could be viewed from a potential utilization point of view, as a specific case of management. This view is aptly summarized by Geesing *et al.* 2004, who mention in their defense of the use of *Prosopis*:

Notwithstanding the unquestionable ecological changes produced by *Prosopis* invasion, where the species have been introduced it is necessary to make the best of a situation that is hardly reversible.

Acknowledgements

Parts of the work reported here have been supported by grants from the Development Marketplace Award, World Bank, Department of Biotechnology, Government of India, Blumoon Foundation, and Rainforest Concern. We acknowledge the support and cooperation of the Karnataka Forest Department in facilitating the field work in south India. Dr. Anil Joshi and artisans from two villages in Natham and Chittor in south India were instrumental in motivating us to consider substituting Lantana for bamboo and canes. Criticisms and suggestions of the work at the BESTNet/DIVERSITAS ecoSERVICES Workshop held at Arizona State University, Tempe, October 2007 greatly helped in formulating and fine tuning the ideas presented in the paper. We gratefully acknowledge the useful comments and criticism offered by Charles Perrings, Hal Mooney, and Mark Williamson.

References

Achhireddy, N.R. and Singh, M. (1984). Allelopathic effects of Lantana (*Lantana camara*) on milkweed vine (*Morrenia odorata*). *Weed Science*, 32, 757–61.

Achhireddy, N.R., Singh, M., Achhireddy, L.L., Nigg, H.N. and Nagy, S. (1985). Isolation and partial characterization of phytotoxic compounds from Lantana (*Lantana camara* L.). *Journal of Chemical Ecology*, 11, 979–88.

Anderson, L.W. (2003). California's Reaction to *Caulerpa taxifolia*: A model for invasive species rapid response actions. In Abstracts: Third International Conference on Marine Bioinvasions, March 16–19, 2003. Scripps Institution of Oceanography La Jolla, California. http://massbay.mit.edu/resources/pdf/MarinePDF/2003/MBI2003abs1.pdf.

Anon. (1962). *Lantana* Linn. (Verbenaceae), In B.N. Sastri, ed. *The Wealth of India: A Dictionary of Raw Materials and Industrial Products*, vol VI, Council of Scientific and Industrial Research, New Delhi, pp. 31–34.

Atkinson, C.T., Dusek, R.J., Woods, K.L. and Iko, W.M. (2000). Pathogenicity of avian malaria in experimentally-infected Hawaii Amakihi. *Journal of Wildlife Diseases*, 36, 197–204.

Atkinson, C.T., Woods, K.L., Dusek, R.J., Sileo, L.S. and Iko, K.W. (1995). Wildlife disease and conservation in Hawaii: pathogenicity of avian malaria (*Plasmodium relictum*) in experimentally infected I'iwi (*Vestiaria coccinea*). *Parasitology*, 111 Suppl., S59–S69.

Bhatt, Y.D., Rawat, Y.S. and Singh, S.P. (1994). Changes in ecosystem functioning after replacement of forest by Lantana shrubland in Kumaun Himalaya. *Journal of Vegetation Science*, 5, 67–70.

Boureima, M., Mayaki, A. and Issa, M. (2001). *Etudes socio-economiques sur la commercialization des produits et sous-produits de la foret de Prosopis et sur la mise en place des marches ruraux dans l'arrondissement N'Guigmi*. Niamey, the Niger Institut national de recherches agronomiques du niger (INRAN) and FAO.

Braithwaite R.W., Lonsdale, W.M. and Estbergs, J.A. (1989). Alien vegetation and native biota in tropical Australia: the impact of *Mimosa pigra*. *Biological Conservation*, 48, 189–210.

Braithwaite, R.W. and Lonsdale, W.M. (1987). The rarity of *Sminthopsis virginiae* in relation to natural and unnatural habitats. *Conservation Biology*, 1, 341–43.

Chavan, S.R. and Nikam, S.T. (1982). Investigation of *Lantana camara* L. (Verbenaceae) leaves for larvicidal activity. *Bulletin of Haffkine Institute*, 10, 21–22.

Cilliers, C.J. (1983). The weed, *Lantana camara* L., and the insect natural enemies imported for its biological control in South Africa. *Journal of the Entomological Society of Southern Africa*, 46, 131–38.

Cronk, Q.C.B. and Fuller, J.L. (1995). *Plant Invaders*. Chapman and Hall, London, pp. 82–87.

Culvenor, C.C.J. (1985). Economic loss due to poisonous plants in Australia. In A.A. Seawright, M.P. Hegarty, L.F. James and R.F. Keeler, eds. *Proceedings of the Australia-U.S.A. Poisonous Plants Symposium*, Brisbane, Australia, pp. 3–13.

D'Antonio, C.M. and Kark, S. (2002). Impacts and Extent of Biotic Invasions in Terrestrial Ecosystems. *Trends in Ecology and Evolution*, 17, 202–04.

Davis, C.J., Yoshioka, E. and Kageler, D. (1992). Biological control of Lantana, prickly pear, and Hamakua Pamakani in Hawaii: a review and update. In C.P. Stone, C.W. Smith and J.T. Tunison, eds. *Alien Plant Invasions in Native Ecosystems of Hawaii: Management and Research*, Cooperative National Park Resources Studies Unit, Hawaii, pp. 411–31.

Day, M.D., Wiley, C.J., Playford, J. and Zalucki, M.P. (2003). *Lantana Current Management Status and Future Prospects*. ACIAR Monograph 102, Canberra.

Dunbar, K.R. and Facelli, J.M. (1999). The impact of a novel invasive species, *Orbea variegata* (African carrion flower), on the chenopod shrublands of South Australia. *Journal of Arid Environment*, 41, 37–48.

Emerson, R.W. (1878). Fortune of the Republic. In *The Complete Works of Ralph Waldo Emerson*. Ams Pr Inc.

Feinsinger, P. (1987). Effects of plant species on each others pollination: Is community structure influenced? *Trends in Ecology and Evolution*, 2, 123–26.

Fensham, R.J., Fairfax, R.J. and Cannell, R.J. (1994). The invasion of *Lantara camara* L. in Forty Mile Scrub National Park, north Queensland. *Australian Journal of Ecology*, 19, 297–305.

Ganeshaiah, K.N. and Uma Shaanker, R. (2001). Impact of invasive species on the diversity, health and productivity of ecosystems: A study in the tropical forests of south India. Report submitted to CIFOR, Indonesia.

Geesing, D., Al-Khawlani, M. and Abba, M.L. (2004). Management of introduced Prosopis species: can economic exploitation control an invasive species? *Unasylva*, 217, 36–44.

Ghazoul, J. (2001). Direct and indirect effect of human disturbance on the reproductive ecology of tropical forest trees. In K.N. Ganeshaiah, R. Uma Shaanker and K.S. Bawa, eds. *Tropical Ecosystems: Structure, Diversity and Human Welfare*, Oxford-IBH Publications, New Delhi, pp. 97–100.

Ghisalberti, E.L. (2000). *Lantana camara* L. (Verbenaceae). *Fitoterapia* 71, 467–86.

Gopal, B. (1987). *Water Hyacinth*. Aquatic Plant Studies, 1. Elsevier Science Publishing Co., New York.

Greathead, D.J. (1968). Biological control of *Lantana*. A review and discussion of recent developments in East Africa. *Proceedings of National Academy of Science*, 14, 167–75.

Greenfield, B. and Nicholson, H. (2007). SA Arid Lands *Opuntia* species management plan, DRAFT. South Australian Arid Lands Natural Resource Management Board.

Gujral, G.S. and Vasudevan, P. (1983). *Lantana camara* L., a problem weed. *Journal of Scientific and Industrial Research*, 42, 281–86.

Gurevitch, J. and Padilla, D. (2004). Are invasive species a major cause of extinctions? *Trends in Ecology and Evolution*, 19, 470–74.

Haider, S.Z. (1989). *Recent Work in Bangladesh on the Utilization of Water Hyacinth*, Commonwealth Science Council/Dhaka University, Dhaka.

Harley, K.L.S. (1973). Biological control of *Lantana* in Australia. In A.J. Wapshere, ed. *Proceedings of the III International Symposium on Biological Control of Weeds*, Montpellier, France, pp. 23–29.

Hegde, R., Suryaprakash, S., Achoth, L. and Bawa, K.S. (1996). Extraction of NTFPs in the forests of BR Hills 1. Contribution to rural income. *Economic Botany*, 50, 243–50.

Hobbs, R.J. and Humphries, S.E. (1994). An integrated approach to the ecology and management of plant invasion. *Conservation Biology*, 9, 761–70.

Hobbs, R.J. and Mooney, H.A. (1986). Community changes following shrub invasion of grassland. *Oecologia*, 70, 508–13.

Holm, L.G., Plucknett, D.L., Pancho, J.V. and Herberger, J.P. (1991). *The World's Worst Weeds: Distribution and Biology*. Krieger Publishing Company, Florida.

Humphries, S.E. and Stanton, J.P. (1992). *Weed Assessment in the Wet Tropics World Heritage Area of north Queensland*. Report to The Wet Tropics Management Agency.

Humphries, S.E., Groves, R.H. and Mitchell, D.S. (1991). Plant invasions of Australian ecosystems: a status review and management directions. *Kowari*, 2, 1–134.

Jain, R., M. Singh and Dezman, D.J. (1989). Qualitative and quantitative characterization of phenolic compounds from *Lantana* (*Lantana camara*) leaves. *Weed Science*, 37, 302–07.

Jenkins, C.D., Temple, S.A., Van Riper, C. and Hansen, W.R. (1989). Disease-related aspects of conserving the endangered Hawaiian crow. In J.E. Cooper, ed. *Disease and Threatened Birds. ICBP Technical Publication Number 10*. International Council for Bird Preservation, Cambridge, U.K., pp. 77–87.

Jensen, A. M. and Hajej, M.S. (2001). The road of hope: control of moving sand dunes in Mauritania. *Unasylva*, 207, 31–36.

Joshi, A.P. (2002). *Lantana*. Himalayan Environment Studies and Conservation Organisation (HESCO), Dehra Dun.

Jules, E.S., Kauffman, M.J., Ritts, W.D. and Carroll, A.l. (2002). Spread of an invasive pathogen over a variable landscape: A nonnative root rot on port orford cedar. *Ecology*, 83, 3167–81.

Julien, M.H. and Griffiths, M.W. (1998). *Biological Control of Weeds: A World Catalogue of Agents and their Target Weeds*. Fourth Edition, CAB International, Wallingford.

Julien, M.H., Hill, M.P., Center T.D. and Jianquig, D. (eds). (2001). *Biological and Integrated Control of Water Hyacinth Eichhornia crassipes*. Proceedings PR102 2001.

Kannan, R., Aravind N.A., Joseph, G., Uma Shaanker, R. and Ganeshaiah, K.N. Invasive weed aids livelihoods: *Lantana* as a substitute for bamboo in South India. *Conservation and Society*, In Press.

Kannan, R., Aravind, N.A., Joseph, G., Ganeshaiah, K.N. and Uma Shaanker, R. (2008). Lantana Craft: A Weed for a Need. *Biotech News*, 3, 9–11.

Katabazi, B.K. (1983). The tsetse fly *Glossina fuscipes* in the sleeping sickness epidemic area of Busoga, Uganda. *East African Medical Journal*, 60, 397–01.

Lamb, D. (1991). Forest regeneration research for reserve management: some questions deserving answers. In N. Goudberg, M. Bonell and D. Benzaken, eds. *Tropical Rainforest Research in Australia: present status and future directions for the Institute for Tropical Rainforest Studies*, pp. 177–81. Institute for Tropical Rainforest Studies.

Lamb, R. (1982). Some effects of *Lantana camara* on community dynamics of eucalypt woodland, pp. 304. *Proceedings of the 52nd ANZAAS Congress*, Townsville.

Larsson, P. (2004). *Introduced Opuntia spp. in Madagascar, problems and opportunities*. SLU External Relations, SLU. Minor field studies/Swedish University of Agricultural Sciences, SLU External Relations. Vol. 285.

Levine, J.M., Vila, M., D'Antonio, C.M., Dukes, J.S., Grigulis, K. and Lavorel, S. (2003). Mechanisms underlying the impacts of exotic plant invasions. *Proceedings of Royal Society of London, B*, 270, 775–81.

Loyn, R.H. and French, K. (1991). Birds and environmental weeds in south-eastern Australia. *Plant Protection Quarterly*, 6, 137–49.

Luken, J.O. and Thieret, J.W. (eds). (1997). *Assessment and management of plant invasions*. Springer-Verlag, Inc., New York.

Martin, P.H. (1999). Norway maple (*Acer platanoides*) invasion of a natural forest stand: understorey consequences and regeneration pattern. *Biological Invasions*, 1, 215–22.

Matthews, S. (2004). *Tropical Asia invaded: the growing danger of invasive alien species*. The Global Invasive Species Programme (GISP) Secretariat, Cape Town, South Africa. Available at: www.gisp.org/downloadpubs/gispAsia.pdf

Mauchamp, A., Aldaz, I., Ortiz, E. and Valdebenito, H. (1998). Threatened species, a re-evaluation of the status of eight endemic plants of the Galápagos. *Biodiversity and Conservation*, 7, 97–107.

Mbulamberi, D.B. (1990). Recent outbreaks of human trypanosomiasis in Uganda. *Insect Science and its Application*, 11, 289–92.

McGarry, D., Shackleton, C.M., Fourie, S. *et al.* (eds). (2005). *A Rapid Assessment of the Effects of Invasive Species on Human Livelihoods, Especially of the Rural Poor*. pp. 210. South African National Biodiversity Institute, Kirstenbosch.

McNeely, J. (2000). Invasive species: a costly catastrophe for native biodiversity. *Land Use and Water Resources Research*, 2, 1–10.

Mersie, W. and Singh, M. (1987). Allelopathic effect of *Lantana* on some agronomic crops and weeds. *Plant Soil*, 98, 25–30.

Mooney, H.A. and Hobbs, R.J. (2000). *Invasive Species in a Changing World*. Island Press, Washington, DC.

Morton, O. (1994). *Marine algae of Northern Ireland*. Ulster Museum, Belfast.

Murali, K.S. and Siddapa Setty, R. (2001). Effect of weeds *Lantana camara* and *Chromelina odorata* growth on the species diversity, regeneration and stem density of tree and shrub layer in BRT sanctuary. *Current Science*, 80, 675–78.

Murali, K.S., Uma Shankar, Uma Shaanker, R., Ganeshaiah, K.N. and Bawa, K.S. (1996). Extraction of NTFPs in the forests of BR Hills 2. Impact of NTFPs extraction on regeneration, population structure and species composition. *Economic Botany*, 50, 251–69.

Mwangi, E. and Swallow, B. (2005). Invasion of *Prosopis juliflora* and local livelihoods: Case study from the lake Baringo area of Kenya. ICRAF Working Paper– no. 3. World Agroforestry Centre, Nairobi.

National Academy of Sciences. (1976). *Making Aquatic Weeds Useful: Some Perspectives for Developing Countries*, National Academy of Sciences, Washington DC.

National Centre for Human Settlements and Environment. (1987). *Documentation of Forest and Rights. Volume 1*. National Centre for Human Settlements and Environment, New Delhi.

Navarro, L.A. and Phiri, G. (eds). (2000). Water hyacinth in africa and the middle east. A Survey of Problems and Solutions, International Development Research Centre (IDRC), Ottawa.

Neser, S. and Cilliers, C.J. (1989). Work towards biological control of *Lantana camara*: perspectives. In E.S. Delfosse, ed. *Proceedings of the VII International Symposium on Biological Control of Weeds*, Rome, Italy, pp. 363–69.

Nigam, J.N. (2002). Bioconversion of water-hyacinth (*Eichhornia crassipes*) hemicellulose acid hydrolysate to motor fuel ethanol by xylose–fermenting yeast. *Journal of Biotechnology*, 97, 107–16.

Nolan, W.J. and Kirmse, D.W. (1974). The Papermaking Properties of Water Hyacinth, *Hyacinth Control Journal*, 12, 90.

Okoth, J.O. and Kapaata, R. (1987). A study of the resting sites of *Glossina fuscipes fuscipes* (Newstead) in relation to *Lantana camara* thickets and coffee and banana plantations in the sleeping sickness epidemic focus, Busoga, Uganda. *Insect Science and its Application*, 8, 57–60.

Parker I.M., Simberloff, D., Lonsdale, W.M. *et al.* (1999). Impact: toward a framework for understanding the ecological effects of invaders. *Biological Invasions*, 1, 3–19.

Parsons, W.T. and Cuthbertson, E.G. (2001). *Noxious Weeds of Australia*, 2nd edition, CSIRO Publishing, Collingwood.

Perkins, R.C.L. and Swezey, O.H. (1924). The introduction into Hawaii of insects that attack *Lantana*. *Bulletin of the Experiment Station of the Hawaiian Sugar Planters' Association* 16, 1–83.

Perrings, C. (2005). *Biological Invasions and Poverty*. [online] URL: http: www.gisp.org/publications/brochures/invasivesandpoverty.pdf

Perrings, C., Williamson, M., and Dalmazzone, S. (2000). *The Economics of Biological Invasions*. Edward Elgar, Cheltenham, UK/Northampton, USA.

Pimentel, D., Lach, L., Zuniga, R. and Morrison, D. (2000). Environmental and Economic Costs of Nonindigenous Species in the United States. *BioScience*, 50, 53–65.

Pimentel, D., McNair, S., Janecka, J. *et al.* (2001). Economic and environmental threats of alien plant, animal, and microbe invasions. *Agriculture, Ecosystems and Environment*, 84, 1–20.

Sanders, R.W. (1987). Taxonomic significance of chromosome observations in Caribbean species of *Lantana* (Verbenaceae). *American Journal of Botany*, 74, 914–20.

Sanders, R.W. (2001). The genera of Verbenaceae in the southeastern United States. *Harvard Papers in Botany*, 5, 303–58.

Schmitz D.C., Simberloff, D., Hofstetter, R.H., Haller, W. and Sutton, D. (1997). The ecological impact of non-indigenous plants. In D. Simberloff, D.C. Schmitz and T.C. Brown, eds. *Strangers in Paradise: Impact and Management of Nonindigenous Species in Florida*, Island Press, Washington, D.C., pp. 39–61.

Secretariat of the Convention on Biological Diversity. (2005). *Handbook of the Convention on Biological Diversity Including its Cartagena Protocol on Biosafety*, 3rd edition. Montreal, Canada.

Shackleton, C.M., McGarry, D., Fourie, S., Gambiza, J., Shackleton, S.E. and Fabricius, C. (2007). Assessing the Effects of Invasive Alien Species on Rural Livelihoods: Case Examples and a Framework from South Africa. *Human Ecology*, 35, 113–27.

Sharma, O.P. (1994). Plant toxicosis in north–western India. In M.S. Colgate and P.R. Dorling, ed. *Plant associated Toxins: Agricultural, Phytochemical and Ecological Aspects*, pp. 19–24. CAB International, Wallingford.

Sharma, O.P. and Sharma, P.D. (1989). Natural products of the *Lantana* plant- the present and prospects. *Journal of Scientific and Industrial Research*, 48, 471–78.

Sharma, O.P., Makkar, H.P.S. and Dawra, R.K. (1988). A review of the noxious plant *Lantana camara*. *Toxicon*, 26, 975–87.

Shaukat, S.S., Siddiqui, I.A., Ali, N.I. and Zaki, M.J. (2001). Biological and chemical control of soilborne fungi and effect of these on growth of mungbean. *Pakistan Journal of Biological Sciences*, 4, 1240–3.

Siges, T.H., Hartemink, A.E., Hebinck P., and Allen, B.J. (2005). The Invasive shrub *Piper aduncum* and rural livelihoods in the Finschhafen area of Papua New Guinea. Human *Ecology*, 33, 875–92.

Simberloff, D., (1998). A Global Threat to Biodiversity and Stability. In P.H. Raven, ed. *Nature and Human Society*, National Research Council (U.S.), Board on Biology, Tania Williams, pp. 325–34.

Singh, M. and Achhireddy, N.R. (1987). Influence of *Lantana* on growth of various citrus rootstocks. *HortScience*, 22, 385–86.

Singh, P. (1976). *Lantana* weed and the *Lantana* lacebug. *Indian Forester*, 102, 474–76.

Thakur, M.L., Ahmad, M., and Thakur, R.K. (1992). *Lantana* weed (*Lantana camara* var. *aculeata* Linn.) and its possible management through natural insect pests in India. *Indian Forester*, 118, 466–86.

Tu, M., Hurd, C., and Randall, J.M. (2001). *Weed Control Methods Handbook: Tools and Techniques for Use in Natural Areas*. Wildland Invasive Species Program, The Nature Conservancy.

Uma Shaanker, R. and Ganeshaiah, K.N. (2006). Bioresources for Human Well-Being: Promises and Challenges. Abstract. Renewable Resources and Biofefineries Conference, 6th–8th September, 2006, University of York, UK.

Uma Shaanker, R., Ganeshaiah, K.N., Smitha Krishnan, et al. (2004a). Livelihood gains and ecological costs of non timber forest product dependence: assessing the roles of dependence, ecological knowledge and market structure in three contrasting human and ecological settings in south India. *Environmental Conservation*, 31, 242–53.

Uma Shaanker, R., Ganeshaiah, K.N., Srinivasan, K., Ramanatha Rao, V. and Hong, L.T. (2004b). *Bamboos and Rattans of the Western Ghats: Population biology, Socio-economics and Conservation strategies*. ATREE, IPGRI and UAS, Bangalore.

Van Riper III, C.S., Van Riper, S.G., Goff, M.L. and Laird, M. (1986). The epizootiology and ecological significance of malaria in the Hawaiian land birds. *Ecological Monographs*, 56, 327–44.

Van Wilgen, B.W. and Van Wyk, E. (1999). Invading alien plants in South Africa: Impacts and solutions. In, D. Eldridge and D. Freudenberger, eds., *People and Rangelands: Building the Future*. Proceedings of the VI International Rangeland Congress, Townsville, Australia, pp. 566–71.

Vitousek P.M. and Walker L.R. (1989). Biological invasions by *Myrica faya* in Hawaii: plant demography, nitrogen fixation, ecosystem effects. *Ecological Monographs*, 59, 247–65.

Vitousek, P. M., Walker, L.R., Whiteaker, L.D., Mueller-Dombois D., and Matson, P.A. (1997). Biological invasion by *Myrica faya* alters ecosystems development in Hawaii. *Science*, 238, 802–04.

Vivrette, N.J. and Muller, C.H. (1977). Mechanism of invasion and dominance of coastal grassland by *Mesembryanthemum crystallinum*. *Ecological Monographs*, 47, 301–18.

Walker, B.H. and Steffen, W. (1997). An overview of the implications of global change for natural and managed terrestrial ecosystems. *Conservation Ecology*, 1, 2. Available from the Internet. [online] URL:http://www.consecol.org/voll/iss2/art2.

Warner, R.E. (1968). The role of introduced diseases in the extinction of the endemic Hawaiian avifauna. *Condor*, 70, 101–20.

Wikelski, M., Foufopoulos, J., Vargas, H. and Snell, H. (2004). Galápagos Birds and Diseases: Invasive Pathogens as Threats for Island Species. *Ecology and Society* 9, 5. [online] URL: http://www.ecologyandsociety.org/vol9/iss1/art5/

Williamson, M. (1996). *Biological Invasions*. Chapman and Hall, London

Willson, B.W. (1968). Insects on trial in fight against *Lantana*. *Queensland Agricultural Journal*, (December), 748–51.

Wootton, L.S., Halsey, S.D., Bevaart, K., McGough, A., Ondreicka, J. and Patel, P. (2005). When invasive species have benefits as well as costs: managing Carex kobomugi (Asiatic sand sedge) in New Jersey's coastal dunes. *Biological Invasions*, 7, 1017–1027.

CHAPTER 15

Prevention: Designing and Implementing National Policy and Management Programs to Reduce the Risks from Invasive Species

Reuben P. Keller and David M. Lodge

15.1 Introduction

Human trade and travel have facilitated the establishment of many species beyond their native range (Mack et al. 2000). Although most non-native species have positive or negligible impacts, many become invasive and cause net harm (Lodge *et al.* 2006). Invasive species are recognized as one of the greatest threats to biodiversity worldwide (Sala *et al.* 2000), cause large economic costs to countries across the globe (Pimentel 2002), and have carried with them many diseases of humans, animals and plants (Jones *et al.* 2008). As levels of international trade and travel continue to grow, so does the potential for the introduction and spread of invasive species (Leprieur *et al.* 2008).

In response to the threats from invasive species many countries have adopted the twin goals of preventing the introduction of new invaders and of controlling the spread and population size of those already established. These policy goals have been formalized through local, state, and national legislation, and through the ratification of international treaties. The 168 signatory countries to the international Convention on Biological Diversity (CBD) have agreed to, as far as possible, "prevent the introduction of, control or eradicate those alien species which threaten ecosystems, habitats or species" (Article 8(h)). Progress towards these goals differs greatly by country, as does the level of resources allocated to achieving them. Recogniz-ing that a large proportion of the world's countries have signed this convention, and that the impacts of invasive species are large across the globe, we assume that the goals encompassed by Article 8(h) are shared by most countries of the world.

Our aim in this chapter is to describe and critically evaluate the "state-of-the-art" for national policy and management efforts designed to prevent the arrival of invasive species. The broad framework for achieving prevention is widely agreed upon and consists of identifying and managing the vectors that transport live species. More specific recommendations are given by the International Union for the Conservation of Nature in its policy and management guidelines for nations working to satisfy their obligations under the CBD (IUCN 2000). Additionally, these elements appear in the programs already adopted by some countries, including Australia (http://www.daff.gov.au/ba), New Zealand (http://www.biosecurity.govt.nz/) and South Africa (http://www.info.gov.za/acts/2004/a10-04/index.html), as well as those recommended for the United States (Lodge *et al.* 2006).

15.1.1 Economic and ecological context for invasive species prevention

Invasion occurs as a process, with risks present at multiple stages (Fig. 15.1; Lodge *et al.* 2006). To become invasive, a species must first enter a

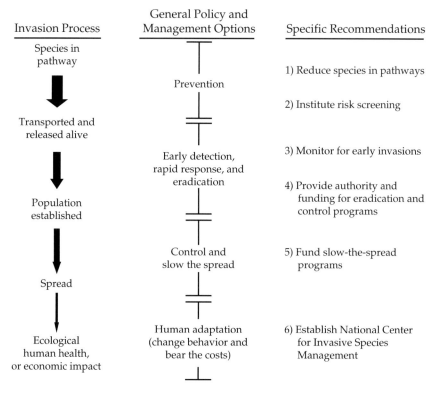

Figure 15.1 The biological invasion process (left column) and the corresponding options for management (right column). Declining arrow thickness represents the declining number of species reaching each step (i.e. not all species pass through each filter). From Lodge et al. 2006.

vector and survive transport. Next, a species is termed established if it is released and begins reproducing beyond human cultivation. Finally, if a species spreads and causes negative impacts it is termed invasive (Kolar & Lodge 2001). Management options differ for species at each stage in the process. The challenge for managers and policy-makers is to implement programs that will prevent invasive species from transiting the entire process.

Although the goal of preventing the arrival, establishment and spread of species is straightforward, the measures required to achieve it are not. Effectively preventing the introduction of invasive species requires targeted management of the many vectors that deliver live organisms. This will often be costly, and may interfere with valuable economic, social, or recreational activities. For example, the ballast water of ships has transported many of the world's most damaging invasive aquatic species (Ricciardi 2006). Effective ballast water treatment would prevent many future invasions (Drake & Lodge 2004), but currently available technologies for achieving this are expensive and, if implemented, would increase the costs of international shipping. Such tradeoffs exist for management of many other vectors and lead naturally to a cost-benefit approach for determining what management and policy actions are rational. For example, it would not be rational to prevent the arrival of invasive species if prevention efforts would cost more, or lead to greater losses, than would have been caused by the invasive species. In contrast, if a single species presents extremely high risks as an invader it may be rational to entirely eliminate its vector of transport, even though other species in that vector may present a low risk. In the following we explore the state of the current science that surrounds and supports risk assessment and risk management programs for invasive species.

15.2 Prevention

Human trade and travel create many vectors that transport live organisms (Hulme *et al.* 2008). Lodge *et al.* (2006) classified these vectors as either commerce in live organisms (intentional) or transport related (unintentional) (Fig. 15.2). Vectors of intentional introduction include the pet and aquarium trades, the nursery plants trade, and the live food trade (Keller & Lodge 2007). Species involved in commerce are pre-selected for desirable attributes. In contrast, species introduced by transportation related vectors can be seen as the accidental consequence of other human activities. These vectors include the ballast water and hulls of ships, and the parasites and diseases often transported with other species (Drake & Lodge 2007a, b; Jones *et al.* 2008). Unintentionally introduced species do not arrive with any expectation of benefits to society, but they do have a nonzero risk of becoming invasive and causing harm.

Different policy and management frameworks are appropriate for commerce in live organisms relative to transport-related vectors. For commercial species, allowing importation will be appropriate if the expected economic and social benefits exceed the expected costs. Thus, benefits and costs must be considered on a species-specific basis, meaning that species level risk assessment and risk management decisions will be required. In contrast, species in transportation-related vectors are not introduced with the expectation of net benefits, and the most common management goal is to reduce the number of species and individual organisms being transported. An example of the latter is ballast water exchange, which aims to reduce the number of viable organisms in ballast tanks by exchanging coastal water with deep ocean water during ship transit. Because ports are in coastal areas, the organisms taken on in the deep ocean should pose a lower risk of invasion when released. We consider risk analysis for these two categories of vectors in greater detail below.

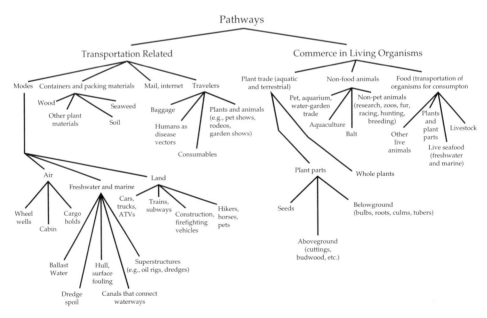

Figure 15.2 A selection of the anthropogenic vectors that transport species across the globe. Vectors are either transportation related, which introduce species as unintentional products of other movement, or commerce related, which introduce selected species for economic benefits. From Lodge *et al.* 2006.

15.2.1 Risk assessment for prevention of introductions via commerce in live organisms

Intentional species introductions, such as those of ornamental and food species, are carried out for the economic gain of exporters, importers and retailers, and to meet consumer demand. Most of these species bring overall benefits to the importing country and do not become invasive (Smith *et al.* 1999). A proportion of intentionally introduced species do, however, become invasive (Jeschke & Strayer 2005), and it is desirable to prevent the introduction of these species. The goal of risk assessment in this context is to discriminate between those species that will and will not cause net harm so that risk management can be targeted at preventing the arrival of the highest risk species (Keller *et al.* 2007a).

In many countries, risk assessment is already coupled to strong prevention programs for invasive species that threaten human health, agricultural crops, livestock, and forestry. Recognizing the large risk posed to human health by the SARS virus, or the risk that foot and mouth disease poses to agricultural production, is not difficult. The market costs that arise from such invasions can be relatively easily quantified so that prevention programs are justified based on purely economic terms. In contrast, species that pose risks to the environment and/or economy, but that don't pose a known direct risk to markets, have rarely been managed to prevent introduction. Indeed, in the majority of countries, risk assessments to determine environmental impacts are only conducted after a species is already established and causing harm. In the US, for example, at least 9 out of 16 taxa banned for introduction by the federal Lacey Act were already present in the country when listed, and at least 7 of those were already established (Fowler *et al.* 2007). Because the eradication of established invaders is usually not feasible, risk assessment and management conducted after a species is established has less potential for reducing impacts than if the risk assessment were conducted prior to introduction. Only with recent concerns that imported species also pose a risk to biodiversity, ecosystem function, and ecosystem services have a few countries begun to employ risk assessments to try to keep out a wider range of invasive species.

The proactive prevention approach to managing invasive species is now widely accepted as the most cost-effective way to reduce total impacts of invasive species (Lodge *et al.* 2006). Its use is recommended by the IUCN (2000), and it is slowly being adopted by more countries. Australia and New Zealand, which have strong proactive approaches to preventing the arrival of invasive species, have mandated for almost a decade that the risk of most taxa proposed for introduction should be assessed prior to first importation. Species assessed as having a high risk are prohibited from importation. Although this approach creates both environmental and economic benefits compared to the alternative of allowing all species for introduction (Keller *et al.* 2007a), most countries still allocate few resources to pre-import risk assessment. One reason for this slow uptake of policies is that many countries lack the scientific capacity and/or the government institutions to rigorously assess species risk and monitor the trades to ensure that high risk species are not being imported. Despite this, many countries that do have both the scientific and institutional capacity have not yet made strong investments in invasive species prevention.

15.2.1.1 DESIGNING SPECIES-SPECIFIC RISK ASSESSMENT TOOLS

Programs for lowering the risk of invasions from commerce in living organisms should have a number of characteristics. First, risk management decisions should be rapid enough to meet the needs of the trades proposing to introduce the species. The Australian weed risk assessment, which is used to assess all new plant introductions, generally requires less than 2 days to provide a result for a given species (Gordon *et al.* 2008). A system giving quick responses reduces costs to the parties paying for the assessment and will minimally disrupt commerce.

Second, the accuracy for predicting the identity of both invasive and non-invasive species must be acceptable to stakeholders. The required level of accuracy is a policy decision, and will depend on a nation's aversion to the lost benefits from false positives (i.e. non-invasive species that are predicted to be invasive and kept out) and the invasion costs of false negatives (i.e. invasive species that are predicted to be benign or beneficial and allowed for import) (Hewitt & Hayes 2002). Additionally, many policy-makers will wish to avoid costly surprise outcomes—species that cause far higher impacts than could have been predicted. This may rationally lead to a high required accuracy for correctly predicting invaders, possibly at the expense of a lower accuracy for correctly predicting non-invaders (Horan *et al.* 2002).

Third, consistency inspires confidence (NRC 2002). This means that different risk assessors should arrive at the same conclusions for any given species. If this condition is not met then the accuracy of the risk assessment system will be called into question, and those affected by the outcomes are unlikely to find it acceptable.

Fourth, systems that are transparent and well documented inspire confidence because all interested parties can determine how a conclusion was reached (NRC 2002). This condition ensures that the scientific support for the system can reasonably be queried, and that the validity of methods and predictions is open to scrutiny.

15.2.1.2 RISK ASSESSMENT TOOLS CURRENTLY IN USE

Ideally, a risk assessment would output the separate probabilities that a given species will cause different levels of impact. Policy-makers could then make informed decisions about whether or not to allow a species for introduction, based on their aversion to different levels of risk and the values they place on different economic and environmental factors (Hewitt & Hayes 2002). With currently available tools, however, this level of detail is rarely available. In most cases it is simply not possible to accurately predict how a particular species will act in its non-native range, how the recipient ecosystem will respond to the new species, and how society will value the resulting changes. Thus, in order to achieve something practical, simplifications must be accepted for the inputs and outputs of risk assessment.

Risk assessments for invasive species range from entirely qualitative to highly quantitative, and each differs in the extent to which it meets the conditions outlined above (Keller and Drake 2009). At the extreme qualitative end are risk assessments based solely on expert opinion gathered in unstructured ways. Although data limitations may sometimes make this the only type of risk assessment possible, it suffers from the twin problems of low consistency (i.e. different experts may arrive at different predictions) and a lack of transparency (i.e. the reasoning behind a prediction will be difficult to document). We are not aware of an expert opinion system that has been rigorously tested for accuracy.

Also at the qualitative end of the spectrum are risk assessment systems based on expert literature review. The logic of this approach is that scientists can read the published literature about a species to determine the risks posed by introduction. Literature review is probably the most widely used risk assessment method around the globe, and most of the invasive species regulations at both state and federal levels in the US are based on this approach. Literature reviews may be more defensible than expert judgment approaches when the sources used to arrive at a prediction are well documented. They may still be inconsistent, however, because different assessors will weight alternative types of impacts, and the information from different sources (e.g. peer reviewed science journals vs. government reports, experimental vs. observational studies), in different ways. Additionally, literature review risk assessments are often time consuming to complete (months to years). This final point probably goes a long way towards explaining why so few species are proactively regulated in jurisdictions that rely on this approach. It is simply not possible to assess the number of species being introduced using risk assessments that take months to years to complete.

Some literature review risk assessment systems have been developed to include an ordinal scoring component. One example is the USDA Generic Risk Assessment for Non-indigenous Aquatic Species (Orr 2003), which requires assessors to predict the probability and consequences of establishment as low, medium, or high. These rankings are combined to give a single prediction of invasiveness, again on a scale of low, medium, and high. Although this adds a crude quantitative element to the risk assessment, it would not reduce the amount of time required to reach a conclusion. Additionally, it does not address the issue of repeatability because definitions of terms like low, medium, and high are inherently subjective.

Quantitative risk assessment approaches have the potential to best satisfy the multiple criteria of rapidity, accuracy, consistency, and transparency (Keller and Drake 2009). The most widely used quantitative risk assessment tools are known as "trait-based" (Keller and Drake 2009), and have arisen from the long history of ecological work looking for traits associated with successful establishment and negative impacts outside of a species' native range (e.g. Baker 1974). In the last couple of decades, methods for identifying these traits and using them to produce risk assessment tools have improved greatly. A trait-based risk assessment begins with a list of species that have been introduced to the region of interest, and information about whether each species has or has not become invasive. Thus, the range of impacts that non-native species can have is simplified to a binary outcome (i.e. invasive or not). Data on the traits of each species are then gathered and analyzed to find patterns that accurately discriminate between invasive and non-invasive species. If consistent patterns are found, the traits of species not yet introduced can be assessed to determine whether they fit the profile associated with previous invaders. The assumption of trait-based risk assessment is that invasive species in the future will be similar to those that have been invasive in the past. Thus, if a species that shares a suite of traits with previous invaders is proposed for introduction, it is assumed to present a high risk (for reviews, see Kolar and Lodge 2001; Rejmánek et al. 2005; Hayes and Barry 2008; Keller and Drake 2009).

Trait-based risk assessment tools are generally focused on limited taxonomic groups and well defined regions or ecosystems. This controls for variation in the associations between traits and outcomes across each of these scales. Currently available tools include risk assessments for fish invaders in the North American Great Lakes (Kolar & Lodge 2002) and California (Marchetti et al. 2004), mollusks in the North American Great Lakes and the 48 contiguous United States (Keller et al. 2007b), birds in New Zealand (Veltman et al. 1996) and Australia (Duncan et al. 2001), and woody plants in the US (Reichard & Hamilton 1997). These risk assessment tools generally have true-positive rates (i.e. correctly identify invaders) of 80–95 per cent. True negative (i.e. correctly identify non-invaders) rates tend to be lower (Lonsdale and Smith 2001).

The most prominent trait-based tool for risk assessment of intentionally introduced species is the Australian Weed Risk Assessment (WRA) (Pheloung 1995). This has been applied as a mandatory screening tool for introductions of new plant species to Australia since 1997, has been adopted in a modified form by New Zealand, and has good accuracy in a range of settings around the world (Gordon et al. 2008). The Australian WRA tool consists of 49 questions about the natural history, biogeography, ecology, and invasion history of the plant being assessed. Each question is assigned a numerical weight, and the outcome is a score that is translated into a prediction that the species will be invasive or benign, or that further information is required for a reliable assessment to be made.

The weights applied to each question in the Australian WRA were assigned by expert opinion during the original development of the tool in the 1990s. Statistical approaches developed since would likely improve the consistency, transparency, and accuracy of the WRA. Logistic regression or machine-learning (e.g. Categorical and Regression Trees) algorithms, for example, are able to identify the patterns in trait data that most consistently discriminate between invasive and non-invasive species. These tools discover the variable weights that allow for greatest discrimination between the two groups of species. Thus, quantitative approaches are less susceptible to the preconceived opinions and biases of experts. An additional benefit is that these approaches often identify that just a small number of variables are required for accurate prediction. For example, Caley et al. (2006a) applied categorical tree analysis to the original training data used to develop the Australian WRA. They concluded that the answers to just four questions are sufficient to give similar, although slightly lower, accuracy to the full risk assessment. Policy-makers could determine whether the additional accuracy from the full tool justifies the extra resources required to produce predictions.

Creating and testing trait-based risk assessment tools requires biological and statistical expertise. In many cases, especially for taxonomic groups and regions that are well studied, tools could be created by agency scientists or graduate students quite quickly (<1 year). Once developed, a statistically-based risk assessment tool can give a result for a species proposed for importation in 1–2 days. Many of the tools already developed may be applicable to multiple regions, which has turned out to be the case for freshwater mollusks in the US (Keller et al. 2007b), and terrestrial plants in numerous regions (Gordon et al. 2008). Thus, it may not be necessary to develop new tools for each combination of taxonomic group and target region.

A final approach to species-specific risk assessment combines intensive literature review and other quantitative tools. While this intensive approach may improve accuracy, this comes at a substantial cost in time (>6 months to assess each species) and expertise. The range of analyses for this approach usually begins with an extensive literature review, and could also include some or all of the following: environmental niche modeling to determine the potential range of the species (Herborg et al. 2009); population modeling to predict how rapidly a population would grow under specified conditions (Sakai et al. 2001); modeling to predict how quickly the species will expand its geographic range (Muirhead et al. 2009), and the trait-based tools described above. The specific analyses included, and how the result of each component is weighted, entail considerably more decisions and more opportunities for loss of consistency and transparency relative to the simple trait-based systems described above. Thus the overall accuracy will depend largely on the skill of the assessor at selecting and correctly using appropriate tools. The upshot of this approach is that a richer and potentially more accurate result is possible, and this may be required when assessing the risks from a species that holds potential for both large benefits and large damages. In such cases, careful consideration of the economic and ecological impacts of an introduction may be justified, and importers and consumers may be willing to wait for the response and bear the cost of the risk assessment.

Mandrak and Cudmore's (2004) risk assessment for Asian carps (Grass Carp *Ctenopharyngodon idella*; Bighead Carp *Hypophthalmichthys nobilis*; Silver Carp *Hypophthalmichthys molitrix*; Largescale Silver Carp *Hypophthalmichthys harmandi*; Black Carp *Mylopharyngodon piceus*) in Canada is a good example of the intensive approach. This assessment used highly quantitative environmental niche models to predict the regions in Canada where each species could survive. This was combined with a qualitative literature review, leading to a final determination of risk. This determination of risk has far higher geographic resolution than that offered by trait-based approaches. For example, regions of Canada where some of the species are unlikely to survive were identified, raising the possibility of management approaches that include trade in the species in some regions but not others. This additional detail is beneficial, but as mentioned above comes at a substantial cost.

In conclusion, we note that the trait-based species-level decision-making process of the Australian WRA is widely viewed as a model system for other nations (IUCN 2000; Lodge et al. 2006; Gordon et al. 2008). In Australia and New Zealand, where this tool is used as a mandatory screen for new plant introductions, species predicted to be invasive are placed on a black list and banned from introduction, while species predicted to be benign are placed on a white list and allowed. Similar processes have been proposed for the US (NISC 2001; Lodge et al. 2006) and the approach is prominent in the IUCN (2000) guidelines for national policy development. Trait-based approaches are recommended because they make rapid species assessments possible, and have high consistency, high reported accuracy (but see next section), and high transparency. Other methods are either less defensible (e.g. expert opinion) or much slower (e.g. intensive approach, literature survey) and may not satisfy the needs of policy-makers and importers. Importantly, the time taken to perform literature surveys and the intensive approach (e.g. Mandrak & Cudmore 2004, Kolar et al. 2005) make it infeasible to assess all species proposed for importation.

15.2.1.3 HOW ACCURATE ARE RISK ASSESSMENT TOOLS?

For scientific purposes, the best way to determine the accuracy of a risk assessment tool would be to assess a large number of species that have not been introduced to the region of interest, and then introduce a randomly drawn subset of species from the pool of potential future introductions. After a time lag to allow species to overcome any barriers to population growth, the outcome of introductions (invasive or not) would be compared to the original predictions to determine their accuracy (Lonsdale & Smith 2001). Because such an approach would be unethical and would in any case not provide guidance soon enough to inform urgent management needs, scientists have developed a number of other approaches to estimate accuracy. Each has its drawbacks, and on balance they may collectively inflate apparent accuracies. We discuss some of these issues in the following, while Smith *et al.* (1999), Londsale and Smith (2001), Caley *et al.* (2006b), Williamson (1999, 2006) and Keller *et al.* (2007a) provide more details.

Expert opinion and literature review-based risk assessment tools are rarely, if ever, assessed for accuracy in a rigorous way. Indeed, because they are usually conducted in small numbers for species already established, and conducted by people already familiar with the species in question, tests of accuracy would not be very informative. In contrast, quantitative risk assessment tools are routinely assessed for accuracy. The general method for doing this is to use some form of cross-validation. When data for a large enough number of species are available, data for a subset of the species are held aside, the remaining species used to create the model, and the model tested on the hold out species. When the number of species for which data are available is limited (as it often is), the model can be created many times, each time with a small number (sometimes only one species) of randomly chosen species held out. The accuracy of the model is then calculated as the percentage of times that the model correctly predicts the true invasiveness of species held out. Cross-validation is common because the computational power required is readily available and because it tests each model on an effectively independent dataset (i.e. the hold outs). We don't know of a better method than cross-validation to assess the accuracy of quantitative risk assessment tools, but in the following we describe a number of reasons that this approach may overestimate accuracy.

First, these tools are created on the assumption that future invaders will have similar impacts to past invaders, and that the traits associated with invasion will be constant over time (Williamson 2006). In reality, species or the invaded ecosystem may be quite different in the future, because of the introduced species themselves and perhaps because new trade routes open, bringing species with novel combinations of traits or impacts. This will make future use of the tool less accurate than suggested by cross-validation.

Second, risk assessment tools are often built using species datasets that contain a higher proportion of invaders than the actual proportion of introduced species that have become invasive (Lonsdale & Smith 2001). Data are often more readily available for invaders than non-invaders. Ideally a model will be trained on a training dataset that contains the same proportion of invaders as the proportion of all introduced species in that taxon and region that have become invasive. The effect of having a higher proportion of invaders in the training dataset is to make the model better at correctly identifying invaders, but at the expense of poorer performance with non-invaders. This bias may be acceptable, but it should be recognized and, where possible, quantified.

Third, we note the general issue that the proportion of introduced species that become invasive (often referred to as the base rate of invasion) can be very low, depending on the taxonomic group and region in question. When the base-rate is low, the proportion of non-invasive species tested by the risk assessment will be very high, so that even small error rates will lead to a large number of non-invasive species that are misidentified as invasive. In contrast, because there are very few invasive species, most of them will be correctly identified as such. This means the vast majority of species predicted to be invasive may, in fact, pose a low risk. Many non-invaders will be kept from trade, where they would otherwise create benefits. This issue is explored in detail by Smith *et al.* (1999) and

Williamson (2006). A more recent analysis (Keller *et al.* 2007a) has shown that, despite the large proportion of non-invaders incorrectly identified as invasive, risk assessment can yield net economic benefits for the importing country.

As a result of these issues, reported accuracies of many quantitative risk assessment tools are probably over estimated. How much this accuracy is inflated is not known, meaning that the best approach is to be aware of the issues, to avoid those that can be avoided, and to be clear about those that cannot be avoided.

15.2.1.4 VECTOR ELIMINATION FOR INTENTIONALLY INTRODUCED SPECIES

We note here that if a vector of intentional introduction presents large risks of invasion but few benefits it may be rational to avoid the costs of species risk assessment and simply disallow the introduction of all species through that vector. This could conceivably occur for vectors where the importing nation already has native species, or widely established non-indigenous species, that could satisfy demand, e.g. earthworms for fish bait in the US (Keller *et al.* 2007c). In this case, the marginal benefits of introducing new non-native species will be low, while the risks of invasion and subsequent damage from these new species may be significant.

15.2.2 Risk assessment for prevention of transportation-related introductions

In contrast to intentional introductions, species that hitchhike on transportation vectors are not expected to provide direct benefits to humans. They do, however, have a nonzero probability of becoming invasive. Many of the world's most damaging invaders have been introduced accidentally. The Zebra Mussel (*Dreissena polymorpha*), for example, is native to the Ponto-Caspian Basin in Eastern Europe and has been spread across much of Western Europe and North America in ballast water or attached to the hulls of ships and recreational boats (Bossenbroek *et al.* 2001). The Emerald Ash Borer (*Agrilus planipennis*) was discovered in the US in 2002 after being introduced as a contaminant of wood packing crates. It is rapidly spreading, and as it does it causes close to 100 per cent mortality of ash trees (BenDor *et al.* 2006). Other vectors of accidental introduction are shown in Fig. 15.2.

The number and magnitude of transportation related vectors makes preventing accidental introductions an enormous challenge. Shipping, one of the strongest vectors, accounts for most of the international movement of cargo, with the global merchant shipping fleet now including at least 94,000 vessels of 100 gross tonnes or larger (IMO 2007). Individual countries face huge challenges in monitoring this vector partly because of its sheer magnitude, and partly because it can introduce species in a number of ways, including through ballast water, hull-fouling, and as contaminants of traded goods. India had total imports of 2.5million shipping containers during 2005–06 (http://www.ipa.nic.in/oper4d_2006.htm). The port of Shanghai processed an order of magnitude more containers during the same period. Each container is potentially vectoring viable populations of invasive species, but globally very few containers are ever inspected. Even New Zealand, which has only around 300,000 containers arriving per year and is considered to be a world leader in inspections, inspects just 17 per cent of incoming cargo containers (Everett 2000).

As well as contaminants of containers, ships can introduce species in the ballast water that is carried to maintain safe ship operating weight. When cargo is loaded, ballast is discharged, along with any organisms in the ballast tanks. Endresen *et al.* (2004) have estimated that, globally, 3500 million tonnes of ballast water are discharged by ships every year. Drake and Lodge (2007b) identified 93 unique taxa from partial sampling of the ballast tanks of 41 ships entering the North American Great Lakes between 2000 and 2002. The volume of ballast water carried, and the number of species entrained, pose large invasion risks to all nations with shipping ports.

The potential for hitchhiking species to arrive with or on human travelers is also large. Over 67 million passengers, along with their baggage, passed through Heathrow International Airport in London, UK, during 2006. Each traveler poses an independent risk of transporting seeds of invasive species attached to their shoes, diseased fruit in their baggage, or non-native species in other ways (Fig. 15.2). Inspecting every piece of baggage and cleaning every pair of shoes is not currently the policy of any country, and the logistical and resource challenges to doing so would be immense.

The diversity of vectors of accidental introduction means that many different approaches will be required to reduce risks. Because these vectors can rarely be eliminated, the general goal of risk management is usually to reduce the number of organisms transported to a level that leaves an acceptably low risk of invasion. As for commerce in living organisms, this level of risk is a policy decision. Achieving the risk threshold set for each vector, however, will require risk management tools designed specifically for that vector. Because risk management will often be expensive, risk assessment of the threats posed by each vector can be used to determine where the greatest reductions in invasion risk can be achieved with available resources. As for intentional introductions, risk assessment for vectors will be most informative when based on quantified risks. Despite this, such analyses have rarely been conducted. Most countries instead have a largely haphazard approach to selecting which vectors of accidental species introduction should be addressed. Although this has led to many useful programs, it does not have the rigor that would come from quantitative analyses and is unlikely to be the most efficient way to spend resources.

Because the diversity of accidental vectors is so great, and because risk management is usually vector-specific, we do not attempt an exhaustive examination of the types of risk assessment and risk management programs being employed. Instead, in the following we describe three illustrative approaches to reducing the invasion risks from vectors of unintentional introduction of both aquatic and terrestrial organisms.

First, because pests and diseases can be transported with introduced live plants, most plants imported to the US are required to be accompanied by a phytosanitary certificate. This program operates under the US Plant Protection Act and is consistent with US obligations as a signatory to the WTO's Agreement on the Application of Sanitary and Phytosanitary Measures (Hedley 2004). Phytosanitary certificates must be filled out by plant health inspectors in the exporting country and detail that the plants are not contaminated with invasive diseases or pests. Exports from the US to Europe must be similarly assessed, this time by US inspectors. Although some organisms inevitably slip through, this program aims to ensure that they are few enough in number that the risk of establishment and invasion is acceptably low.

Second, the ISPM 15 standard for treatment of wood and wood packaging has been widely adopted around the world. This standard was developed in response to invasions, for example, by Emerald Ash Borer (*Agrilus planipennis*). Countries that have signed this standard agree to treat wood packaging material (e.g., wooden pallets) prior to export to ensure that it does not contain pests (https://www.ippc.int/IPP/En/default.jsp).

Finally, the International Maritime Organization (IMO 2004) has produced a convention that, once ratified by enough countries, will require all ships to treat ballast water to ensure that an acceptably low number of species are transported. For a number of years efforts to reduce ballast invasions have concentrated on ballast water exchange, which requires ships to flush and re-fill ballast tanks on the high seas on the assumption that any organisms taken up there are unlikely to survive when released at coastal ports. The success of ballast water exchange has been questioned (Ricciardi 2006, Costello *et al*. 2007a), however, supporting adoption of the higher standards in the IMO convention (IMO 2004). See Chapter 16 'Burgiel *et al*. (this volume)' for more discussion of international agreements that aim to prevent unintentional introductions.

15.3 Preventing invasive species: opportunities for improved risk assessments

15.3.1 Quantitative vs. qualitative approaches

Risk assessment is conducted to estimate the probability that certain events will occur under specified conditions. In the case of programs for preventing the arrival of invasive species, risk assessment should aim to determine the probability that different levels of impact will arise from the introduction of a species, or from different types of vector management. Probability estimation is thus central to risk assessment. Surprisingly, however, most risk assessment systems do not incorporate explicit probability estimations, do not quantitatively estimate uncertainty around probability estimates, and do not appropriately multiply probabilities together for different stages of invasion when these are estimated independently. For example, the US Generic Risk Assessment for Aquatic Organisms (Orr 2003) requires assessors to estimate the probability of a species traversing each stage of invasion as low, medium, or high. These categories are not defined quantitatively, and it is inevitable that different assessors will treat these categories differently. Each assessor is performing probability estimation, but is doing so under their interpretation of the categories. If these categories were defined numerically (e.g. low = 0–10 per cent, medium = 11–50 per cent, high = 51–100 per cent) the risk assessment tool would be more consistent (i.e. different assessors would be more likely to get the same results) and it would be easier to question the reasoning that leads to predictions.

Data for precise probability estimation in risk assessment is rarely available. For example, the full costs of invasive species are rarely known and we lack knowledge about how interacting economic and ecological systems respond to different types and levels of control. This means that estimated probabilities will often have large bounds of uncertainty. Quantifying these bounds will give policymakers and managers the best possible information on which to base their decisions.

15.3.2 What is an appropriate risk threshold?

Risk management determines appropriate responses to the probabilities of given events occurring. Because management actions are discrete but probabilities occur on a continuous scale, it is usually necessary to establish thresholds of risk that trigger different management actions. As discussed above, these thresholds are policy rather than scientific decisions. Thresholds will differ based on levels of aversion to different outcomes. Under WTO, a country is free to establish their risk threshold as long as it is applied consistently across the countries from which imports are received, and across the full range of products imported.

We summarize three considerations that lead many to argue that thresholds for action to prevent the arrival of an invasive species (e.g. restrict import of a species; require actions to reduce the number of organisms entrained in a vector of accidental introduction) should be low. First, the damages from an invasive species are generally high compared to the benefits from the introduction of a non-invasive species (Keller *et al.* 2007a). The value of preventing introduction of an invader is thus usually greater than the lost value from accidentally banning the introduction of a non-invasive species. Second, invasive species are rarely eradicated. The costs of a false negative (e.g. allowing import of an invasive species) can be expected to recur over a long time period, leading to high cumulative impacts (Lodge *et al.* 2006). Finally, acceptable thresholds for other comparable risks are low. For example, the US EPA generally considers that an activity or product has an unacceptable risk if it raises by 1 in 1,000,000 an individual's probability of contracting cancer (http://www.epa.gov/OUST/rbdm/sctrlsgw.htm).

15.3.3 Pre-border vs. border prevention

For many vectors, nations can choose to implement prevention programs before organisms reach their border. This offers greater reductions in risk compared to prevention at the border. Additionally, it can offer large economic savings to a government by putting the onus of prevention on importers. The Australian WRA process is an example of pre-border prevention because species predicted to be invasive never enter the vector that would have otherwise transported them to Australia. Assuming reasonable compliance, this type of program will save resources by reducing the need for trained botanists searching plant shipments at the border. Ballast water exchange and the use of phytosanitary certificates are also programs for pre-border prevention, and demonstrate that pre-border prevention is equally applicable to intentional and unintentional vectors of introduction.

Most nations do not have the capacity to monitor the total volume of trade and travelers arriving at their border. Thus, when pre-border prevention methods are not available, prioritization will be required to determine which vectors pose the greatest immediate risks. Additionally, within a vector, it may be advantageous to determine which donor regions pose the greatest threat. Costello et al. (2007b) showed that the likelihood of a given unit of trade resulting in an invasion to San Francisco depends on the origin of the traded goods. Invasions will generally be more likely to occur when the source of the vector is an ecosystem that is similar to the arrival region. These considerations can be included when deciding how to allocate limited resources for border level prevention.

15.3.4 Resource availability

Applying risk assessment tools to vectors of introduction requires a government agency that can administer the tools, along with the infrastructure and personnel to monitor and enforce the regulations created. For many countries these institutional frameworks already exist in the forms of agencies to protect agriculture, natural resources, and human health. In these countries the implementation of trait-based risk assessment tools for screening species could be implemented relatively quickly and with a cost far below that of controlling the invaders that would otherwise establish (Keller et al. 2007a).

In other nations the infrastructure and personnel required to regulate species and vectors does not exist, and/or other issues such as malnutrition, human diseases, and defense are prioritized over preventing invasive species. It is beyond the scope of this chapter to suggest solutions to these issues. In many situations, however, it will be possible to take some positive steps with little investment. For example, a country that is not able to operate effective border quarantine could adopt identical standards to the US phytosanitary certificate program described above. Because any business that exports to the US will have the facilities to achieve this level of plant hygiene, the less developed nation may be able to effectively piggy-back off the US regulations, and thus achieve some protection with little investment.

We also note that nations without resources to fund their own programs will gain relatively more from international and regional agreements. Thus, less developed nations may benefit disproportionately from supporting treaties like the International Maritime Organization's proposed ballast water treatment standards and the ISPM 15 standards for treating wood packaging. International programs for preventing the spread of invasive species are described in more detail in Burgiel et al. Chapter 16 (this volume).

The sheer number of vectors, and the reality of limited resources to address them, means that all nations will need to prioritize the species and vectors that they address. Ideally this prioritization will be based on an assessment of the vector-specific risks. A simple way to do this would be to investigate the recent history of each vector and the species it has introduced. Vectors that have led to large damages could be prioritized over vectors that have caused less damage. Combining this with predictions of future trade and travel patterns to determine which vectors are likely to grow and shrink would make the approach more robust. More sophisticated assessments for prioritizing vectors would likely be based on this approach, but could be more quantitative and include consideration of the risks posed at the species level, and the cost-benefit trade-offs involved in modifying these vectors.

15.3.5 Gaps and overlaps among government agencies

Although functioning government agencies are required for invasive species prevention, they are not sufficient. The US and Canada, for example, have well-resourced government agencies, but still maintain relatively open borders, even for species known to be invasive elsewhere. Agency mandates that leave gaps in vector management or overlap when it comes to invasive species issues contribute to this situation (Kaiser 2006). Concern about environmental invaders has grown greatly in the last couple of decades, but most government agencies pre-date this concern. In many cases there are no agencies that have historically had the mandate to prevent the arrival of environmental invaders. In other cases, multiple agencies have this mandate, leading to a situation where individual agencies are reluctant to act for fear of becoming the agency presumed responsible for all invasion prevention. Australia and New Zealand have addressed this issue by establishing biosecurity agencies that are responsible for coordinating invasive species prevention. The US established the National Invasive Species Council (NISC) in 1999 to serve this purpose. Although NISC is made up of high-ranking officials from the federal agencies that have responsibility for preventing invasions, it has thus far had little success at implementing prevention programs.

15.3.6 Species spread across borders

Species can spread naturally across borders from an invaded to an uninvaded country. Where borders are long, it may be impractical to eliminate this route of invasion. This means that a risk analysis approach, where the threats posed by species are weighted against the costs and options for prevention and control, will be necessary. Ultimately, however, a regional approach may be optimal, whereby jurisdictions in contiguous geographic regions (e.g. Canda, US, and Mexico) establish coordinated policies for preventing invasive species (see Burgiel *et al*. Chapter 16 this volume for more discussion of regional and international approaches).

15.4 Conclusions

Preventing the introduction of invasive species will almost always be less costly than managing and adapting to species once they arrive (Lodge *et al*. 2006; Keller *et al*. 2007a). Although this is widely recognized by biologists and economists, only a handful of countries have embraced this approach in their policies for environmental invaders. As nations work to meet their obligations under the CBD, it is likely that the number of countries with programs for preventing invasive species will grow.

Australia and New Zealand are generally considered to implement the most effective programs for preventing the introduction of invasive species. Because the general issues faced by all countries with respect to invasive species are similar, it should be possible for nations to learn from the experiences of others, and thus to reduce the costs and uncertainty involved with implementing new policies. Tools for risk assessment of species and vectors, and the management options that are appropriate, for example, will be the same for many nations.

Differences among countries are also important of course. For example, different levels of resources to devote to risk assessment and quarantine, and whether a nation is an island or not, will influence policy and management approaches. We have described a number of these considerations, and how they apply to different situations.

For the foreseeable future invasive species will remain a large threat to ecosystems, economies and human health in all countries (Sala *et al*. 2000). Although efforts to prevent invasions will often require large investments to design risk assessment tools, re-arrange or create agencies, or to modify trade practices, the available evidence strongly suggests that such actions will yield net benefits for society (Keller *et al*. 2007a).

References

Baker, H.G. (1974). The evolution of weeds. *Annual Review of Ecology and Systematics*, 5, 1–24.

BenDor, T.K., Metcalf, S.S., Fontenot, L.E., Sangunett, B., and Hannon, B. (2006). Modeling the spread of the emerald ash borer. *Ecological Modelling*, 197, 221–36.

Bossenbroek, J.M., Kraft, C.E., and Nekola, J.C. (2001). Prediction of long-distance dispersal using gravity models: Zebra mussel invasion of inland lakes. *Ecological Applications*, 11, 1778–88.

Caley, P., Lonsdale, W.M., and Pheloung, P.C. (2006a). Quantifying uncertainty in predictions of invasiveness, with emphasis on weed risk assessment. *Biological Invasions*, 8, 1595–1604.

Caley, P., Lonsdale, W.M., and Pheloung, P.C. (2006b). Quantifying uncertainty in predictions of invasiveness. *Biological Invasions*, 8, 277–86.

Costello, C., Drake, J.M. and Lodge, D.M. (2007a). Evaluating the effectiveness of an environmental policy: ballast water exchange and invasive species in the North American Great Lakes. *Ecological Applications*, 17, 655–62.

Costello, C., Springborn, M., McAusland, C., and Solow, A. (2007b). Unintended biological invasions: does risk vary by trading partner? *Journal of Environmental Economics and Management*, 54, 262–76.

Drake, J.M. and Lodge, D.M. (2004). Global hot spots of biological invasions: evaluating options for ballast-water management. *Proceedings of the Royal Society Biological Sciences Series B* (print), 271, 575–80.

Drake, J.M. and Lodge, D.M. (2007a). Hull fouling is a risk factor for intercontinental species exchange in aquatic ecosystems. *Aquatic Invasions*, 2, 121–31.

Drake, J.M. and Lodge, D.M. (2007b). Rate of species introduction in the Great Lakes via ships' ballast water and sediments. *Canadian Journal of Fisheries and Aquatic Sciences*, 64, 530–38.

Duncan, R.P., Bomford, M., Forsyth, D.M., and Conibear, L. (2001). High predictability in introduction outcomes and the geographical range size of introduced Australian birds: a role for climate. *Journal of Animal Ecology*, 70, 621–32.

Endresen, Ø., Beyrens, H.L., Brynestad, S., Andersen, A.B., and Skjong, R. (2004). Challenges in global ballast water management. *Marine Pollution Bulletin*, 48, 615–23.

Everett, R.A. (2000). Patterns and pathways of biological invasions. *Trends in Ecology and Evolution*, 15, 177–78.

Fowler, A.J., Lodge, D.M. and Hsia, J.F. (2007). Failure of the Lacey Act to protect US ecosystems against animal invasions. *Frontiers in Ecology and the Environment*, 5, 353–59.

Gordon, D.R., Onderdonk, D.A., Fox, A.M. and Stocker, R.K. (2008). Consistent accuracy of the Australian weed risk assessment system across varied geographies. *Diversity and Distributions*, 14, 234–42.

Hayes, K.R. and Barry, S.C. (2008). Are there any consistent predictors of invasion success? *Biological Invasions*, 10, 483–506.

Hedley, J. (2004). The International Plant Protection Convention and invasives. In M.L. Miller and R.N. Fabian, eds. *Harmful Invasive Species: Legal Responses*, Environmental Law Institute, Washington, D.C., pp. 185–201.

Herborg, L.-M., Drake, J.M., Rothlisberger, J.D. and Bossenbroek, J.M. (2009). Identifying suitable habitat for invasive species using ecological niche models and the policy implications of range forecasts. In R.P. Keller, D.M. Lodge, M.A. Lewis, and J. Shogren, eds. *Bioeconomics of Invasive Species*. Oxford University Press, New York, U.S.A. In press.

Hewitt, C.L. and Hayes, K.R. (2002). Risk assessment of marine biological invasions. In E. Leppäkoski, S. Gollasch and S. Olenin, eds. *Invasive Aquatic Species of Europe*, Kluwer Academic Publishers, Dordrecht, The Netherlands, pp. 456–66.

Horan, R.D., Perrings, C., Lupi, F., and Bulte, E.H. (2002). Biological pollution prevention strategies under ignorance: the case of invasive species. *American Journal of Agricultural Economics*, 84, 1303–1310.

Hulme, P.E., Bacher, S., Kenis, M., *et al*. (2008). Grasping at the routes of biological invasions: a framework for integrating pathways into policy. *Journal of Applied Ecology*, 45, 403–414.

International Maritime Agency [IMO]. (2004). International Convention for the Control and Management of Ships' Ballast Water and Sediments. [online] URL: http://www.imo.org.

International Maritime Agency [IMO]. (2007). International shipping and world trade: facts and figures. IMO Library Services. [online] URL: http://www.imo.org.

International Union for the Conservation of Nature [IUCN]. (2000). IUCN guidelines for the prevention of biodiversity loss caused by alien invasive species. [online] URL: http://iucn.org.

Jeschke, J.M. and Strayer, D.L. (2005). Invasion success of vertebrates in Europe and North America. *Proceedings of the National Academy of Sciences*, 102, 7198–7202.

Jones, K.E., Patel, N.G., Levy, M.A., *et al*. (2008). Global trends in emerging infectious diseases. *Nature*, 451, 990–94.

Kaiser, B. (2006). On the garden path: an economic perspective on prevention and control policies for an invasive species. *Choices*, 21, 139–42.

Keller, R.P. and Drake, J.M. (2009). Trait based risk assessment for invasive species. In R.P. Keller, D.M. Lodge, M.A. Lewis and J. Shogren, eds. *Bioeconomics of Invasive Species*. Oxford University Press, New York, U.S.A. In press.

Keller, R.P. and Lodge, D.M. (2007). Species invasions from commerce in live organisms: problems and possible solutions. *BioScience*, 57, 428–36.

Keller, R.P., Lodge, D.M., and Finnoff, D.C. (2007a). Risk assessment for invasive species produces net bioeconomic benefits. *Proceedings of the National Academy of Sciences*, 104, 203–207.

Keller, R.P., Drake, J.M. and Lodge, D.M. (2007b). Fecundity as a basis for risk assessment of nonindigenous freshwater molluscs. *Conservation Biology*, 21, 191–200.

Keller, R.P., VanLoon, C., Cox, A.N., Lodge, D.M., Herborg, L.M., and Rothlisberger, J. (2007c). From bait shops to the forest floor: earthworm use, transport and disposal by anglers. *American Midland Naturalist*, 158, 321–28.

Kolar, C.S., and Lodge, D.M. (2001). Progress in invasion biology: predicting invaders. *Trends in Ecology and Evolution*, 16, 199–204.

Kolar, C.S. and Lodge, D.M. (2002). Ecological predictions and risk assessment for alien fishes in North America. *Science*, 298, 1233–1236.

Kolar, C.S., Chapman, D.C., Courtenay Jr., W.R., Housel, C.M., Williams, J.D., and Jennings, D.P. (2005). *Asian carps fo the genus Hypophthalmichthys: a biological synopsis and environmental risk assessment*. Report to U.S. Fish and Wildlife Service.

Leprieur, F., Beauchard, O., Hugueny, B., Grenouillet, G., and Brosse, S. (2008). Null model of biotic homogenization: a test with the European freshwater fish fauna. *Diversity and Distributions*, 14, 291–300.

Lodge, D.M., Williams, S., MacIsaac, H., *et al*. (2006). Biological invasions: recommendations for policy and management. *Ecological Applications*, 16, 2035–54.

Lonsdale, W.M. and Smith, C.S. (2001). Evaluating pest-screening systems—insights from epidemiology and ecology. In R.H. Groves, F.D. Panetta, and J.G. Virtue, eds. *Weed Risk Assessment*, PP 52–60. CSIRO Publishing, Collingwood, Victoria, Australia.

Mack, R.N., Simberloff, D., Lonsdale, W.M., Evans, H., Clout, M., and Bazzaz, F.A. (2000). Biotic invasions: Causes, epidemiology, global consequences, and control. *Ecological Applications*, 10, 689–710.

Mandrak, N.E. and Cudmore, B. (2004). *Risk assessment for Asian carps in Canada*. Department of Fisheries and Oceans, Ottawa, Canada.

Marchetti, M.P., Moyle, P.B., and Levine, R. (2004). Invasive species profiling? Exploring the characteristics of non-native fishes across invasion stages in California. *Freshwater Biology*, 49, 646–661.

Muirhead, J.R., Bobeldyk, A.M., Bossenbroek, J.M., Egan, K.J., and Jerde, C.L. (2009). Estimating dispersal and predicting spread of nonindigenous species. In R.P. Keller, D.M. Lodge, M.A. Lewis, and J. Shogren, eds. *Bioeconomics of Invasive Species*. Oxford University Press, New York, U.S.A. In press.

National Invasive Species Council [NISC]. (2001). *Meeting the Invasive Species Challenge: National Invasive Species Management Plan*. National Invasive Species Council, Washington D.C.

National Research Council [NRC]. (2002). *Predicting Invasions by Nonindigenous Plants and Plant Pests*. National Academy of Sciences, Washington, DC, USA.

Orr, R. (2003). Generic nonindigenous aquatic organisms risk analysis review process. In G.M. Ruiz and J.T. Carlton, eds. *Invasive Species: Vectors and Management Strategies*, Island Press, Washington, DC, pp. 415–38.

Pheloung, P.C. (1995). *Determining the Weed Potential of New Plant Introductions to Australia* Agriculture Protection Board, Perth, Australia.

Pimentel, D., ed. (2002). *Biological Invasions: Economic And Environmental Costs Of Alien Plant, Animal And Microbe Species*. CRC Press, Boca Raton, Florida, USA.

Reichard, S.H. and Hamilton, C.W. (1997). Predicting invasions of woody plants introduced into North America. *Conservation Biology*, 11,193–203.

Ricciardi, A. (2006). Patterns of invasion in the Laurentian Great Lakes in relation to changes in vector activity. *Diversity and Distributions*, 12, 425–433.

Rejmánek, M., Richardson, D.M., Higgins, S.I., Pitcairn, M.J., and Grotkopp, E. (2005). Ecology of invasive plants: state of the art. In H.A. Mooney, R.N. Mack, J.A. McNeely, L.E. Neville, P.J Schei, and J.K. Waage, eds. *Invasive Alien Species: A New Synthesis*, pp. 104–161. Island Press, Washington D.C., USA.

Sakai, A.K., Allendorf, F.W., Holt, J.S *et al*. (2001). The population biology of invasive species. *Annual Review of Ecology and Systematics*, 32, 305–332.

Sala, O.E., Chapin, F.S., Armesto, J.J., *et al*. (2000). Global Biodiversity Scenarios for the Year 2100. *Science*, 287, 1770–1774.

Smith, C.S., Lonsdale, W.M., and Fortune, J. (1999). When to ignore advice: invasion predictions and decision theory. *Biological Invasions*, 1, 89–96.

Veltman, C.J., Nee, S., and Crawley, M.J. (1996). Correlates of introduction success in exotic New Zealand birds. *American Naturalist*, 147, 542–557.

Williamson, M. (1999). Invasions. *Ecography*, 22, 3–12.

Williamson, M. (2006). Explaining and predicting the success of invading species at different stages of invasion. *Biological Invasions*, 8, 1561–68.

CHAPTER 16

Globalization and Bioinvasions: The International Policy Problem

Charles Perrings, Stas Burgiel, Mark Lonsdale, Harold Mooney, and Mark Williamson

16.1 Dimensions of the policy problem

Invasive species control is a public good. Once provided, the benefits it offers in terms of enhanced protection of human, animal, and plant health and the productivity of agriculture, forestry, aquaculture, and fisheries, are available to everyone. Like all public goods it will be undersupplied if left to the market. This makes it a collective responsibility—a legitimate role of government at many different scales. This role involves two different functions. One is the development of broad strategies and supporting institutions, statutes, regulations, or agreements for addressing the problem. This may be either directly through public expenditure on controls or indirectly through incentives to the people whose actions may be the cause of the problem. Broad strategies of this kind define an invasive species policy. A second function involves the implementation of that policy, and specifically the use of public resources to undertake all of the actions described in this volume: inspection and interception at the port of entry, sanitary and phytosanitary measures both along pathways and *in situ*, detection, eradication, and control of harmful species that have been introduced, established, and spread. This is invasive species management.

In this chapter we consider both broad issues of policy and specific challenges to management. The general thrust of the volume is that the risks associated with invasive species are increasing as a result of globalization, but are also changing in ways that require novel approaches. We review the changes that are taking place and explore the implications for both policy and management. Specifically, we argue that the tight linkage between invasive species risks and the closer integration of the world economic system means that the best strategy for dealing with the problem involves global coordination and cooperation, both to internalize what are potentially major externalities of world trade and to address the provision of a global public good. We also argue that the uncertainty which follows from the rapid evolution of the global system—for example, that surrounding emerging zoonotic diseases—demands a strategy that is at once precautionary (that it protects the system whilst allowing learning about the consequences of novelty) and cost effective (in practice, that it does not involve the costs associated with unnecessary restriction of trade or travel). While precaution is a feature of existing instruments, such as the Sanitary and Phytosanitary (SPS) Agreement, we note that it is an option that is effectively only available to a subset of countries and that this defect increases global exposure to invasive species risks. Traditional invasive species management involves a mix of defensive measures such as inspection, interception, detection, eradication, and control. We consider ways of strengthening such measures whilst, at the same time, realizing the potential benefits of a more coordinated international approach.

Countering the movement and spread of invasive species requires coordination across countries to identify common threats, implement prevention or management measures, and ultimately to ensure that new outbreaks do not breach the weakest points in the biosecurity system. Currently, few countries have the resources, the biogeographical conditions, and the institutional, statutory, and regulatory environment to mount effective defenses against invasive species (New Zealand and Australia are commonly cited examples of countries that do satisfy these conditions). For most countries, cooperative efforts at the regional and global level offer a more cost effective solution. Yet it remains the case that the default strategy is still national defense, not international cooperation.

Under the World Trade Organization (WTO) and its constituent agreements, particularly the General Agreement on Tariffs and Trade (GATT) and the SPS Agreement, countries have the right to take actions to protect food safety, animal or plant health (see Appendix 1). There is no mechanism to address the collective risks posed by the international trading system other than the successive renegotiations of the GATT conducted by the World Trade Organization (WTO). The position with respect to human health is rather different. Indeed, the 2005 International Health Regulations (see Appendix 2) administered by the World Health Organization (WHO) mandate both coordinated and cooperative international action to address human health risks. We consider the implications of this model for other invasive species risks. However, we also note that the compartmentalization of the problem—the separation of human from other disease risks, for example—is one reason for the inconsistency in approaches to collective action. Discussions in fora such as the Convention on Biological Diversity (CBD), the International Plant Protection Convention (IPPC), and the UN Food and Agriculture Organization (FAO) have identified many gaps in institutional jurisdiction, but these bodies are not themselves able to fill those gaps.

We note that all of the main international agreements require states to ensure that they are not themselves a source of risk to others. Indeed, this is a central tenet of the IHR (2005). Articles 6 and 7 of the regulations require member countries to notify the WHO of any event that may have implications for international health, while Article 9 requires notification of any international public health risk due to the movement of people, disease vectors, or contaminated goods (Appendix 2). The Convention on Biological Diversity (1993), too, asserts that "States have...the sovereign right to exploit their own resources pursuant to their own environmental policies, *and the responsibility to ensure that activities within their jurisdiction or control do not cause damage to the environment of other States or of areas beyond the limits of national jurisdiction*" (Our italics) (Article 3). Article 14 of the convention goes further, requiring states to notify other of any action or event that is "likely to significantly affect adversely the biological diversity of other States or areas beyond the limits of national jurisdiction." More particularly, where the potential damage to other states is "imminent or grave", states are required both to notify potentially affected states and to "initiate action to prevent or minimize such danger or damage" and to "encourage international cooperation to supplement such national efforts and, where appropriate and agreed by the States or regional economic integration organizations concerned, to establish joint contingency plans" (Article 14d, 14e).

In what follows, we argue for a strategy for dealing with invasive species risks that extends beyond traditional defensive measures. We argue, first, that since invasive species are an externality of trade, transport, and travel that involve public goods, they require collective regulation of international markets through the GATT. A review of the SPS Agreement is on the agenda for the beleaguered Doha round, and provides an opportunity for the introduction of measures to bring conformity between that agreement and the International Health Regulations. Second, we argue for the development of mechanisms to generate and disseminate information on invasive species risks and their impacts. This is needed for three reasons: 1) to enable the more effective use of defensive measures allowed under the SPS Agreement, especially by developing countries; 2) to facilitate the coordination of actions by different international institutions; and 3) to support cooperative international action on invasive species that may

Table 16.1 Areas of missing or overlapping responsibility for invasion pathways identified by the Convention on Biological Diversity

CBD identified pathway gaps		Responsible organizations
Animals that are not plant pests, including pets, aquarium species, live bait, live food	→	OIE, FAO, WTO SPS
Marine biofouling	→	IMO (for ships of large tonnage – i.e. there is a gap for ships of low tonnage)
Civil air transport	→	ICAO, IATA
Aquaculture, mariculture	→	FAO
Conveyances, tourism, emergency aid and development assistance, military activities, and inter-basin water transfers and canals	→	none
Flowers, livestock, specific agricultural goods, timber and other raw materials	→	WTO SPS, IPPC, OIE

cause widespread harm to animals and plants, and hence damage to the ecosystem services they generate. Third, we argue for the development of a global program of investment in building the capacity to identify and respond to trade-related invasive species risks. Capacity building to identify and respond to health risks is already called for under the IHR, but is not funded. A similar requirement is needed under the SPS Agreement, along with the establishment of a global fund for both.

16.2 The institutional environment

The options open to states are strongly structured by the existing institutional landscape. This landscape is quite fractured. Institutions concerned with the impact of invasive species on the general environment, for example, include the CBD and its Cartagena Protocol on Biosafety, the Ramsar Convention on Wetlands of International Importance, and the Convention on Migratory Species (CMS) along with many other multilateral environmental agreements. Institutions concerned with pathways (transport routes) include the International Maritime Organization (IMO), the International Civil Aviation Organization (ICAO), the International Air Transport Association (IATA), and the UN Convention on the Law of the Sea (UNCLOS). Institutions concerned with agriculture, aquaculture, forestry, and fisheries include World Animal Health Organization (OIE), International Plant Protection Convention (IPPC), and the UN Food and Agriculture Organization (FAO) along with its Compliance Agreement and Code of Conduct for Responsible Fisheries. Institutions concerned with trade include the Convention on International Trade in Endangered Species (CITES), but are dominated by the WTO and the GATT, along with supporting agreements such as the SPS Agreement and the Agreement on Technical Barriers to Trade (TBT Agreement). Institutions concerned with human health include the WHO and the International Health Regulations.

Many of these bodies have overlapping mandates. The CBD, FAO, and WTO, for example, have broad and frequently overlapping coverage. Since environmental, agricultural, and trade issues often overlap, there is at least a *prima facie* case for coordinated action. In practice, however, the outcome in any particular case is dominated by the body with the strongest set of rules (frequently the WTO's SPS Agreement). In many other instances, nobody has explicit responsibility. The invasive species risks of tourism, emergency aid and development assistance, military activity, and inter-basin water transfers are all cases in point. The CBD has identified a number of gaps and areas of inconsistency induced by overlapping responsibility at the international level (Table 16.1).

One manifestation of the fragmentation of effort amongst multiple instruments is that many prevention and management efforts have dealt with the global issues piecemeal. For example, the World Animal Health Organization's (OIE) specifically focuses on diseases related to livestock and other commercially valuable animal species, and actions are taken species by species. It neglects both interactions between species, and consequently the wider and longer term effects of species dispersal. There is currently some attempt to develop a more integrated approach, at least in terms of introduction pathways. The IPPC, for instance, has developed an international standard to curtail the movement of pests through solid wood packaging material (the likely source of introductions of the Asian Long-horned Beetle and the Emerald Ash Borer into the US) and has recently indicated an interest in looking at plant pest issues related to the aviation, shipping, and waste management sectors. Similarly, the ICAO has also called for guidance on the role of civil aviation in preventing the movement of invasive species.

The CBD's ninth Conference of the Parties prioritized invasive species, with animal invasive species (those that are not plant pests), hull fouling, civil aviation, tourism, and development assistance as key priorities within this. In the case of animal invasive species the CBD has initiated a process to liaise with the other relevant international institutions to discuss issues of competence, mandate, and types of guidance that might be developed. Such discussions have, however, been slow and cumbersome given (a) the relative inertia of international institutions and (b) that the expansion of organizational mandates ultimately requires agreement by member countries whose representatives are generally not the same as those sitting at the table in parallel international environmental or trade organizations (CBD, 2008). A useful measure would be to give the CBD membership, or at least observer status, on the SPS committee.

The CBD, with support from the Global Invasive Species Programme and IUCN's Invasive Species Specialist Group, has started collecting best practices in the area of animal invasive species with a view to creating tools and guidance for countries. Such information will not only be provided directly to countries for immediate application, but also will be fed into international discussions to decide which institution(s) should be responsible for the problem of animal invasive species. In this way, experience with implementation of management strategies can also serve as an input into international procedures. In the same way, negotiation and adoption of the IMO's International Convention on the Control and Management of Ships' Ballast Water and Sediments was facilitated by a series of national level pilot projects under the Global Ballast Water Programme (GloBallast).

It is not clear whether such exercises will address the problem of overlapping remits and integration. A number of institutions have very focused remits that compromise their ability to deal with invasive species. For example, the IPPC only deals with pests of plants. It can address invasive animals that impact upon agricultural crops or native plant species, but it has no remit to address invasive animals that may impact other animals or human health. Similarly the IMO only addresses shipping and marine vessels of large tonnage. This might be appropriate for ballast water which is associated with large ships, but it means that it has nothing to say about hull fouling issues on smaller vessels that are a significant source of introductions.

Another issue is the legal status attaching to the guidance offered by different agreements. For example, the CBD is perceived as a "soft law" convention, as it does not have any concrete enforcement mechanisms to ensure implementation. Additionally, much of the issue-specific guidance developed under the CBD is in the form of non-binding guidelines, guiding principles, and frameworks. Other international institutions are less limited in what they can do. For example, the IPPC and OIE have developed international standards in their respective areas of expertise that are recognized as compatible with international trade rules under the WTO's SPS Agreement. Thus, a country implementing an IPPC or OIE standard would not have to worry about suits brought by other countries under the WTO's dispute settlement process. The relative status of guidance issued under other international institutions (e.g. IMO, ICAO, and Cartagena Protocol on Biosafety) in the context of the WTO dispute settlement process is less clear.

The CBD does potentially have a mechanism for stronger, more binding agreements through the development of protocols. There may not be much appetite among member states for this, however, and it is significant that the Cartagena Protocol on Biosafety is the only protocol completed since the Convention entered into force in 1994. Nevertheless, a protocol on invasive species provides a mechanism by which the multiplicity of agreements bearing on the different dimensions of the invasive species problem could be coordinated. In the absence of a protocol to the CBD, however, we identify four principles that should guide international policy on invasive species. These are set out in Sections 16.3, 16.4, 16.5, and 16.6.

16.3 Provide the public good at the right geographical scale

Chapter 1 of this volume noted three aspects of globalization that have served to increase the risks of invasive species. One is that as economies have become more open with respect to imports they have become more likely to experience the introduction of potentially invasive species (Dalmazzone 2000; Vilà and Pujadas 2001). A second is that as the speed of transport increases the likelihood that "passenger" species will survive the journey increases (Tatem *et al.* 2006; Ruiz and Carlton 2003; Burgiel *et al.* 2006). A third is related to the development of Regional Trade Agreements (RTAs). As the volume of trade between countries in bioclimatically similar zones increases, and as scrutiny of imports and exports between countries in such zones decreases, the likelihood that introduced species will successfully spread rises (e.g. Perrault *et al.* 2003). This has two implications. On the one hand, since invasive species are an externality of international trade, the solution to the problem involves intervention in international markets and hence in the international trade regime. On the other, control of invasive species is a public good both nationally and internationally, meaning that provision of the appropriate level of control requires international cooperation.

We observed that while Article XX of the GATT allows countries to protect themselves against the invasive species risks of the growth in world trade, it does so under a very restrictive set of conditions. Aside from global threats to human health, invasive species risks are addressed as bilateral trade issues. Moreover, it is assumed that the importing country captures all of the benefits of any measures taken. This may be true in some instances, but in general it is not. Not only is the information generated in the process of inspection and interception a public good that can, for example, inform the decisions taken by other importers of the same commodity group, but the interception of pests or pathogens by one country reduces the likelihood that it will infect/re-infect its trading partners with the same pest or pathogen. Because they are public goods, however, inspection (the knowledge it generates) and interception (the protection it offers the wider community) are international public goods, and will be undersupplied if left to individual countries. The existing trade rules allow for only partial internalization of the most invasive species externalities on a bilateral basis.

One solution to this problem would be to add a level of scrutiny to the international movement of commodities and people. This would add inspection and interception services in areas beyond national jurisdiction, the cost being borne by the exporting country. This would be to provide the global public good directly. A second solution would be to invoke the incremental cost principle (that informs the Global Environment Facility's funding decisions) to pay individual countries to bring their inspection and interception effort up to a level that equates the marginal global costs and benefits of the inspection and interception regime. This would be to deliver the global public good through the actions of individual countries.

There is currently a considerable difference between the treatment of infectious human pathogens and all other potentially invasive species. The international community has done much more to protect global human health than to protect the health of all other species, including those on which people most depend. A significant step towards development of mechanisms to address invasive species risks might be achieved by bringing the International Health Regulations and the Sanitary and Phytosanitary Agreement into conformity with one another. As they stand, the two instruments aim to offer very different levels of protection to the global community. The aims and scope of the IHR are "to prevent, protect against, control and provide a public health response to the international spread of disease in ways that are commensurate with and restricted to public health risks, and which avoid unnecessary interference with international traffic and trade" (IHR 2005:

Article 2, see Appendix 2). The aims of the SPS Agreement, by contrast, are to ensure that the defensive measures taken by members to protect their own human, animal, and plant health do not constitute a barrier to trade (SPS Agreement 1995: Article 2, see Appendix 1). The intent of the two documents is very different. The IHR is focused on protecting the global human health. The SPS Agreement is focused on minimizing the risk that national defensive measures do not unnecessarily restrict trade. The SPS Agreement should be strengthened to offer the same focus on global risks to human, animal, and plant health.

A second difference between the IHR and the SPS Agreement lies in the commitment they make to building the capacity to discharge countries' responsibilities in each case. Once again, the IHR embodies a far more constructive approach. It both imposes reporting obligations on countries, and requires that they develop the capacity "to meet develop, strengthen and maintain...the capacity to detect, assess, notify and report events in accordance with these Regulations" within a period of five years (IHR 2005: Article 5(1)). At the same time, however, it imposes an obligation on the World Health Organization to "assist States Parties, upon request, to develop, strengthen and maintain the capacities referred to."

There is nothing equivalent for the SPS Agreement. Article 9(1) of that agreement refers to a commitment "to facilitate the provision of technical assistance to other Members, especially developing country Members, either bilaterally or through the appropriate international organizations." Article 9(2) then states that "where substantial investments are required in order for an exporting developing country Member to fulfill the sanitary or phytosanitary requirements of an importing Member, the latter shall consider providing such technical assistance as will permit the developing country Member to maintain and expand its market access opportunities for the product involved" (Article 9(2)). A commitment to "consider" bilateral assistance, or to "facilitate" multilateral assistance is much weaker than the obligation on the WHO under the IHR.

The essential point here is that the control of species that threaten the wider community is a public good. At the national level, an inspection and interception policy that protects the residents of a country against invasive pathogens provides benefits that are neither rival nor exclusive. If one person benefits from the protection offered by the policy, it neither affects the costs of the policy nor the benefits it offers to others. As with any public good, however, each person has an incentive to free ride on the efforts of others. The same is true at the international level. If one country is protected by an international inspection and interception regime, it has no implications for the benefits offered to other countries. The difficulty with invasive species control is that it is a very particular kind of public good.

The benefits offered by any inspection and interception regime to all members of the community (whether national or international) are only as good as the benefits offered by the least effective provider—the weakest link in the chain. Sandler (1997) accordingly classifies public goods of this sort as "weakest link" public goods. In the national case, for example, if one quarantine facility fails to contain an invasive pathogen, the fact that all others may do so is irrelevant. If one landowner fails to control an invasive plant, it nullifies the control efforts of all others. Internationally, if one country fails to monitor, detect, or contain an infectious disease, it compromises the control efforts of all other countries (Perrings et al. 2002). This is the motivation for buttressing the capacity of countries to comply with the IHR through the support offered by the WHO. We need something analogous for other invasive species.

Perrings et al. (2002) recommended a number of changes to the current regime in order to address this problem. Specifically, they suggested the need for a coordinated international response to the monitoring problem that would build on the experience of the Center for Disease Control (CDC) in monitoring and reporting on human diseases. They proposed an international body with responsibility for invasive species generally, that would be charged with developing and maintaining a database that would include the species-level data provided by the CDC, the OIE, and the IPPC. Aside from monitoring trends and providing risk assessments and recommendations for action, they argued that such an organization should be able to coordinate responses to invasive species threats in poorer countries.

One way such a body might enhance the operation of the current system is to provide low income countries with the information needed to protect themselves under the current SPS Agreement. That agreement, as we have already noted, is primarily designed to limit the damage to trade of actions taken by member states in protection of their own environment. In fact it explicitly states that: "Sanitary and phytosanitary measures shall not be applied in a manner which would constitute a disguised restriction on international trade" (SPS Agreement, 1995: Article 2(3), see Appendix 2). It allows countries to introduce temporary trade restrictions, but also insists that any sanitary or phytosanitary measure be based on "scientific principles" and that it not persist in the absence of "sufficient scientific evidence". The inability to advance sufficient scientific evidence is one reason why the trade restrictions allowed under the SPS Agreement are overwhelmingly implemented by high and middle income countries, and almost never by low income countries. A CDC-equivalent (in addition to the existing Standards and Trade Development Facility) might provide low income countries with the monitoring and scientific support needed to make the current system work.

Beyond this, we argue that the SPS Agreement be brought into conformity with the IHR in ways that explicitly address the global risks associated with invasive species. Since the SPS Agreement can only be renegotiated within the context of the GATT, this implies extension of the protections of the SPS Agreement (a) to cover global threats, (b) to support development of member states' capacity to implement the terms of the agreement, and (c) to implement global monitoring and risk assessment.

16.4 Precautionary action should be targeted, and support learning

A number of chapters in this volume have confirmed the observation that for most harmful invasive pests and pathogens the cost of prevention is significantly less than the cost of cure, suggesting that for such species the best policy is generally exclusion (Leung *et al.* 2002). However, this does not mean that this is the best policy for all potentially invasive species. Economic analysis of the most appropriate mechanisms for internalizing the costs of invasive species (Costello and McAusland 2003; Mcausland and Costello 2004; Costello *et al.* 2007) indicates that conventional measures to deal with international trade-externalities, such as tariffs, are unlikely to be effective for this problem. One reason for this is that although the vast majority of introduced species have relatively little impact on the host system, the ones that do have an impact tend to have a disproportionately large effect. That is, the distribution of losses and gains associated with invasive species may be something like that shown in Fig. 16.1. Most of the damage is done by a few species in the left tail of the distribution. At the same time, there is considerable uncertainty about

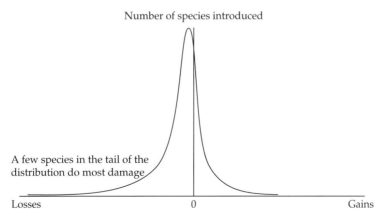

Figure 16.1 Distribution of the net damage costs of invasive species.

which commodity groups and trade routes are most likely to yield species in the left tail. In this case, the optimal strategy for any one country is unlikely to involve use of an instrument that increases the cost of all imports. Instead, it will involve an effort to identify the main sources of risk and to address those directly.

The larger policy question this raises is how to develop and implement the precautionary strategy authorized in the IHR and the SPS Agreement for dealing with the few species that are likely to cause most of the damage. The IHR and the SPS Agreement both permit trade restrictive action to avoid the consequences of the introduction of potentially harmful species despite uncertainty about the risks. Consistent with the precautionary principle they also require investigation to determine the level of risk. The SPS Agreement, for example, states that:

"...members have the right to take sanitary and phytosanitary measures necessary for the protection of human, animal or plant life or health...[but] shall ensure that any sanitary or phytosanitary measure is applied only to the extent necessary to protect human, animal or plant life or health, is based on scientific principles and is not maintained without sufficient scientific evidence."

(Article 2(1,2))

What should trigger a precautionary action? Decision-makers generally treat probabilities close to zero in a particular way. The probability of a very unlikely outcome tends either to be overestimated, or to be set equal to zero—the deviation of the perceived from the actual risk generally depending on the value of the outcome. For example, people asked to estimate the probability of deaths from different causes tend to underestimate deaths from frequent causes, but to overestimate deaths from infrequent causes (Pigeon et al. 1992). McDaniels, Kamlet, and Fischer (1992) described this as a "dread" effect. In the language of decision theory, this implies that people weight the probabilities attaching to different outcomes according to the relative importance of the outcome (Quiggin, 1982). It has also been observed that decision-makers weight probabilities relative to some reference point. Starmer and Sugden (1989) and Tversky and Kahneman (1992), for example, suggested that the weights attaching to the probability of loss tends to be higher than that attaching to gains. The weights may also reflect the confidence that decision-makers have in the probability estimates. In such cases, extreme, unique, rare, and irreversible events with few historical precedents will attract greater weight than they might objectively deserve. In all cases, though, precautionary action tends to be triggered where there is the prospect of significant harm, even if the objective probability of that is very low. Empirically, examples of precautionary action triggered by the prospect of significant harm extend all the way from global human health threats to local animal or plant health threats.

It follows that initiation of precautionary action implies a weaker test than that required by standard scientific proof. Historic examples of such tests include "scientifically based suspicion", "reasonable grounds for concern", and the "balance of evidence" (Harremoës et al. 2001). All are based on data that fail standard tests of scientific proof, but nevertheless provide at least a conditional basis for decisions about the introduction of species. They are "early warning signs" rather than conclusive proof. Precautionary action in response to such early warnings then implies a reversal of the burden of proof. So the decision to exclude commodities of a particular type or from a particular region pending investigation, or the quarantining of introduced species at the port of entry, conditionally assumes that the commodities or introduced species are infected. Subsequent investigation either validates or invalidates that assumption, and the commodities or quarantined species are either admitted or rejected, but the action has protected the host system in the meantime.

It turns out that this approach to the problem is quite consistent with the large and long-standing literature on decision-making in the face of irreversible change. This literature finds that where the consequences of choices are imperfectly understood and there is scope for learning, the information foregone by taking action now is valuable and will generally lead to the action being deferred. The information that can be gained by delay is said to be a "quasi-option value" of the decision (Epstein 1980; Dixit and Pindyck 1994). In the case of invasive species, the quasi-option value of an action such as the quarantining of intercepted species is the value of the information acquired by delaying their introduction to the host system.

Although the "evidence" that triggers the precautionary action authorized under the IHR and the SPS Agreement may be weaker than standard scientific proof, it is still possible to apply scientific principles to its acquisition. Springborn *et al.* in Chapter 10 this volume, show that the optimal strategy for detecting potentially invasive species in imports depends on the degree of uncertainty involved. Where the risks associated with commodity groups, trade routes, vessel types etc is known, the optimal approach will be targeted inspection and interception. Where they are unknown, it will involve random inspection and interception. In both cases the optimal resources committed to detection will depend on the expected value of the damage due to a marginal change in import volumes. In other words, if the probability that a particular commodity group from a particular country will be infected is known to be high, and if the damage of infection is also known to be high, then that commodity group should be subject to targeted inspection. But if the probability that any given commodity group will be infected is unknown, then the group should be inspected randomly at an intensity that reflects the expected damage due to infections. The instrument for internalizing the damage cost in this case is a combination of an inspection/interception charge and insurance against damage, levied by the importing country on the exporting country.

So the trigger for action in the case of species introductions is the prospect of potential harm to the host system: a product of the (weighted) probability of the establishment and spread of the introduced species and the damage caused were that to occur. For known pests and pathogens, where the probability of establishment and spread is close to one and the potential damage is known to be high, exclusion is the appropriate response. For novel species, where neither the likelihood of establishment and spread nor the potential damage caused is known, a precautionary response implies delay in their introduction combined with investigation of their potential interactions with species in the host system. Although the SPS Agreement currently treats the host system as the nation state and limits admissible precautionary action to the protection of national borders, the IHR focuses on global threats to health and mandates action at that level. Bringing the SPS Agreement into conformity with the IHR would enable precautionary action at that level, but also enable investigation of the potential consequences of introductions across a range of systems.

16.5 Many invasive species problems may be most effectively managed at the regional scale

The role of international organizations and agreements is complemented by a network of regional institutions and initiatives that help mediate between the international and national levels. On the one hand, these regional bodies can help identify national priorities and serve as a first step for developing methods to address key threats or species. For example, the IPPC's international standard on solid wood packaging material originated from regional work by the North American Plant Protection Organization. On the other hand, these regional bodies can also take international guidance and priorities and support their development and application at the national level. The Invasives Information Network under the Inter-American Biodiversity Information Network (IABIN I3N) has, for example, helped develop and disseminate not only information tools but also national policy models throughout the Americas and beyond. These regional networks may also become important means for countries to communicate amongst themselves about their experiences and to coordinate on common priorities or threats. Finally, regional bodies can also help build and support capacity, particularly in areas such as island regions, where it is unlikely that single countries will ever be able to develop independent, fully-fledged biosecurity systems. For these and other reasons, coordination at the regional scale may be more feasible than coordination at the global scale.

A major trend in the world trading system has been the proliferation of bilateral and regional trade agreements (RTAs), and especially the proliferation of South-South RTAs (World Bank 2005). There are currently around 230 RTAs in force accounting for around 40 per cent of world trade. This number is expected to grow rapidly. Indeed, as we noted in Chapter 1, more than 420 RTAs had been notified to the WTO up to December 2008 and it is expected that 400 will be implemented by 2010 (http://www.wto.org/english/tratop_e/region_e/region_e.htm).

The development of RTAs is relevant to the problem of invasive species for two main reasons. The first is that RTAs can potentially lead to increased volumes of trade and lower levels of inspection between countries in which bioclimatic conditions are broadly similar, and hence in which the risk that introduced species will establish, naturalize, and spread is high. There is, for example, some evidence that the North American Free Trade Association (NAFTA) has facilitated the spread of species within the free trade area (Perrault *et al*. 2003). Secondly, the promotion of trade in agricultural products between regions in which resources for the detection and control of potentially invasive species are weak must be a concern. Against this is the fact that many RTAs explicitly address environmental issues, in recognition of the fact that different patterns of trade involve different environmental risks. Indeed, cooperation within RTAs may be an important part of the solution to biological invasion externalities and the free rider problems attaching to the control of non-indigenous species. Schiff and Winters (2002) argue that if there are economies of scale or transboundary externalities, regional cooperation can provide the answer.

In many cases, the environmental agreements within RTAs are designed to force compliance with environmental laws. So, for example, the NAFTA Commission for Environmental Cooperation exists to ensure that member states do not seek a trade benefit or attract inward investment by failing to comply with environmental laws. Similarly, the environmental chapter of the US–Singapore Free Trade Agreement requires both countries to enforce their environmental laws, and includes fines for non-compliance (World Bank 2005). A second role of the environmental agreements is to harmonize environmental standards. The Southern Common Market (MERCOSUR), for example, includes an environmental working group charged with eliminating the use of environmental barriers to trade, promoting "upward harmonization" of environmental management systems and securing cooperation on shared ecosystems. Indeed, many of the main South-South RTAs—MERCOSUR, the Andean Pact, the Common Market for Eastern and Southern Africa (COMESA), the Southern African Development Community (SADC), the ASEAN Free Trade Area (AFTA), and the Caribbean Community (CARICOM)—include agreements on standards (World Bank, 2005).

The existence of RTAs makes it possible for invasive species risks to be managed at the level of the group. There are several options. First, prevention and control efforts can focus on the main hubs for moving goods and people into and out of a region. Quarantine and pre-border controls focused on these strategic points can benefit not just that country but the entire region. The entry points into a region with the lowest level of invasive species controls (whether monitoring, inspection, management, or prevention) are also likely to be the most vulnerable to new introductions, and thereby a potential source for further spread of established invasive species across the region. It is, however, possible for regional invasive species control efforts to focus on these entry points. More particularly, regional initiatives and institutions can play a role in training and capacity development, exchanging experiences across countries and regions, and providing regional centers of expertise for risk assessment. In addition, regional initiatives allow countries to identify and collaborate on common priorities, whether that is controlling the spread of an established invasive species, preventing the introduction of particularly harmful species, or promoting regional funding and capacity-building priorities.

Existing examples of regional strategies and actions plans on invasive species include the European Strategy on Invasive Alien Species under the Bern Convention, the regional Guidelines for Invasive Species Management in the Pacific, and the Caribbean Regional Invasive Species Intervention Strategy. A particularly robust example of regional cooperation can be found in the Pacific where a constellation of four different initiatives and organizations is collaborating to support both national and region-wide efforts. The institutions involved include the South Pacific Regional Environment Programme (SPREP), the Secretariat of the Pacific Community (SPC), the Pacific Invasives Initiative (PII), and the Pacific Invasives Learning Network (PILN). SPREP is the Pacific region's major intergovernmental organization charged with protecting and managing the environment. SPREP can convene the region's governments to help coordinate strategies and policies to address the environmental aspects of invasive species and has initiated work around regional strategies and funding programs. SPREP grew out of the SPC and the two bodies maintain close ties and coordination in their activities. The SPC is the Pacific's oldest regional intergovernmental organization and serves as a technical advisory and implementation body to assist with social, land, and marine resource issues. Its work on invasive species focuses on quarantine, agriculture, and aquaculture related issues, including management, training and policy/legislation with some overlap on the environmental side.

The Pacific Invasives Initiative (PII) is a non-governmental organization comprised of a partnership of eight organisations that mounts demonstration projects, from the initial planning stages to implementation to exchange of lessons learned. PII therefore has a very technical orientation on invasive species eradication, control and management, which includes ensuring that skills related to project development and implementation are shared across the Pacific. The Pacific Invasive Learning Network (PILN) is a partnership of several national, regional and international organizations which serves approximately 14 inter-agency and multi-stakeholder teams from Pacific Island countries and territories. PILN's focus is on developing in-country capacity through networking and cross-team collaboration, skill and resource sharing, links to technical expertise, and information exchange. Most other regions also have bodies that serve some of these functions, although the breadth covered by the Pacific is rare.

There are many examples where regional cooperation would have been advantageous. Failure to contain the spread of *Caulerpa taxifolia*, an aquatic weed released/escaped to the Mediterranean from an oceanographic museum in Monaco in the early to mid-1980s, has had widespread repercussions. *C. taxifolia* now touches all the major shores of the eastern Mediterranean and has also moved into the Tyrrhenian and Adriatic Seas (Meinesz *et al.* 2001). Our point is that the existence of RTAs and other regional institutions, while potentially increasing risk, at least makes it possible to secure collaboration in the management of invasive species that pose regional threats.

16.6 Pay special attention to the impacts of invasive species on the poor

Another important dimension of the invasive species problem is that it tends to bear most heavily on people who are least able to take precautionary preventive action or to deal with the consequences. The vulnerability of people to emerging infectious diseases, like the vulnerability of agriculture to livestock and crop diseases, tends to be greatest in low income countries. In the 1990s, Oerke *et al.* (1994) suggested the impact of invasive agricultural pests on yields may be around 50 per cent in the poorest countries. Examples of pests and pathogens that have had particularly severe effects on crop yields in the world's poorest region, Sub-Saharan Africa, include Witchweed (*Striga hermontheca*), Grey Leaf Spot (*Circosporda zeae-maydis*), Cassava Mealybug (*Phenacoccus manihoti*), the Cassava Green Mite (*Mononychellus tanajoa*), and the Large Grain Borer (*Prostephanus truncatus*) (Rangi 2004). The last of these was apparently introduced from south and central America during the 1970s and is now established in east, central, south, and west Africa. Since it primarily affects grain in storage, it has potentially severe consequences for those who rely on the ability to hold stocks of grain as insurance against crop failure (Farrell and Schulten 2002).

In this volume, Perrings, Fenichel, and Kinzig show that vulnerability to the effects of known harmful animal pathogens is greatest amongst the poorest countries. Indeed, there are many examples of the failure of farmers to cope with the effects of invasive species. Borggaard *et al.* (2003), for example, note that Cogon Grass (*Imperata cylindrica*) has invaded shifting cultivation plots in many

South and Southeast Asian countries. However, since shifting cultivators are amongst the poorest members of those societies, they have been unable to undertake the control measures needed to eradicate it (Johnson and Shilling 2003). They have no option but to live with the costs it imposes. Moreover this is true of a large number of invasive pests and pathogens affecting people, livestock, crops, and wildlife.

The problem is all the harder to solve when invasive species control is a public good, and those who undertake control measures are not able to capture the benefits of their actions. There are a few examples of collective actions that have been effective in limiting the damage due to invasive species, but these are exceptional. The South African "Working for Water" campaign, which seeks to control invasive *Pinus*, *Hakea* and *Acacia* species in the fynbos, is the best known example of these (van Wilgen and Richardson Chapter 13 this volume; Turpie and Heydenrych 2000; le Maitre *et al*. 2002). But few others have been as effective in either their environmental or their social goals.

Where invasive species have effectively displaced functionally similar native species, the people who have come to depend on the invasive species will not generally support its eradication. Indeed, it is not always obvious that eradication or control is the optimal strategy in such cases. An example of this is siam weed (*Chromoleana odorata*), introduced into Ghana in the 1960s and spread to approximately 60 per cent of the land area. While it has had major ecological impacts there is little local support for its eradication due to its value as a source of fuel, fiber, building materials, and medicinal products (Rangi 2004). Mesquite (*Prosopis juliflora*) in the semi-arid Nama and karoo biomes of South Africa is similarly highly valued for its capacity to provide a more reliable source of fuel and fiber than many native species in dry conditions (GISP 2004). In Chapter 14 this volume, Shaanker *et al*. make the case for exploitation of *Lantana camara* in India.

In such cases, the marginal damage associated with the further spread of the species in question may be small while the costs of eradication may be very large indeed, making an adaptive response the best option. But this is not true where the marginal damage of further spread is very large—as with many pathogens, for example. In such cases, the limited capacity of poor countries and poor people to control the spread of disease involves costs both to themselves and to others, and requires a proportionate international response. We have already noted the difference between the IHR and the SPS Agreement in this respect. Two issues require urgent international attention.

The first of these concerns the information available to countries to protect themselves under the SPS Agreement. At present, low income countries see the Agreement primarily as one of a number of instruments available to restrict their exports. There is in fact little doubt that the measures taken under the SPS Agreement do provide trade protection. Large numbers of countries are ineligible to supply certain markets with a range of animal products and food crops because of restrictions based on threats to plant and animal health (Sumner 2003). The complaints lodged by developing countries over the use of the SPS Agreement show this to be persistent source of frustration. High income countries are seen as excluding agricultural products from low income countries on the basis of the disease risks they involve, including both animal diseases, such as foot and mouth disease and bovine spongiform encephalopathy, and a range of plant pests and pathogens (Jaffee and Henson 2005).

The other side of the same coin is that low income countries themselves make disproportionately little use of the protections offered by the SPS Agreement, in part because of the difficulty they have in marshalling the scientific evidence required to exclude harmful pests and pathogens. The establishment of a CDC-like mechanism to monitor and evaluate invasive species risks, and to provide countries with the information they need to use the provisions of the Agreement, would be one step in this direction.

A second issue requiring attention concerns the risks posed to the global community by the introduction of novel species, especially emergent zoonotic diseases that have their origins in low latitudes (Jones *et al*. 2008). Part of the problem here lies in the asymmetry between the IHR and the SPS Agreement. Zoonotic diseases that affect both humans and other species are addressed differently depending on which host is being considered. But part of the problem also lies in the capacity of low income countries to undertake the monitoring and surveillance needed to identify problem pests and pathogens before they are introduced to the global system. Bringing the SPS Agreement into conformity with the IHR would address the first issue. The second really does require the resources needed to develop the in-country capacity required to monitor and evaluate the risks. What is needed for this to happen is a resource equivalent to the WHO funds committed to building the capacity to comply with the terms of the IHR, to be committed to the identification of other emerging risks.

16.7 Conclusions

The rapid increase in the number of harmful invasive species in most countries indicates that current attempts to mitigate those risks within national borders—defensive measures for inspection and interception at the port of entry, and detection and eradication of new introductions beyond the port of entry—are not keeping up with the rate of harmful species introductions. While it is certainly possible to enhance the effectiveness of such measures (a number of the contributions to this volume address that issue), it is not sufficient to address the problem. Since the risks facing importers depend on the capacity of exporters to undertake internal sanitary and phytosanitary measures, and since many emergent pests and pathogens originate in areas where that capacity is limited, there is a need to build both capacity and commitment in exporting countries to reduce the risks they pose to others. Bilateral trade agreements can be useful in raising sanitary and phytosanitary standards in exporting countries, and the standards emerging out of the work of the three agencies that support the SPS Agreement—the Codex Alimentarius Commission for food safety, the International Office of Epizootics for animal health, and the International Plant Protection Convention for plant health—provide a benchmark for that. Where achievement of particular sanitary and phytosanitary standards is a condition on exporters, they have an incentive to commit resources to the problem. But more needs to be done. We have observed that while bilateral and regional trade agreements have the potential to improve standards, they necessarily neglect the costs or benefits imposed on third parties. So even if a bilateral agreement internalizes the external effects of trade within the two countries involved, it ignores the effects imposed on third parties. Given the "small world" nature of the global trade network, and the fact that many introductions involve re-infection from third parties, this is not acceptable.

Our central conclusions follow from this. Bringing the SPS Agreement into conformity with the IHR is necessary to recognize and respond to the global nature of many trade-related invasive species risks. While some invasive species issues are highly localized in their effects, and therefore best dealt with at a local level, many others are not. This is particularly true of infectious diseases spread through international trade, but the risks associated with many insects or mites distributed through the trade system have some of the same characteristics. There are two main elements involved in bringing the SPS Agreement into conformity with the IHR. One is the focus on global as distinct from country-level risk. The other is the global support that is offered to individual countries to comply with the terms of the regulations. The establishment of a mechanism to monitor, evaluate and disseminate information on emerging invasive species risks is one part of this. The establishment of a fund to replicate the role of the World Health Organization in building local capacity and in coordinating and implementing rapid responses is another.

Currently, international investment in the management of invasive species is dominated, on the one hand, by coordinated actions in response to particular threats such as SARS, AIDS, or avian flu and, on the other, by bilateral or multilateral conservation and development projects that include invasive species control (Perrings 2007). Lending for invasive species control remains a very small proportion of World Bank lending for environmental and natural resource management (ENRM) projects, and ENRM lending has also fallen substantially, both in absolute terms and as a percentage of total lending. While there is no easy way to extract the invasive species element within ENRM lending, it is safe to say that it too has been declining.

At the same time there are a number of regional initiatives to invest in invasive species control. In Sub-Saharan Africa, for example, the Programme Area on Prevention, Control and Management of Invasive Alien Species of the New Partnership for Africa's Development (NEPAD) generates funds "to minimize the impact of invasive alien species on the African continent's people, economies and ecological systems" using a number of regional trade agreements as the implementing agencies. These include the East African Community, the Southern Africa Development Community, and the Common Market for Eastern and Southern Africa. The plan focuses on improvement in the capacity to undertake risk assessments, awareness raising and information provision, as well as development of the institutional capacity to manage invasive species (UNEP, 2003). While it is worth supporting such regional initiatives—indeed, developing the potential of the Regional Trade Agreements is an essential component of any international invasive species management strategy—it does not take away from the importance of developing a global capacity to deal with those species that pose a global risk.

The two most urgent aspects of the invasive species problem that need to be addressed both relate to the public good nature of the problem. The fact that the control of many pests and pathogens is a "weakest link" public good requires measures to raise the capacity of the countries least able to monitor, evaluate, and manage invasive species risks. This is what gives value to the development of a mechanism to build capacity in low income countries, and to finance control efforts in those countries. The fact that global monitoring, assessment, evaluation, and dissemination is a "best shot" public good supports the notion that wealthier countries should invest in the development of a CDC-like mechanism to generate the information needed by all countries. While this should build on the monitoring undertaken in specific areas by existing bodies, its primary value would be in the integration of different kinds of data.

The present reality is that countries individually incur very large costs in terms of output lost to pests and pathogens, compromised human, animal, and plant health, impacts on ecological functioning and ecosystem services, along with the costs of inspection, interception, eradication, control, and other sanitary and phytosanitary efforts. But while individual country expenditure on invasive species is large—larger than that on almost any other environmental problem—it is significantly less than it should be. Because countries are authorized by the existing institutional and legal structure to neglect many of the external costs of their actions, they do not take these into account in developing their own invasive species management strategies. What the chapters in this volume show is that, even leaving aside human pathogens, the consequences of neglecting impacts of invasive species on the global public good are potentially extremely serious. At the same time, however, they suggest that the options for addressing the problem are within reach. While there are no easy routes to international cooperation and coordination, the historical record in this area suggests that the prospect of extreme losses offer a powerful incentive. A succession of invasive human, animal, and plant pathogens over the past 600 years has shown us just how bad the effects can be. There can be no greater incentive to develop a cooperative international strategy for the mitigation of these risks.

References

Borggaard, O.K., Gafur, A., and Peterson, L. (2003). Sustainability appraisal of shifting cultivation. *Ambio.* **32**, 118–23.

Burgiel, S., Foote, G. Orellana M., and Perrault, A. (2006). *Invasive Alien Species and Trade: Integrating Prevention Measures and International Trade Rules*. Center for International Environmental Law, Defenders of Wildlife and TNC, Washingon D.C.

Convention on Biological Diversity. (2008). In-depth Review of Ongoing Work on Alien Species that Threaten Ecosystems, Habitats or Species: Addendum—Preliminary Report of Expert Workshop on Best Practices for Pre-import Screening of Live Animals in International Trade (9–11 April 2008, South Bend, Indiana, USA) UNEP/CBD/COP/9/INF/31/Add.1 (5 May 2008).

Costello, C. and McAusland, C. (2003). Protectionism, Trade and Measures of Damage from Exotic Species Introduction. *American Journal of Agricultural Economics* **85**(4), 964–975.

Costello, C., Springborn, M., McAusland, C., and Solow, A. (2007). Unintended biological invasions: Does risk vary by trading partner? *Journal of Environmental Economics and Management* **54**, 262–76.

Dalmazzone, S. (2000). Economic factors affecting vulnerability to biological invasions. In C. Perrings, M. Williamson, and S. Dalmazzone (eds). *The Economics of Biological Invasions*, pp. 17–30. Edward Elgar, Cheltenham.

Dixit, A.K. and Pindyck, R.S. (1994). *Investment Under Uncertainty*, Princeton University Press, Princeton.

Epstein, L.G. (1980). Decision making and the temporal resolution of uncertainty. *International Economic Review* **21**, 269–83.

Farrell, G. and Schulten, G.M.M. (2002). Large grain borer in Africa: a history of efforts to limit its impact. *Integrated Pest Management Review* **7**, 67–84.

Global Invasive Species Programme (GISP). (2004). *Africa invaded: the growing danger of invasive alien species*, GISP, Cape Town.

Harremoës, P., Gee, D., MacGarvin, M., Stirling, A., Keys, J., Wynne, B., and Guedes Vaz S., eds (2001). *Late Lessons from Early Warnings: The Precautionary Principle*, Environment Issue Report No 22, European Environment Agency, Copenhagen.

IHR (2005) See World Health Organization (2005).

Jaffee, S.M. and Henson, S. (2005). Agro–food exports from developing countries: the challenges posed by standards. In M.A. Aksoy and J.C. Beghin, eds. *Global Agricultural Trade and Developing Countries*, pp. 91–114. World Bank, Washington, D.C.

Johnson, E.R.R.L. and Shilling, D.G. (2003). *Cogon grass: Imperata cylindrica (L.) Palisot*. Plant Conservation Alliance Alien Working Group, National Park Service, Washington D.C.

Jones, K.E., Patel, N.G., Levy, M.A., Storeygard, A., Balk, D., Gittleman, J.L., and Daszak P. (2008). Global trends in emerging infectious diseases. *Nature* **451**(21), 990–94.

Le Maitre, D.C., van Wilgen, B.W., Gelderblom, C.M., Bailey, C., Chapman, R.A., and Nel, J.A. (2002). Invasive alien trees and water resources in South Africa: case studies of the costs and benefits of management. *Forest Ecology and Management* **160**, 143–59.

Leung, B., Lodge, D.M., Finnoff, D., Shogren, J.F., Lewis, M.A., and Lamberti, G. (2002). An ounce of prevention or a pound of cure: bioeconomic risk analysis of invasive species. *Proceedings of the Royal Society of London, Biological Sciences* **269**(1508), 2407–13.

McAusland, C. and Costello, C. (2004). Avoiding invasives: trade related policies for controlling unintentional exotic species introductions. *Journal of Environmental Economics and Management* **48**, 954–77.

McDaniels, T., Kamlet, M., and Fischer, G. (1992). Risk perception and the value of safety. *Risk Analysis* **12**(4), 495–503.

Meinesz, A., Belsher, T., Thibaut, T., Antolic, B., Mustapha, K.B., Boudouresque, C-F., Chiaverini, D., Cinelli, F., Cottalorda, J-M., Djellouli, A., El Abed, A., Orestano, C., Grau, A.M., Ivesa, L., Jaklin, A., Langar, H., Massuti-Pascual, E., Peirano, A., Tunesi, L., de Vaugelas, J., Zavodnik, N., and Zuljevic, A. (2001). The Introduced Green Alga Caulerpa Taxifolia Continues to Spread in the Mediterranean. *Biological Invasions* **3**(2), 201–210.

Oerke, E.-C., Dehne, H.-W., Schönbeck, F., and Weber, A. (1994). *Crop Production and Crop Protection: Estimated Losses in Major Food and Cash Crops*, Elsevier, Amsterdam.

Perrault A., Bennett, M., Burgiel, S., Delach A., and Muffett, C. (2003). *Invasive Species, Agriculture and Trade: Case Studies from the NAFTA Context*, North American Commission for Environmental Cooperation, Montreal.

Perrings, C., Williamson, M., Barbier, E.B., Delfino, D., Dalmazzone, S., Shogren, J., Simmons, P., and Watkinson, A. (2002). Biological invasion risks and the public good: an economic perspective. *Conservation Ecology* **6**(1), 1. [online] URL: http://www.consecol.org/vol6/iss1/art1

Perrings, C. (2007). Pests, pathogens and poverty: biological invasions and agricultural dependence. In A. Kontoleon, U. Pascual, and T. Swanson, (eds) *Biodiversity Economics: Principles, Methods and Applications*, pp. 133–65. Cambridge University Press, Cambridge.

Pigeon, N., Hood, C., Jones, D., Turner, B., and Gibson, R. (1992). Risk perception. In *Risk: Analysis, Perceptions and Management. Report of a Royal Society Study Group*, London, The Royal Society, London.

Quiggin, J. (1982). A theory of anticipated utility. *Journal of Economic Behaviour and Organisation* **3**(4), 323–43.

Rangi, D.K. (2004). *Invasive alien species: agriculture and development, Proceedings of a global synthesis workshop on b,iodiversity loss and species extinctions: managing risk in a changing world*, UNEP, Nairobi.

Ruiz, G.M. and Carlton, J.T. (2003). Invasion vectors: a conceptual framework for management. In G.M. Ruiz and J.T. Carlton, eds. *Invasive Species: Vectors and Management Strategies*, pp. 461–2. Island Press, Washington D.C.

Sandler, T. (1997). *Global Challenges*, Cambridge, Cambridge University Press.

Schiff, M. and Winters, L.A. (2003). *Regional Integration and Development*, World Bank, Washington, D.C.

Starmer, C. and Sugden, R. (1989). Violations of the independence axiom in common ratio problems: an experimental test of some competing hypotheses. *Annals of Operational Research* **19**, 79–102.

Sumner, D.A., ed. (2003). *Exotic Pests and Diseases: Biology and Economics for Biosecurity*, Iowa State Press, Ames, Iowa.

Tatem, A.J., Hay, S.I., and Rogers, D.J. (2006). Global traffic and disease vector dispersal. *Proceedings of the National Academy of Sciences* **103**(16), 6242–47.

Turpie, J.K. and Heydenrych, B.J. (2000). Economic consequences of alien infestation of the Cape Floral Kingdom's Fynbos vegetation. In C. Perrings, M. Williamson, and S. Dalmazzone, eds. *The Economics of Biological Invasions*, pp. 152–82. Edward Elgar, Cheltenham.

Tversky, A. and Kahneman, D. (1992). Advances in prospect theory: cumulative representation of uncertainty. *Journal of Risk and Uncertainty* **5**(4), 297–323.

United Nations Environment Programme (UNEP). (2003). *Action Plan of the Environment Initiative of the New Partnership for Africa's Development* (NEPAD), UNEP, Nairobi.

Convention on Biological Diversity (1993). *The Convention on Biological Diversity*, United Nations, New York.

Vila, M. and Pujadas, J. (2001). Land use and socio-economic correlates of plant invasions in European and North African countries. *Biological Conservation* **100**: 397–401.

World Bank. (2005). *Global Economic Prospects*, World Bank, Washington D.C.

World Health Organization (2005). *International Health Regulations* (*2005*), WHO Press, Geneva.

APPENDIX 1

Agreement on the Application of Sanitary and Phytosanitary Measures (1995), Articles 1–11

Article 1: General provisions

1. This Agreement applies to all sanitary and phytosanitary measures which may, directly or indirectly, affect international trade. Such measures shall be developed and applied in accordance with the provisions of this Agreement.
2. For the purposes of this Agreement, the definitions provided in Annex A shall apply.
3. The annexes are an integral part of this Agreement.
4. Nothing in this Agreement shall affect the rights of Members under the Agreement on Technical Barriers to Trade with respect to measures not within the scope of this Agreement.

Article 2: Basic rights and obligations

1. Members have the right to take sanitary and phytosanitary measures necessary for the protection of human, animal or plant life or health, provided that such measures are not inconsistent with the provisions of this Agreement.
2. Members shall ensure that any sanitary or phytosanitary measure is applied only to the extent necessary to protect human, animal or plant life or health, is based on scientific principles and is not maintained without sufficient scientific evidence, except as provided for in paragraph 7 of Article 5.
3. Members shall ensure that their sanitary and phytosanitary measures do not arbitrarily or unjustifiably discriminate between Members where identical or similar conditions prevail, including between their own territory and that of other Members. Sanitary and phytosanitary measures shall not be applied in a manner which would constitute a disguised restriction on international trade.
4. Sanitary or phytosanitary measures which conform to the relevant provisions of this Agreement shall be presumed to be in accordance with the obligations of the Members under the provisions of GATT 1994 which relate to the use of sanitary or phytosanitary measures, in particular the provisions of Article XX(b).

Article 3: Harmonization

1. To harmonize sanitary and phytosanitary measures on as wide a basis as possible, Members shall base their sanitary or phytosanitary measures on international standards, guidelines or recommendations, where they exist, except as otherwise provided for in this Agreement, and in particular in paragraph 3.
2. Sanitary or phytosanitary measures which conform to international standards, guidelines or recommendations shall be deemed to be necessary to protect human, animal or plant life or health, and presumed to be consistent with the relevant provisions of this Agreement and of GATT 1994.
3. Members may introduce or maintain sanitary or phytosanitary measures which result

in a higher level of sanitary or phytosanitary protection than would be achieved by measures based on the relevant international standards, guidelines or recommendations, if there is a scientific justification, or as a consequence of the level of sanitary or phytosanitary protection a Member determines to be appropriate in accordance with the relevant provisions of paragraphs 1 through 8 of Article 5 (2) Notwithstanding the above, all measures which result in a level of sanitary or phytosanitary protection different from that which would be achieved by measures based on international standards, guidelines or recommendations shall not be inconsistent with any other provision of this Agreement.
4. Members shall play a full part, within the limits of their resources, in the relevant international organizations and their subsidiary bodies, in particular the Codex Alimentarius Commission, the International Office of Epizootics, and the international and regional organizations operating within the framework of the International Plant Protection Convention, to promote within these organizations the development and periodic review of standards, guidelines and recommendations with respect to all aspects of sanitary and phytosanitary measures.
5. The Committee on Sanitary and Phytosanitary Measures provided for in paragraphs 1 and 4 of Article 12 (referred to in this Agreement as the "Committee") shall develop a procedure to monitor the process of international harmonization and coordinate efforts in this regard with the relevant international organizations.

Article 4: Equivalence

1. Members shall accept the sanitary or phytosanitary measures of other Members as equivalent, even if these measures differ from their own or from those used by other Members trading in the same product, if the exporting Member objectively demonstrates to the importing Member that its measures achieve the importing Member's appropriate level of sanitary or phytosanitary protection. For this purpose, reasonable access shall be given, upon request, to the importing Member for inspection, testing and other relevant procedures.
2. Members shall, upon request, enter into consultations with the aim of achieving bilateral and multilateral agreements on recognition of the equivalence of specified sanitary or phytosanitary measures.

Article 5: Assessment of risk and determination of the appropriate level of sanitary or phytosanitary protection

1. Members shall ensure that their sanitary or phytosanitary measures are based on an assessment, as appropriate to the circumstances, of the risks to human, animal or plant life or health, taking into account risk assessment techniques developed by the relevant international organizations.
2. In the assessment of risks, Members shall take into account available scientific evidence; relevant processes and production methods; relevant inspection, sampling and testing methods; prevalence of specific diseases or pests; existence of pest- or disease-free areas; relevant ecological and environmental conditions; and quarantine or other treatment.
3. In assessing the risk to animal or plant life or health and determining the measure to be applied for achieving the appropriate level of sanitary or phytosanitary protection from such risk, Members shall take into account as relevant economic factors: the potential damage in terms of loss of production or sales in the event of the entry, establishment or spread of a pest or disease; the costs of control or eradication in the territory of the importing Member; and the relative cost-effectiveness of alternative approaches to limiting risks.

4. Members should, when determining the appropriate level of sanitary or phytosanitary protection, take into account the objective of minimizing negative trade effects.
5. With the objective of achieving consistency in the application of the concept of appropriate level of sanitary or phytosanitary protection against risks to human life or health, or to animal and plant life or health, each Member shall avoid arbitrary or unjustifiable distinctions in the levels it considers to be appropriate in different situations, if such distinctions result in discrimination or a disguised restriction on international trade. Members shall cooperate in the Committee, in accordance with paragraphs 1, 2 and 3 of Article 12, to develop guidelines to further the practical implementation of this provision. In developing the guidelines, the Committee shall take into account all relevant factors, including the exceptional character of human health risks to which people voluntarily expose themselves.
6. Without prejudice to paragraph 2 of Article 3, when establishing or maintaining sanitary or phytosanitary measures to achieve the appropriate level of sanitary or phytosanitary protection, Members shall ensure that such measures are not more trade-restrictive than required to achieve their appropriate level of sanitary or phytosanitary protection, taking into account technical and economic feasibility.(3)
7. In cases where relevant scientific evidence is insufficient, a Member may provisionally adopt sanitary or phytosanitary measures on the basis of available pertinent information, including that from the relevant international organizations as well as from sanitary or phytosanitary measures applied by other Members. In such circumstances, Members shall seek to obtain the additional information necessary for a more objective assessment of risk and review the sanitary or phytosanitary measure accordingly within a reasonable period of time.
8. When a Member has reason to believe that a specific sanitary or phytosanitary measure introduced or maintained by another Member is constraining, or has the potential to constrain, its exports and the measure is not based on the relevant international standards, guidelines or recommendations, or such standards, guidelines or recommendations do not exist, an explanation of the reasons for such sanitary or phytosanitary measure may be requested and shall be provided by the Member maintaining the measure.

Article 6: Adaptation to regional conditions, including pest- or disease-free areas and areas of low pest or disease prevalence

1. Members shall ensure that their sanitary or phytosanitary measures are adapted to the sanitary or phytosanitary characteristics of the area—whether all of a country, part of a country, or all or parts of several countries—from which the product originated and to which the product is destined. In assessing the sanitary or phytosanitary characteristics of a region, Members shall take into account, inter alia, the level of prevalence of specific diseases or pests, the existence of eradication or control programmes, and appropriate criteria or guidelines which may be developed by the relevant international organizations.
2. Members shall, in particular, recognize the concepts of pest- or disease-free areas and areas of low pest or disease prevalence. Determination of such areas shall be based on factors such as geography, ecosystems, epidemiological surveillance, and the effectiveness of sanitary or phytosanitary controls.
3. Exporting Members claiming that areas within their territories are pest- or disease-free areas or areas of low pest or disease prevalence shall provide the necessary evidence thereof in order to objectively demonstrate to the importing Member that

such areas are, and are likely to remain, pest- or disease-free areas or areas of low pest or disease prevalence, respectively. For this purpose, reasonable access shall be given, upon request, to the importing Member for inspection, testing and other relevant procedures.

Article 7: Transparency

Members shall notify changes in their sanitary or phytosanitary measures and shall provide information on their sanitary or phytosanitary measures in accordance with the provisions of Annex B.

Article 8: Control, inspection and approval procedures

Members shall observe the provisions of Annex C in the operation of control, inspection and approval procedures, including national systems for approving the use of additives or for establishing tolerances for contaminants in foods, beverages or feedstuffs, and otherwise ensure that their procedures are not inconsistent with the provisions of this Agreement.

Article 9: Technical assistance

1. Members agree to facilitate the provision of technical assistance to other Members, especially developing country Members, either bilaterally or through the appropriate international organizations. Such assistance may be, inter alia, in the areas of processing technologies, research and infrastructure, including in the establishment of national regulatory bodies, and may take the form of advice, credits, donations and grants, including for the purpose of seeking technical expertise, training and equipment to allow such countries to adjust to, and comply with, sanitary or phytosanitary measures necessary to achieve the appropriate level of sanitary or phytosanitary protection in their export markets.
2. Where substantial investments are required in order for an exporting developing country Member to fulfil the sanitary or phytosanitary requirements of an importing Member, the latter shall consider providing such technical assistance as will permit the developing country Member to maintain and expand its market access opportunities for the product involved.

Article 10: Special and differential treatment

1. In the preparation and application of sanitary or phytosanitary measures, Members shall take account of the special needs of developing country Members, and in particular of the least-developed country Members.
2. Where the appropriate level of sanitary or phytosanitary protection allows scope for the phased introduction of new sanitary or phytosanitary measures, longer time-frames for compliance should be accorded on products of interest to developing country Members so as to maintain opportunities for their exports.
3. With a view to ensuring that developing country Members are able to comply with the provisions of this Agreement, the Committee is enabled to grant to such countries, upon request, specified, time-limited exceptions in whole or in part from obligations under this Agreement, taking into account their financial, trade and development needs.
4. Members should encourage and facilitate the active participation of developing country Members in the relevant international organizations.

Article 11: Consultations and dispute settlement

1. The provisions of Articles XXII and XXIII of GATT 1994 as elaborated and applied

by the Dispute Settlement Understanding shall apply to consultations and the settlement of disputes under this Agreement, except as otherwise specifically provided herein.
2. In a dispute under this Agreement involving scientific or technical issues, a panel should seek advice from experts chosen by the panel in consultation with the parties to the dispute. To this end, the panel may, when it deems it appropriate, establish an advisory technical experts group, or consult the relevant international organizations, at the request of either party to the dispute or on its own initiative.
3. Nothing in this Agreement shall impair the rights of Members under other international agreements, including the right to resort to the good offices or dispute settlement mechanisms of other international organizations or established under any international agreement.

APPENDIX 2

International Health Regulations (2005) Articles 2, 5–13

Article 2: Purpose and scope

The purpose and scope of these Regulations are to prevent, protect against, control and provide a public health response to the international spread of disease in ways that are commensurate with and restricted to public health risks, and which avoid unnecessary interference with international traffic and trade.

Article 5: Surveillance

1. Each State Party shall develop, strengthen and maintain, as soon as possible but no later than five years from the entry into force of these Regulations for that State Party, the capacity to detect, assess, notify and report events in accordance with these Regulations, as specified in Annex 1.
2. Following the assessment referred to in paragraph 2, Part A of Annex 1, a State Party may report to WHO on the basis of a justified need and an implementation plan and, in so doing, obtain an extension of two years in which to fulfil the obligation in paragraph 1 of this Article. In exceptional circumstances, and supported by a new implementation plan, the State Party may request a further extension not exceeding two years from the Director-General, who shall make the decision, taking into account the technical advice of the Committee established under Article 50 (hereinafter the "Review Committee"). After the period mentioned in paragraph 1 of this Article, the State Party that has obtained an extension shall report annually to WHO on progress made towards the full implementation.
3. WHO shall assist States Parties, upon request, to develop, strengthen and maintain the capacities referred to in paragraph 1 of this Article.
4. WHO shall collect information regarding events through its surveillance activities and assess their potential to cause international disease spread and possible interference with international traffic. Information received by WHO under this paragraph shall be handled in accordance with Articles 11 and 45 where appropriate.

Article 6: Notification

1. Each State Party shall assess events occurring within its territory by using the decision instrument in Annex 2. Each State Party shall notify WHO, by the most efficient means of communication available, by way of the National IHR Focal Point, and within 24 hours of assessment of public health information, of all events which may constitute a public health emergency of international concern within its territory in accordance with the decision instrument, as well as any health measure implemented in response to those events. If the notification received by WHO involves the competency of the International Atomic Energy Agency (IAEA), WHO shall immediately notify the IAEA.
2. Following a notification, a State Party shall continue to communicate to WHO timely, accurate and sufficiently detailed public health information available to it on the

notified event, where possible including case definitions, laboratory results, source and type of the risk, number of cases and deaths, conditions affecting the spread of the disease and the health measures employed; and report, when necessary, the difficulties faced and support needed in responding to the potential public health emergency of international concern.

Article 7: Information-sharing during unexpected or unusual public health events

If a State Party has evidence of an unexpected or unusual public health event within its territory, irrespective of origin or source, which may constitute a public health emergency of international concern, it shall provide to WHO all relevant public health information. In such a case, the provisions of Article 6 shall apply in full.

Article 8: Consultation

In the case of events occurring within its territory not requiring notification as provided in Article 6, in particular those events for which there is insufficient information available to complete the decision instrument, a State Party may nevertheless keep WHO advised thereof through the National IHR Focal Point and consult with WHO on appropriate health measures. Such communications shall be treated in accordance with paragraphs 2 to 4 of Article 11. The State Party in whose territory the event has occurred may request WHO assistance to assess any epidemiological evidence obtained by that State Party.

Article 9: Other reports

1. WHO may take into account reports from sources other than notifications or consultations and shall assess these reports according to established epidemiological principles and then communicate information on the event to the State Party in whose territory the event is allegedly occurring. Before taking any action based on such reports, WHO shall consult with and attempt to obtain verification from the State Party in whose territory the event is allegedly occurring in accordance with the procedure set forth in Article 10. To this end, WHO shall make the information received available to the States Parties and only where it is duly justified may WHO maintain the confidentiality of the source. This information will be used in accordance with the procedure set forth in Article 11.
2. States Parties shall, as far as practicable, inform WHO within 24 hours of receipt of evidence of a public health risk identified outside their territory that may cause international disease spread, as manifested by exported or imported:
   (a) human cases;
   (b) vectors which carry infection or contamination; or
   (c) goods that are contaminated.

Article 10: Verification

1. WHO shall request, in accordance with Article 9, verification from a State Party of reports from sources other than notifications or consultations of events which may constitute a public health emergency of international concern allegedly occurring in the State's territory. In such cases, WHO shall inform the State Party concerned regarding the reports it is seeking to verify.
2. Pursuant to the foregoing paragraph and to Article 9, each State Party, when requested by WHO, shall verify and provide:
   (a) within 24 hours, an initial reply to, or acknowledgement of, the request from WHO;
   (b) within 24 hours, available public health information on the status of events referred to in WHO's request; and
   (c) information to WHO in the context of an assessment under Article 6, including relevant information as described in that Article.

3. When WHO receives information of an event that may constitute a public health emergency of international concern, it shall offer to collaborate with the State Party concerned in assessing the potential for international disease spread, possible interference with international traffic and the adequacy of control measures. Such activities may include collaboration with other standard-setting organizations and the offer to mobilize international assistance in order to support the national authorities in conducting and coordinating on-site assessments. When requested by the State Party, WHO shall provide information supporting such an offer.
4. If the State Party does not accept the offer of collaboration, WHO may, when justified by the magnitude of the public health risk, share with other States Parties the information available to it, whilst encouraging the State Party to accept the offer of collaboration by WHO, taking into account the views of the State Party concerned.

Article 11: Provision of information by WHO

1. Subject to paragraph 2 of this Article, WHO shall send to all States Parties and, as appropriate, to relevant intergovernmental organizations, as soon as possible and by the most efficient means available, in confidence, such public health information which it has received under Articles 5 to 10 inclusive and which is necessary to enable States Parties to respond to a public health risk. WHO should communicate information to other States Parties that might help them in preventing the occurrence of similar incidents.
2. WHO shall use information received under Articles 6 and 8 and paragraph 2 of Article 9 for verification, assessment and assistance purposes under these Regulations and, unless otherwise agreed with the States Parties referred to in those provisions, shall not make this information generally available to other States Parties, until such time as:

   (a) the event is determined to constitute a public health emergency of international concern in accordance with Article 12; or
   (b) information evidencing the international spread of the infection or contamination has been confirmed by WHO in accordance with established epidemiological principles; or
   (c) there is evidence that:
      (i) control measures against the international spread are unlikely to succeed because of the nature of the contamination, disease agent, vector or reservoir; or
      (ii) the State Party lacks sufficient operational capacity to carry out necessary measures to prevent further spread of disease; or
   (d) the nature and scope of the international movement of travellers, baggage, cargo, containers, conveyances, goods or postal parcels that may be affected by the infection or contamination requires the immediate application of international control measures.
3. WHO shall consult with the State Party in whose territory the event is occurring as to its intent to make information available under this Article.
4. When information received by WHO under paragraph 2 of this Article is made available to States Parties in accordance with these Regulations, WHO may also make it available to the public if other information about the same event has already become publicly available and there is a need for the dissemination of authoritative and independent information.

Article 12: Determination of a public health emergency of international concern

1. The Director-General shall determine, on the basis of the information received, in particular from the State Party within whose territory an event is occurring, whether an event constitutes a public health emergency

of international concern in accordance with the criteria and the procedure set out in these Regulations.

2. If the Director-General considers, based on an assessment under these Regulations, that a public health emergency of international concern is occurring, the Director-General shall consult with the State Party in whose territory the event arises regarding this preliminary determination. If the Director-General and the State Party are in agreement regarding this determination, the Director-General shall, in accordance with the procedure set forth in Article 49, seek the views of the Committee established under Article 48 (hereinafter the "Emergency Committee") on appropriate temporary recommendations.

3. If, following the consultation in paragraph 2 above, the Director-General and the State Party in whose territory the event arises do not come to a consensus within 48 hours on whether the event constitutes a public health emergency of international concern, a determination shall be made in accordance with the procedure set forth in Article 49.

4. In determining whether an event constitutes a public health emergency of international concern, the Director-General shall consider:
   (a) information provided by the State Party;
   (b) the decision instrument contained in Annex 2;
   (c) the advice of the Emergency Committee;
   (d) scientific principles as well as the available scientific evidence and other relevant information; and
   (e) an assessment of the risk to human health, of the risk of international spread of disease and of the risk of interference with international traffic.

5. If the Director-General, following consultations with the State Party within whose territory the public health emergency of international concern has occurred, considers that a public health emergency of international concern has ended, the Director-General shall take a decision in accordance with the procedure set out in Article 49.

Article 13: Public health response

1. Each State Party shall develop, strengthen and maintain, as soon as possible but no later than five years from the entry into force of these Regulations for that State Party, the capacity to respond promptly and effectively to public health risks and public health emergencies of international concern as set out in Annex 1. WHO shall publish, in consultation with Member States, guidelines to support States Parties in the development of public health response capacities.

2. Following the assessment referred to in paragraph 2, Part A of Annex 1, a State Party may report to WHO on the basis of a justified need and an implementation plan and, in so doing, obtain an extension of two years in which to fulfil the obligation in paragraph 1 of this Article. In exceptional circumstances and supported by a new implementation plan, the State Party may request a further extension not exceeding two years from the Director-General, who shall make the decision, taking into account the technical advice of the Review Committee. After the period mentioned in paragraph 1 of this Article, the State Party that has obtained an extension shall report annually to WHO on progress made towards the full implementation.

3. At the request of a State Party, WHO shall collaborate in the response to public health risks and other events by providing technical guidance and assistance and by assessing the effectiveness of the control measures in place, including the mobilization of international teams of experts for on-site assistance, when necessary.

4. If WHO, in consultation with the States Parties concerned as provided in Article 12, determines that a public health emergency of international concern is occurring, it may offer, in addition to the support indicated in paragraph 3 of this Article, further assistance to the State Party, including an assessment

of the severity of the international risk and the adequacy of control measures. Such collaboration may include the offer to mobilize international assistance in order to support the national authorities in conducting and coordinating on-site assessments. When requested by the State Party, WHO shall provide information supporting such an offer.

5. When requested by WHO, States Parties should provide, to the extent possible, support to WHO-coordinated response activities.
6. When requested, WHO shall provide appropriate guidance and assistance to other States Parties affected or threatened by the public health emergency of international concern.

Index

A

Abronia umbellata (Verbena) 59
Acacia 164, 167, 191–3, 197, 246
  cyclops 149–50
  longifolia (Golden Wattle) 192
  mearnsii (Black Wattle) 166, 192
  paradoxa 196
  saligna 190
Acridotheres tristis (Common Myna) 186
Aedes albopictus 8
aesthetic value of landscape 172–3
Africanized honeybee 58, 167
Agreement on the Application of Sanitary and Phytosanitary Measures *see* Sanitary and Phytosanitary (SPS) Agreement
agricultural GDP per capita 52
Agricultural Quarantine Inspection Monitoring (AQIM), US 128, 143
agricultural systems 2
  impact of invasive species 8–9, 163–6
Agrilus planipennis (Emerald Ash Borer) 228, 238
air quality regulation 171
alien species 19–20, 161
  habitat studies 67
  mechanisms of impact 162–3
  native–alien relationship 68–9
  quantification of impacts 161–2
  representation of 67
  scale importance 68
  species pool 66
  see also invasive species
Angiostrongylus cantonensis (Rat Lungworm) 168
animal pathogens 2, 43
  inspection and interception model 44–54
  introduced disease growth rates 46–8
  List A pathogens 43, 47–53
  List B pathogens 47–51
Anopheles gambiae 8
anthropogenic disturbance 4–5, 20
  habitat invasion and 71, 72
Aonidiella aurantii (Red Scale) 186
Apis mellifera (Honeybee) 58, 167, 187
Aplexa marmorata 187
archaeophytes 62, 66
  habitat invasions 68, 69, 74–5
  spread 62–3
Argentine Ant 186, 190
Arundo donax 171
ASEAN Free Trade Area (AFTA) 8, 244
Asian Longhorned Beetle 238
Assessing large-scale risks for biodiversity with tested methods (ALARM) 75
Australia 3, 5, 8, 11, 42, 58, 129, 163–4, 166, 168, 170–2, 184, 190–1, 196, 203, 208, 220, 223, 225–6, 230, 232, 236
Australian Pine 168
Australian Weed Risk Assessment (WRA) 225–6, 230
Avena fatua 163
avian influenza virus 8
avian malaria 170
Azolla filiculoides (Red Water Fern) 187, 191

B

ballast water 228
Batrachochytrium dendrobatidis (amphibian chytrid) 5, 26
beavers 166, 171
Bighead Carp 226
BIOCLIM model 31
biodiversity
  invasive species impact on 202
  South Africa 189–90
  resilience and 5–6, 42
biological control 87
Black Carp 226
Black Wattle 192
Blue Gum Eucalyptus 171
bounding box 37
bovine spongiform encephalopathy (BSE) 9
Bromus tectorum 170–1
Brown Tree Snake 100
Brush-tailed Possum 164, 168–70
Bush crickets 58

C

Cactoblastis cactorum 208
Campuloclinium macrocephalum (Pompom Weed) 189, 196
Canada 172, 226, 231
Cane toad 5, 58
carbon dioxide
  concentrations 37
  storage capacity 167
Carex kobomugi 206
Caribbean Community (CARICOM) 244
Cassava Green Mite 245
Cassava Mealybug 245
Castor canadensis (North American Beaver) 166, 171
Casuarinas equisetifolia (Australian Pine) 168
Caulerpa taxifolia 11–12, 245
Centaurea (knapweeds) 163
  maculosa 168
  solstitialis (Yellow Star Thistle) 148, 163, 166, 172
Center for Disease Control (CDC) 240–1

262 INDEX

Ceratitus capitata (Mediterranean Fruit Fly) 186
Cervus dama 186
Chaunus marinus (Cane toad) 5, 58
Cherax spp. 187
Chile 57
China 53–4
Chondrilla juncea 163
Chromolaena odorata (Triffid Weed) 189–90, 197, 246
Circosporda zeae-maydis (Grey Leaf Spot) 245
citrus canker 127
climate 30
  distribution relationships 37–9
  forcing of 35
  invasion relationships 22–6
  observations 31–3
  regulation 167
  use of uncertain climate information 35–7
  see also climate envelope models; general circulation models (GCMs)
climate change 3, 4, 19
  invasion interactions 23–6
climate envelope models 21, 30–1, 37
  climate data 31–3
  limitations 30–1
Coddling Moth 186
Cogon Grass 245–6
colony collapse disorder, honeybees 167
Columba livia (pigeon) 172–3, 186
Comb Jelly 164
Common Carp 167–8
Common Market for Eastern and Southern Africa (COMESA) 8, 244, 248
Common Myna 186
competitive exclusion 6
concave benefits 83
Conocephalus discolor (Bush cricket) 58
control 10–12, 101–2
  biological control 87
  decision models 112–24
  expected benefits and 193–7
  failure 207–8
    utilization of uncontrollable invasive species 214
  model 102–8
  spatial aspects 146, 150–5
  tropical human-dominated landscapes 203
  see also management options

Convention on Biological Diversity (CBD) 236–9
Convention on International Trade in Endangered Species (CITES) 186, 237
Convention on Migratory Species (CMS) 237
convex damages 83–4, 87
  nonconvex damages 87, 88, 90–3
Coqui Frog 173
Cortaderia jubata (pampas grass) 168
Corvus splendens (crow) 186
costs
  anticipated damages 84–8
  cost-effective management 163–4
  inspection and interception effort 44–6
  of invasive species 2, 7–9, 42, 161–2, 205
  prevention costs 44
  see also economic significance
crop pathogens 2
Ctenopharyngodon idella (Grass Carp) 226
cultural heritage 172–3
cultural services 162, 171–3
  see also ecosystem services
Cydia pomonella (Codling Moth) 186
Cyprinus carpio (common carp) 167–8
Cytisus scoparius (Scotch Broom) 172
Czech Republic 58, 61–3, 66–8, 70–1, 74–5

D
damages *see* costs
detection strategies 10, 100–1, 102
  model 102–8
  see also inspection and interception
disease regulation services 168–9
disease transmission *see* animal pathogens; human pathogens
dispersal
  inspection and interception effort and 44–6
  see also invasions; spread
distributions 19–20
  climate relationships 37–9
  dynamic nature of 20
  future distributions 20–1
dread effect 242
Dreissena polymorpha (Zebra Mussel) 168, 169, 228
Drosophila subobscura 56–8

E
Eastern Bluebird 172
Echium plantagineum 163

Economics costs of bioinvasions
  costs of inspection and interception, eradication, control, damage etc 248
Economic indicators
  GDP 7, 8, 42, 48, 51–3, 192, 202
  income 2, 13, 43, 49, 52, 149, 163, 203, 206, 208, 212–14, 241, 245–8
  investment 10, 45, 84, 148, 162, 191, 196–7, 203–5, 208, 223, 231–2, 237, 240, 244, 247
Economic instruments for bioinvasion externalities
  taxes, charges, subsidies etc 83, 174
economic significance 2, 7–9, 42, 83, 202–3
  optimal policy choices 84–97
  South Africa 191–3
  spatial aspects 146–55
  see also costs; ecosystem services; policy options
ecosystem services 162–73, 202
  cultural services 162, 171–3
    aesthetic value 172–3
    cultural heritage 172–3
    recreation and tourism 172
  provisioning services 162–6
    food, fiber and fuel 163–5
    fresh water 166
    medicine 166
  regulating services 162, 166–71
    air quality 171
    climate 167
    erosion 168
    fire and flooding 170–1
    pests and disease 168–70
    pollination 167
    water purification 167–8
  research and policy recommendations 173–4
Eichhornia crassipes (Water Hyacinth) 164, 168, 187, 207
Eleutherodactylus coqui (Coqui Frog) 173
Emerald Ash Borer 228, 238
emissions 35
ensembles of opportunity (EOEs) 36
Environmental and natural resource management 247
eosinophilic meningoencephalitis 168
eradication 10–12, 101
erosion
  impact, South Africa 190
  regulation 168

Eucalyptus 166, 167, 171, 192
   globulus 171
Eurasian Water Milfoil 172
European Union 8, 75
European Wasp 186
Eurphorbia esula (Leafy Spurge) 163
Externality
   bioinvasions as an externality of trade 12
   taxes 83, 174
   tariffs 12, 43, 84, 145, 236, 241
   charges 110, 240, 243–5
   property rights 36
extinctions 162

F

Fallow Deer 186
fiber 164–6
Fire Ant 170, 172
fire-risk
   impact, South Africa 190
   regulation 170–1
firewood 164
first best rule 113, 117–24
   comparison with second best 119–24
Fisher–KPP–Skellam model 56, 59
flood-risk regulation 171
Flowerpot Snake 186
flu *see* influenza
Food and Agriculture Organization (FAO) 12, 236–7
food security 163–6
foot and mouth disease 2
forecasting 30–1
   probabilistic projections 36
   scenarios 36
   see also climate envelope models; global climate models (GCMs)
Four o'clock 61
freshwater ecosystems, South Africa 190–1
   see also water
fuel 164–6
fynbos ecosystem, South Africa 6, 149–50, 191
   see also South Africa

G

General Agreement on Tariffs and Trade (GATT) 43, 236
generalized cross-validation errors (GCVEs) 31–2
Generic Risk Assessment for Aquatic Organisms 230
Glasshouse Whitefly 148

general circulation models (GCMs) 33–5, 39
globalization 3, 7, 42, 235–6
   dispersal of species 6–8, 12–3, 238
   correlation of risk 240–1, 248
   measures (trade%GDP) 7
   integration of global economy 3
   international agreements 236–7
Golden Apple Snail 164, 167
Golden Wattle 192
government agencies 231–2
Grass Carp 226
'greedy' approach to inspection resource allocation 132, 134
Grey Leaf Spot 245
Grey Squirrel 58, 184
Gypsy Moth 149, 170, 172

H

habitat 66
   alien species representation 67
   habitat-based mapping of invasions 75–6
   invasibility 69–75
   level of invasion 66–9, 70–5
   most invaded habitats 67–8
habitat fragmentation 4
   cost of 9
habitat loss 4
Hakea 197, 246
Heliotropium eurpaeum 163
Hemidactylus mabouia (Common House Gecko) 186
Hemitragus jemlahicus (Himalayan Thar) 186
Himalayan balsam 58
HIV AIDS 1, 6
Honeybee 58, 167, 187
House Sparrow 186
human pathogens 1–2
hydrologic services 166
Himalayan Thar 186
Hypophthalmichthys
   harmandi (Largescale Silver Carp) 226
   molitrix (Silver Carp) 226
   nobilis (Bighead Carp) 226

I

Impatiens (balsams) 60
   glandulifera (Himalayan balsam) 58, 60
Imperata cylindrica (Cogon Grass) 245–6
India 8, 42, 184, 190, 203–4, 207–14, 228, 246

India, *Lantana* case study 209–14
infinite-horizon problems 112
influenza
   avian influenza virus 8
   Spanish flu 1, 12
inspection and interception programs (IIPs) 43, 102
   allocation of inspection resources 127, 130–43
   allocation method and results 132–42
   benefits of inspection 128
   dispersal and 44–6
   globalization and 239–41
   optimal effort model 44–54
   risk-based targeted inspection 127–30
   see also management options; prevention
institutional environment 237
insurance hypothesis 6
intentional introductions
   risk assessment 224–8
   vector elimination 228
Intergovernmental Panel on Climate Change (IPCC) 4, 20, 33, 35–6
international agreements 236–7
International Air Transport Association (IATA) 237
International Civil Aviation Organization (ICAO) 237–8
International Development Research Centre (IDRC) 217
International Health Regulations (IHR) 239–40, 246–8, 256–60
   targeted precautionary action 241–3
International Maritime Organization (IMO) 228–9, 231, 237–8
International Plant Protection Convention (IPPC) 238, 243
International Union for Conservation of Nature (IUCN) 190, 220, 223, 236, 238
invasions 220–1
   climate relationships 22–6
   dynamics 111–12, 204–7
   forecasting 30–1
   habitat-based mapping 75–6
   invasibility 69–75
   level of invasion 69–75
   most invaded habitats 67–8
   propagule pressure 75
   scale importance 68
   spatial aspects 145–55
   stages of 56, 220–1
   see also spread

Invasive Alien Species 30, 42–54,
    161–75, 183–5, 188, 190, 198,
    208, 244, 248
invasive species 1, 21–2, 183
    as pollutants 83
    benefits of 27, 205–6
    ecological significance 3–6
        impact on biodiversity 189–90,
            202
    economic significance 2, 7–9, 42,
        161–2, 183, 202–3
        see also costs
    global dimension of problem 12–13
    impacts on the poor 245–7
    risks 42–3
    see also alien species; ecosystem
        services
Invasives Information Network 243

J
Jointed Cactus 189, 192
jump spread 59
Juniperus pinchotii (Redberry Juniper)
    164

K
knapweeds 163
kudzu 168, 171

L
Lake Tahoe 172
Lake Victoria 164
Lantana 172, 207
    camara 166, 168, 192, 208–14, 246
    ecosystem impacts 209, 211
    invasion and spread 208–9, 211
    management 209–14
    utilization of 212–14
Large Grain Borer 245
Largescale Silver Carp 226
Leafy Spurge 163
learning 127–9, 242–3
    trade-related risk, model 127–32
level of invasion 69–75
    relative importance of determining
        factors 74–5
    vs. invasibility 70–4
Linepithema humile (Argentine Ant)
    186, 190
literature review risk assessment
    224–5, 226
Lodgepole pine 58–9
low income countries 245–7
Lymantria dispar (Gypsy Moth) 149,
    170, 172
Lymnaea columella 187

M
malaria 5, 8
    avian 170
management options 9–12, 100,
    235–6
    cost-effective management 163–4
    globalization and 235–7, 247–8
    institutional environment 237
    invasion dynamics and 204–7
    Lantana 209–14
    mitigation vs. adaptation 10, 43
    model 102–8
    regional scale 243–5
    spatially heterogeneous systems
        150–5
    targeted precautionary action
        241–3
    tropical human-dominated
        landscapes 203
    uncontrollable invasive species
        214
    see also control; detection
        strategies; inspection and
        interception programs; policy
        options; prevention
Manorina melanocephala (Noisy
    Miner) 170
marine ecosystems 4
medicine 166
Mediterranean Fruit Fly 186
Mediterranean Mussel 187
Melaleuca quinquenervia 164, 166,
    171
Melia azedarach 166
Mesquite 189, 246
Metrioptera roeselii (Bush cricket) 58
Mexico... 1, 6, 148, 164, 232
Millennium Ecosystem Assessment
    (MA) 3–4, 6
mitigation 9, 11
    vs. adaptation 10, 43
Mnemiopsis leidyi (Comb Jelly) 164
Mononychellus tanajoa (Cassava
    Green Mite) 245
Mountain Bluebird 58
multi-armed bandits (MAB) 132–4
Multilateral environmental
    agreements
    We need the individual
        agreements listed... 8, 12,
        236, 237
    coordination of international
        policy... 237
    cooperation between nation
        states... 243
Mylopharyngodon piceus (Black Carp)
    226

'myopic' approach to inspection
    resource allocation 132, 134
Myriophyllum
    aquaticum (Parrots Feather) 187
    spicatulum (Eurasian Water Milfoil)
        172
Mytilus galloprovincialis
    (Mediterranean Mussel) 187

N
Nasella Tussock Grass 189
National Academy of Sciences
    (NAS) 12, 207
National Invasive Species Council
    (NISC)... 10, 100–1, 232
National Research Council
    (NRC)... 224
National Science Foundation
    (NSF)... 125
native species 19–20
    invasion susceptibility and 22
    native–alien relationship 68–9
natural hazards regulation 170–1
NCEP model 31
neophytes 62, 66
    habitat invasions 67–8, 69, 73–4
    spread 61–2
New Partnership for Africa's
    Development (NEPAD)
    248
New Zealand... 58, 164, 170, 189,
    196, 220, 223, 225–6, 228,
    232, 236
Nile Perch 164
Noisy Miner 170
non-timber and minor forest
    products (NTFPs) 203
nonconvex damages 87, 88, 90–3
North American Free Trade
    Association (NAFTA) 244

O
Office of Technology Assessment
    (OTA)... 8, 88, 100, 161, 163,
    172–3
Onopordum spp. 163
Opuntia 207, 208
    aurantiaca (Jointed Cactus) 189,
        192
    ficusindica 208
Orconectes rusticus (Rusty Crayfish)
    172
ordinary least squares regression
    (ols) 62
Oryctolagus cuniculus (rabbit) 58,
    164

overfishing 4
Oxybaphus nyctagineus (Four o'clock) 61

P
Pacific Invasive Learning Network (PILN) 245
Pacific Invasives Initiative (PII) 245
Pampas grass 168
Parage aegeria (Speckled Wood butterfly) 58
Parrots Feather 187
Parthenium hysterophorus (Parthenium Weed) 190
Passer domesticus (House Sparrow) 186
pathogens *see* animal pathogens; crop pathogens; human pathogens
perturbed-physics ensembles (PPEs) 36
pest regulation services 168–9
Phenacoccus manihoti (Cassava Mealybug) 245
Phylloxera 2
Physa acuta 187
Pigeon 172–3, 186
pigs, feral 170, 172, 186
Pinus 164, 166, 192, 193, 197, 246
  contorta (Lodgepole pine) 58–9
  pinaster 149–50
Piper aduncum 206
Pistia stratiodes (Water Lettuce) 187
plague 1
Plasmodium relictum (avian malaria) 170
Podalyria calyptrata 190
Policy
  inspection and interception ... 235, 243, 248
  quarantine ... 231–2, 240, 244–5
  eradication ... 162–3, 173, 186, 196, 205, 207, 235, 245–8
  adaptation ... 207, 211, 232, 246
  International ... 235–48
policy options 9, 83–97, 100, 220, 235
  aggressive policy 95
  decision models 112–24
  ecosystem services importance 173–4
  first best rule 113, 117–24
  marginal considerations 83
  optimal policy choices 84–97, 101, 110–11
  prevention of abuse 96–7
  second best rule 111–17, 119–25
  see also management options
pollination services 167
pollution 83
  air quality regulation 171
  invasive species as pollutants 83
Pomacea canaliculata (Golden Apple Snail) 164, 167
Pompom Weed 189, 196
prevention 100–1, 102, 220, 222–32
  costs 44
  economic and ecological context 220–2
  gaps and overlaps among government agencies 231–2
  model 102–8
  pre-border vs. border prevention 230–1
  quantitative vs. qualitative approaches 229–30
  risk assessment 223–32
  see also inspection and interception programs (IIPs); management options
probabilistic projections 36
probability estimation 229–30
Procambarus clarkii 187
Prosopis 166, 189, 192, 207–8
  glandulosa 167
  juliflora (Mesquite) 189, 246
Prostephanus truncatus (Large Grain Borer) 245
provisioning services 162–6
  see also ecosystem services
public goods 235, 240, 248
  free-riding ... 240
  geographical scale 239–41
  invasive species control as an international public good ... 13, 239–41, 248
  invasive species control as a local public good ... 239
Pueraria lobata (kudzu) 168, 171

Q
quasi-option value 242–3

R
rabbit 58, 164
Rabbit hemmorhagic disease (RCD) ... 46
Ramphotyphlops braminus (Flowerpot Snake) 186
Rat Lungworm 168
rats 11, 184
recreation 172
Red Scale 186
Red Water Fern 187, 191
Red-eared Slider Turtle 186
Redberry Juniper 164
regional networks 243–5
Regional Trade Agreements (RTAs) 8, 239, 243–5
regulating services 162, 166–71
  see also ecosystem services
resilience
  biodiversity and 5–6, 42
  native species pool and 22
  reduction in 5–6, 162
resource allocation for inspection 127, 130–43
  allocation method and results 132–42
  inspection allocation decision problem 130–2
  learning model of trade-related risk 129–30
resource availability
  for risk assessment 231
  habitat invasion and 70–4
  see also resource allocation for inspection
Rhododendron ponticum 148
rinderpest 2
ripple spread 59
risk
  learning model of trade-related risk 127–32
  targeted inspection 127–43
risk assessment 223–32
  accuracy of tools 227–8
  appropriate risk threshold 230
  designing species-specific tools 223–4
  quantitative vs. qualitative approaches 229–30
  resource availability 231
  tools currently in use 224–6
  trade in live organisms 223–8
  transportation-related introductions 228–9
Risks of bioinvasions ... 25, 69, 76, 228–9
  risk and uncertainty ... 242
  risk assessment ... 66, 76, 223–32, 240–1, 244, 248, 252
  risk aversion ... 142
  risks of diseases ... 252
  risk management ... 222–3, 229–30
roads 5

Ruffe fish 172
Russian mustard 61
Rusty Crayfish 172

S

Salt Cedar 166
Salvinia 187
San Francisco Bay 7
Sanitary and Phytosanitary (SPS) Agreement 43–6, 235, 236, 251–5
  benefits 44
  conformity with International Health Regulation 239–41, 246–8
  targeted precautionary action 241–3
  see also inspection and interception programs (IIPs)
scenarios 36
Scientific Committee on Problems in the Environment (SCOPE)... 184
Scotch Broom 172
Scurius carolinensis (Grey Squirrel) 58, 184
sea lamprey 172
second best policies 111–17
  comparison with first best 119–24
Secretariat of the Pacific Community (SPC) 245
Senecio jacobaea (Tansy Ragwort) 163
shipping 228
Sialia
  currucoides (Mountain Bluebird) 58
  mexicana (Western Bluebird) 58
  sialis (Eastern Bluebird) 172
Siam Weed 246
Silver Carp 226
Sirex noctilio (Wood Wasp) 186
Sisymbrium volgense (Russian mustard) 61
Solanum elaegnifolium (satansbos) 189
Solenopsis invicta (Fire Ant) 170
South Africa 183–98
  control related to expected benefits 193–7
  fynbos ecosystem 6, 149–50
  impacts of invasive species 188–97
    biodiversity 189–90
    economic assessments of 191–3
    erosion 190
    estimates of future impacts 193
    fire regimes 190

freshwater ecosystems 190–1
grazing resources 189
human health and safety 190
water resources 188–9
invasive species 183–7
  amphibians 186
  birds 186
  freshwater biota 187
  mammals 184–6
  marine organisms 187
  plants 184
  reptiles 186
  terrestrial invertebrates 186–7
pathways of invasion 187–8
Southern African Development Community (SADC)... 244
South Pacific Regional Environment Programme (SPREP) 245
Southern Common Market (MERCOSUR) 244
Spanish flu 1, 12
spatial aspects of biological invasions 145–55
  spatially differentiated studies 146, 148–50
  spatially explicit studies 146, 149–50
  spatially heterogeneous system model 150–4
  spatially implicit studies 146–8
species distributions *see* distributions
species redundancy 6
Speckled Wood butterfly 58
Sphaeroma quoyanum 168
sports fishing 172
spread 56
  across borders 232
  control of 146
  jump spread 59
  ripple spread 59
  spatial aspects 146–55
  time to complete spreading 61–3
  variation between species 59–61
  variation within species 56–9
  see also invasions
Standard International Trade Classification (SITC)... 48
Starling 172, 186
Stipa trichotoma (Nasella Tussock Grass) 189
Striga hermontheca (Witchweed) 245
Sturnus vulgaris (Starling) 172, 186
Sus scrofa (pig) 170, 172, 186

T

Tamarix ramosissima (Salt Cedar) 166

Tansy Ragwort 163
Tarebia granifera 187
targeted precautionary action 241–3
  inspection 127–43
Technical Barriers to Trade (TBT)... 237, 251
Theory of Biotic Acceptance 69
Theory of Fluctuating Resource Availability 74
tourism 172
Trachemys scripta (Red-eared Slider Turtle) 186
trade 7–8, 42, 52–3, 127
  agreements... 8, 247
  growth... 53
  inspection and interception programs (IIP) 43–54
  learning model of trade-related risk 127–32
  link to General Agreement on Tariffs and Trade (GATT) and the Sanitary and Phytosanitary Agreement (SPS)... 12
  optimal policy choices 84–97
  Regional Trade Agreements (RTA)... 8, 53, 239, 243, 247–8
  restrictions... 241
  risk assessment 223–9
  routes/pathways... 7, 8, 10, 11, 42, 52, 227, 242–3
  tariffs... 43, 84
  volumes... 48, 52, 98
  volumes and introductions... 2, 3, 8, 11
trait-based risk assessment 225–6
transportation-related introductions 228–9
travellers as vectors 228–9
Trialeurodes vaporariorum (Glasshouse Whitefly) 148
Trichosurus vulpecula (Brush-tailed Possum) 164, 168–70
Triffid Weed 189–90, 197
Tropical House Gecko 186
tropical human-dominated landscapes 203
tuberculosis 6

U

uncertainty 10, 39, 101–2, 129
  emissions 35
  use of uncertain climate information 35–7
  see also forecasting

UN Convention on the Law of the Sea (UNCLOS)... 237
United Kingdom... 2, 3, 8, 9, 36, 42, 67, 228
United Nations Environment Program (UNEP)... 12, 248
United States... 12, 67, 220, 225
United States Environmental Protection Agency (EPA)... 230
United States Department of Agriculture... 127, 197

V

Varroa destructor 186–7

Vespula germanica (European Wasp) 186
Vulpia spp. 163

W

water
  ecosystem services 166, 173–4
  invasive species impacts 188–9
  purification 167–8
Water Hyacinth 164, 168, 187, 207
Water Lettuce 187
Western Bluebird 58
Witchweed 245
Wood Wasp 186
World Animal Health Organization (OIE) 43, 46–9, 237–8

World Bank environmental and natural resource management (ENRM) lending 247
World Health Organization (WHO) 236
World Trade Organization (WTO) 236
WORLDCLIM 32

Y

Yellow Star Thistle 148, 163, 166, 172

Z

Zebra Mussel 168, 169, 228
zoonotic diseases 246–7